Topics in Applied Physics   Volume 45

W0079139

Topics in Applied Physics Founded by Helmut K. V. Lotsch

Hydrodynamic Instabilities

and the

Transition to Turbulence

Edited by H. L. Swinney and J. P. Gollub

With Contributions by
F. H. Busse   P. A. Davies   R. C. Di Prima   J. P. Gollub
J. M. Guckenheimer   D. D. Joseph   O. E. Lanford
S. A. Maslowe   H. L. Swinney   D. J. Tritton
E. D. Yorke   J. A. Yorke

Second Edition

With 82 Figures

Springer-Verlag
Berlin Heidelberg GmbH

Professor *Harry L. Swinney*, PhD

The University of Texas at Austin, Department of Physics,
Austin, TX 78712, USA

Professor *Jerry P. Gollub*, PhD

Department of Physics, Haverford College,
Haverford, PA 19041, USA   and

Department of Physics, University of Pennsylvania,
Philadelphia, PA 19104 USA

ISBN 978-3-540-13319-3   ISBN 978-3-540-38449-6 (eBook)
DOI 10.1007/978-3-540-38449-6

Library of Congress Cataloging in Publication Data. Main entry under title: Hydrodynamic instabilities and the transition to turbulence. (Topics in applied physics; v. 45) Includes bibliographies and index. 1. Hydrodynamics. 2. Stability. 3. Turbulence. I. Swinney, H.L., 1939–. II. Gollub, J.P., 1944–. III. Title: Hydrodynamic instabilities. IV. Series. QA911.H9 1985 532′.5 85-2851

The use of registered names, trademarks, etc. in this publication does not imply, even in the absence of a specific statement, that such names are exempt from the relevant protective laws and regulations and therefore free for general use.

2153/3130-543210

Preface to the Second Edition

In the four years that have elapsed between the first and second editions of this book, much progress has been made in understanding hydrodynamic instabilities and the transition to turbulence. For example, the strange attractors discussed theoretically by Lanford in Chap. 2 have been convincingly observed in experiments on weakly turbulent flows, and several "universal" routes to chaos have been identified in theoretical and experimental studies. Many other noteworthy advances have been made using quite different theoretical methods. For example, the evolution of convection patterns has been studied using two-dimensional model equations.

Brief descriptions of these and other developments, along with numerous added references, are included in this second edition. We hope that the reduced cost of this edition in paperback will make it accessible to many additional scientists and students in the various fields to which it is relevant, especially physics, mathematics, and engineering.

We appreciate the assistance of our contributors, and the support of the National Science Foundation Fluid Mechanics Program.

We dedicate this book to the memory of our colleague and friend, Richard C. DiPrima (9 August 1927–10 September 1984), whose contributions to hydrodynamic stability theory will long be remembered.

Austin and Haverford, February 1985   *H. L. Swinney · J. P. Gollub*

Preface to the First Edition

Although much of the universe is filled with fluids in turbulent motion, the processes by which turbulence develops are poorly understood. When a fluid is driven away from thermal and mechanical equilibrium, it will often undergo a sequence of instabilities, each of which leads to a change in the spatial or temporal structure of the flow. The nature of these instabilities, which sometimes lead to turbulence, is the subject of this volume.

Hydrodynamic instabilities and turbulence have been extensively studied for more than a century, but the research has been primarily concerned with either the first instability that occurs with increasing Reynolds number or with turbulence at very large Reynolds number. The transition from laminar to turbulent flow has until recently been largely beyond the reach of both theory and experiment. This situation has been changed dramatically by the use of computers in laboratory experiments and in numerical analyses of nonlinear systems. While past experiments were primarily photographic or measured time-averaged quantities, recent experiments using computers and modern optical and cryogenic techniques have distinguished between many different dynamical regimes of flows undergoing transition. Numerical studies of nonlinear models have also revealed entirely unexpected results, such as chaotic behavior in a system with only three variables. Another development of great potential importance is the application of new mathematical concepts from the qualitative theory of differential equations, sometimes known as dynamical systems theory, to the transition to turbulence problem. More traditional methods such as bifurcation theory and stability analysis also continue to contribute major new insights.

This book is a collaboration between physicists, mathematicians, and fluid dynamicists, each of whom is a recognized leader in the field. The various chapters include: introductions to the relationship between dynamical systems theory and turbulence (Chaps. 2 and 4); a review of hydrodynamic stability and bifurcation theory (Chap. 3); three case studies – convection, rotating fluids, and shear flows (Chaps. 5–7); a review of the many types of instabilities that occur in geophysics (Chap. 8); and a discussion of instabilities and chaotic behavior in nonhydrodynamic systems (Chap. 9).

Although not all of the book is strictly introductory, the authors have tried to make the majority of it accessible to physicists, mathematicians, engineers, and graduate students who do not have significant background in fluid dynamics and advanced mathematics. It is our hope that it will provide an introduction to the literature of this rapidly developing field.

We owe special thanks to D. D. Joseph for his encouragement and advice in this endeavor, and to our contributors for their efforts to communicate with clarity to a new audience. We also acknowledge the support of the National Science Foundation.

Austin and Haverford, October 1980   *H. L. Swinney · J. P. Gollub*

Contents

Contributors

Busse, Friedrich H.
  Department of Earth and Space Science and
  Institute of Geophysics and Planetary Physics, University of California
  Los Angeles, CA 90024, USA

Davies, Peter A.
  Department of Civil Engineering, University of Dundee
  Dundee DD1 4HN, UK

DiPrima, Richard C. (deceased)

Gollub, Jerry P.
  Department of Physics, Haverford College
  Haverford, PA 19041, USA   and

  Department of Physics, University of Pennsylvania
  Philadelphia, PA 19104 USA

Guckenheimer, John M.
  Department of Mathematics, Cornell University
  Ithaca, NY 14853, USA

Joseph, Daniel D.
  Department of Aerospace Engineering and Mechanics
  University of Minnesota,
  Minneapolis, MN 55455, USA

Lanford, Oscar E.
  I.H.E.S., F-91440 Bures-sur-Yvette, France

Maslowe, Sherwin A.
  Mathematics Department, McGill University, 805 Sherbrooke Street West
  Montreal H3A 2K6, Canada

Swinney, Harry L.
   The University of Texas at Austin, Department of Physics
   Austin, TX 78712, USA

Tritton, David J.
   School of Physics, The University
   Newcastle upon Tyne NE1 7RU, England

Yorke, Ellen D.
   Department of Physics, University of Maryland Baltimore County
   Catonsville, MD 21228, USA

Yorke, James A.
   Institute for Physical Science and Technology
   University of Maryland, College Park, MD 20742, USA

1. Introduction

H. L. Swinney and J. P. Gollub

The problem of explaining the origin of turbulent flows has been recognized for more than a century. Early discussions of hydrodynamic stability included an article on the instability of fluid jets by *Rayleigh* [1.1] in 1879 and a paper by *Reynolds* [1.2] in 1883 on "direct" and "sinuous" flow in pipes. Much is now known about a variety of hydrodynamic instabilities [1.3–5]. In addition, strongly turbulent flows have been rather well characterized using statistical methods [1.6]. However, a convincing and quantitative explanation of the origin of chaotic fluid motion remains elusive.

It is natural to ask why this problem has proven to be so difficult, and what recent progress justifies the writing of this book. A number of the experimental difficulties that have hindered progress toward understanding the transition problem are described in Sect. 1.1. On the theoretical side, the basic difficulty is the intractability of the nonlinear hydrodynamic equations; although they are believed to be correct even for strongly turbulent flows, the mathematical challenge of obtaining explicit solutions is formidable, and solutions are not generally unique, except at low Reynolds number.

In the past decade both theory and experiment have undergone profound development. This progress, which provided the principal motivation for writing this book, is described briefly in the following sections and in detail in the chapters of this book.

1.1 Experimental Difficulties and Advances

An operational definition of turbulence has been lacking, so that the concept of the onset of turbulence has been ill defined. Until recently, the practical definition has been the appearance of apparent randomness in photographs of flows containing materials which permit visualization of streamlines or other features. However, this approach omits the possibility of complex flow patterns that are nevertheless highly ordered. The qualitative appearance of a flow is not necessarily a reliable indicator of its fundamental behavior.

A second experimental difficulty has been the absence of sensitive and quantitative experimental techniques capable of measuring the time dependence of the fundamental dynamical variables: the velocity field, temperature field, vorticity, etc. The prediction and measurement of time-averaged fields (e.g., velocity profiles) is of limited use in achieving a fundamental understanding of the onset of turbulence.

A third difficulty has been an inadequate degree of control over boundary conditions and other experimental parameters. Flows which are near a hydrodynamic instability are extremely sensitive to external perturbations. Moreover, precision control is essential to successful discrimination between random flow produced by internal dynamics and random flow produced by external influences.

Finally, experimentation has been hindered by the lack of an adequate conceptual framework. Since much of the theoretical work on turbulence has concentrated on the strongly turbulent regime rather than the transition region, experimenters have had to be largely empirical in their work; clear hypotheses permitting validation of well-defined models of transition to turbulence have been lacking.

A number of noteworthy developments during the past five years have qualitatively advanced the art of experimentation on instabilities and turbulence. One of these new methods is the use of laser Doppler techniques [1.7], in which the frequency shift of light scattered from a moving fluid provides remote measurements of the instantaneous local velocity. Laser probes have very high spatial and temporal resolution, respond linearly to a component of the local velocity, do not perturb the flow significantly, and can be used to measure the velocity at several locations simultaneously.

Local measurements can also be made by the older technique of hot wire anemometry [1.8], in which the heat transport from a small resistively heated wire (or film) is used to infer the local velocity perpendicular to the wire. In some circumstances (especially enclosed geometries) this probe significantly disturbs the flow, and it also does not have a linear response. An important complement to the quantitative local measurements obtained by laser Doppler velocimetry and hot wire anemometry are the qualitative instantaneous global pictures that can be obtained by the old technique of flow visualization, which has been greatly improved in recent years. Thus the development of new techniques and the refinement of older ones have permitted a striking improvement in studies of transition phenomena in some systems [1.9].

Another example of a qualitative advance in experimentation has been the application of cryogenic techniques to the study of heat transport in convecting fluids [1.10]. Temperature resolution of about 1 part in 10^7 can be attained at low temperatures. Containers with a heat capacity far lower, but a thermal conductivity far higher, than the fluid they contain are easily constructed for use at low temperatures. Radiative heat transport is negligible at low temperatures. For these and other reasons, fluctuations in the heat transport resulting from instabilities or turbulence in liquid helium can be measured with great precision.

Finally, the most important experimental development has been the application of laboratory computers to data collection and the control of experiments. By collecting measurements of velocity, heat flux, etc., over a long period of time and then using fast Fourier transform techniques, the spectra of these dynamical variables can be easily measured. Such spectra form an excellent method of distinguishing between qualitatively different hydrodynamic flows. A spectrum containing a single peak and its harmonics indicates a periodic oscillation in the flow. A spectrum containing peaks at several incommensurate frequencies is known as a quasiperiodic flow, in the sense that a given segment of the time history comes close to recurring after a sufficiently long (but finite) time. Finally, a spectrum containing broadband noise is said to be nonperiodic. The onset of nonperiodicity in a flow may be a useful operational definition of the onset of turbulence.

1.2 Hydrodynamic Stability and Bifurcation

In recent years the theoretical developments in the study of instabilities and turbulence have been as profound as the developments in experimental methods. As we remarked at the beginning of this introduction, the subject of hydrodynamic stability is an old one. The basic technique is linear stability analysis [1.3–5], in which the effect of a small fluctuation away from a solution to the equations is examined as a function of a parameter such as the Reynolds number. The analysis is usually in terms of normal modes of the system, which constitute a complete set of eigenfunctions of the problem. The hydrodynamic equations are linearized in the amplitude of the fluctuation, and the linearized equations are solved to determine whether the fluctuation decays or grows in time. In this way the boundaries of a stable region in parameter space can be determined. Investigation of the development of secondary flow and the onset of higher instabilities requires a *nonlinear* stability analysis, a subject that *Chandrasekhar* [1.3] discussed in only a few pages of his classic (1961) book on stability. The seminal ideas on nonlinear stability analysis date from 1960 [1.11, 12], but the real flowering of the subject has occurred in the past decade as large computers have become widely available. Results of nonlinear studies of particular flows are discussed in Chaps. 5–7.

Another approach to the problem of hydrodynamic stability is bifurcation theory, which classifies and characterizes the types of bifurcations (splitting of solutions) that can occur in nonlinear systems. A great deal can be learned by consideration of the bifurcation properties of simplified models, which correspond well to certain hydrodynamic cases. The subject of hydrodynamic stability and bifurcation is discussed extensively in Chap. 3 of this volume by *Joseph*, with particular attention to recent developments in this field. Although the rigorous results are largely confined to the bifurcation of steady solutions into other steady solutions or into time-periodic solutions, *Joseph* also discusses the problem of turbulence from the standpoint of bifurcation theory.

1.3 Dynamical Systems

Recent developments in the qualitative theory of ordinary differential equations (dynamical systems theory) have provided a new and stimulating point of view about the problem of the onset of turbulence. New insights in this field have been achieved primarily through detailed computer studies of the dynamics of simple nonlinear models. The basic observation is that relatively simple systems of three or more coupled nonlinear first-order equations often have chaotic solutions. These solutions (sometimes called strange attractors) are much more irregular than solutions of deterministic equations were believed to be until recently. This observation has generated the hypothesis that the transition to turbulence can be qualitatively explained by models that are highly simplified in comparison with the full hydrodynamic equations. *Lanford* (Chap. 2) discusses the mathematics of dynamical systems and the hypothesis that turbulence may be viewed as a strange attractor. This point of view is further elucidated by *Yorke* and *Yorke* in Chap. 4; they also discuss several concrete examples of chaotic dynamical systems, including the Lorenz equations and the cascades of bifurcations exhibited by one-dimensional maps.

While the dynamical systems point of view has not yet achieved much quantitative predictive power, it has perhaps produced some conceptual enlightenment and has stimulated some highly precise experimentation.

1.4 Convection, Rotation, and Shear Flows

Progress in understanding instabilities and the onset of turbulence has been made primarily by focussing attention on a small number of relatively simple hydrodynamic systems: Rayleigh-Bénard convection, the flow between concentric rotating cylinders, and parallel shear flows. The first of these, the flow of a fluid confined to a horizontal layer with an imposed vertical temperature gradient, forms the subject of Chap. 5 by *Busse*. This topic has been one of the primary testing grounds for *nonlinear* theories of hydrodynamic stability. One important method is the introduction of series expansions of the velocity and temperature fields in powers of a small parameter. Such perturbation expansions can be used to determine the form and stability of the new steady-state solutions which evolve from an instability. Nonlinear theory can also be used to examine the stability of the new steady state with respect to other types of disturbances. From both theoretical analysis and experiments, it is now clear that a uniform fluid layer undergoes quite a variety of qualitatively distinct instabilities, resulting in complex temporal and spatial behavior. Busse's chapter also includes a brief review of recent experiments in this field. While the early instabilities are reasonably well understood, the transition to turbulence can occur in a variety of ways which depend sensitively on geometrical factors. Much of this behavior is not currently understood.

Another experimental system of major importance in studies of instabilities and transition is the flow of a fluid confined to the annulus between concentric rotating cylinders; this instability is discussed by *Di Prima* and *Swinney* in Chap. 6. The best known instability in this system is the Taylor instability, which results in a toroidal vortex flow when the inner cylinder reaches a critical angular velocity. At higher rotation rates, the Taylor vortex flow becomes unstable to transverse oscillations, and travelling waves appear in the flow. Eventually the flow becomes turbulent in a process that is not well understood.

The subject of parallel shear flows is reviewed by *Maslowe* in Chap. 7. These flows, which include boundary layers, free shear layers, and flows in channels and pipes, are of great practical importance. Maslowe discusses the linear stability of these flows using a normal mode approach, the time evolution of perturbations which are not normal modes, nonlinear theories, and experiments on shear flows. The mathematical ideas in this field can be elegant, as Maslowe clearly shows.

1.5 Instabilities in Geophysics and Nonhydrodynamic Systems

Instabilities and turbulence have an importance that transcends laboratory systems and engineering applications. Much of the universe is filled with fluids in turbulent motion, and instabilities are quite common in planetary geophysics, oceanography, atmospheric science, and astrophysics. In Chap. 8, *Tritton* and *Davies* discuss the physical mechanisms underlying these instabilities. They focus on several common situations: stratified shear flows (shear flows in the presence of a density gradient); shear flows in rotating systems; instabilities arising from differential diffusion in multicomponent systems (such as salt water); and baroclinic instabilities, in which the density stratification is not vertical. These instabilities are important in meteorology and atmospheric dynamics.

Finally, in Chap. 9 *Guckenheimer* reviews the subject of chaotic dynamics in systems other than Newtonian fluids. One example concerns the problem of erratic field reversals in magnetohydrodynamic systems. Another is the complicated time evolution of certain chemically reacting systems. Population dynamics in ecological systems can be modeled by systems of differential or difference equations having complex periodic and nonperiodic solutions. Models of forced nonlinear electrical oscillators also show complicated dynamics. Guckenheimer's chapter concludes with a thoughtful discussion of the difficulty of making quantitative comparisons between models and experiments for systems with chaotic dynamics.

1.6 Summary

The subject of hydrodynamic instabilities and the transition to turbulence has strong connections to fundamental mathematical problems and diverse applications. It is our hope in assembling this volume that these connections will come through clearly.

This book is not concerned at all with the structure of strongly turbulent flows. For a summary of this fascinating and active field, we refer the reader to the recent volumes edited by *Bradshaw* [1.13], and *Libby* and *Williams* [1.14].

References

1.1   Lord Rayleigh: On the stability of jets. Proc. London Math. Soc. **X**, 4 (1879)
1.2   O. Reynolds: An experimental investigation of the circumstances which determine whether the motion of water shall be direct or sinuous, and of the law of resistance in parallel channels. Phil. Trans. R. Soc. **174**, 935 (1883)
1.3   S. Chandrasekhar: *Hydrodynamic and Hydromagnetic Stability* (Clarendon Press, Oxford 1961)
1.4   C. C. Lin: *The Theory of Hydrodynamic Stability* (University Press, Cambridge 1955)
1.5   D. D. Joseph: *Stability of Fluid Motions I and II* (Springer, Berlin, Heidelberg, New York 1976)
1.6   A. S. Monin, A. M. Yaglom: *Statistical Fluid Mechanics*, 2 vols. (MIT Press, Cambridge 1971, 1975)
1.7   T. S. Durrani, C. A. Greated: *Laser Systems in Flow Measurement* (Plenum, New York 1977)
1.8   A. E. Perry, G. L. Morrison: A study of the constant temperature hot wire anemometer. J. Fluid. Mech. **47**, 577–599 (1971)
1.9   P. R. Fenstermacher, H. L. Swinney, J. P. Gollub: Dynamical instabilities and the transition to chaotic Taylor vortex flow. J. Fluid Mech. **94**, 103–128 (1979)
1.10 G. Ahlers: Low temperature studies of the Rayleigh-Bénard instability and turbulence. Phys. Rev. Lett. **33**, 1185–1188 (1974)
1.11 J. T. Stuart: On the non-linear mechanics of wave disturbances in stable and unstable flows. Part 1. The basic behaviour in plane Poiseuille flow. J. Fluid Mech. **9**, 353–370 (1960)
1.12 J. Watson: On the non-linear mechanics of wave disturbances in stable and unstable parallel flows. Part 2. The development of a solution for plane Poiseuille flow and for plane Couette flow. J. Fluid Mech. **9**, 371–389 (1960)
1.13 P. Bradshaw (ed.): *Turbulence*, 2nd ed., Topics in Applied Physics, Vol. 12 (Springer, Berlin, Heidelberg, New York 1978)
1.14 P. A. Libby, F. Williams (eds.): *Turbulent Reacting Flows*, Topics in Applied Physics, Vol. 44 (Springer, Berlin, Heidelberg, New York 1980)

2. Strange Attractors and Turbulence

O. E. Lanford

With 1 Figure

The objective of this chapter is to provide an elementary introduction to some aspects of the modern qualitative theory of differential equations and its possible relation to the theory of turbulence.

2.1 Basic Principles

The idea that the qualitative theory of differential equations might have something to do with turbulence was forcefully argued in a paper by *Ruelle* and *Takens* [2.1] published in 1971, and the resulting theory has become widely known as the Ruelle-Takens theory of turbulence.

Before going any further, we need to issue a disclaimer. Turbulence is a complicated and many-faceted phenomenon. Notably, turbulent flows have striking and varied spatial organizations, and the onset of a time-dependent turbulence is frequently preceded by the development of fascinatingly complex stationary spatial patterns. The ideas we are going to describe have nothing to say about these matters; they concentrate almost entirely on the temporal structure of turbulent flows. Secondly, this circle of ideas has not – yet at least – contributed to the prediction of practical parameters such as turbulent viscosities and the like; the objective is rather to clarify the mathematical structure underlying the physical phenomena. Finally, in spite of its mathematical character, the Ruelle-Takens approach is still mathematically speculative in the sense that it is based on some concrete mathematical conjectures about the Navier-Stokes equations, conjectures which are so far supported only by indirect evidence, not by any solid and precise analysis of the equations themselves.

That said, let us now proceed to summarize the fundamental picture to be developed here. It is

The mathematical object which accounts for turbulence is an attractor or a few attractors, of reasonably small dimension, imbedded in the very-large-dimensional state space of the fluid system. Motion on the attractor depends sensitively on initial conditions, and this sensitive dependence accounts for the apparently stochastic time dependence of the fluid.

We shall explain what the words mean and describe what phenomena can be understood on the basis of this picture. We begin by outlining, in very broad and general terms, some of the physical and mathematical presuppositions of this approach:

1) Turbulence is to be understood within the framework of the Navier-Stokes equations. This hypothesis is open to criticism on mathematical grounds since there is no general global existence theorem for solutions of the initial value problem for the Navier-Stokes equation. Although no examples are known, it is conceivable that solutions may become singular so that the equations break down (meaning perhaps that physical principles beyond those of classical fluid dynamics need to be introduced to obtain a complete theory). While the possible importance of breakdown of Navier-Stokes equations for the theory of turbulence cannot be ruled out at our present stage of mathematical understanding, we shall proceed to develop a view of turbulence which ignores such breakdowns.

2) The fluid system is regarded as a mechanical system with friction. Mathematically, its motion is viewed as governed by a first-order (in time!) differential equation on a *state space*. The state space is the analog of the phase space for a Hamiltonian system of a finite number of degrees of freedom; a point of this space means a complete specification of the instantaneous state of the fluid, i.e., a specification of the velocity at each point together perhaps with a more or less complete specification of the thermodynamic state of the fluid there.

3) Although the state space for the fluid system is infinite dimensional, the point of view to be developed is that the phenomena underlying turbulence are essentially finite dimensional. The intuition is that there is some finite set of essential "modes" or degrees of freedom which effectively govern the behavior, while the remaining infinitely many degrees of freedom simply respond passively. These intuitive ideas can be supported by the following plausibility argument: The infinite dimensionality of the state space arises from the possibility of exciting disturbances of arbitrarily small spatial dimensions, but the frictional mechanisms of viscosity (and, in some cases, thermal conductivity) become arbitrarily strong in damping out disturbances as the size of the disturbances becomes arbitrarily small. The problem of making precise technical sense of this intuition remains, nevertheless, obscure.

4) Because of the presence of friction, the fluid systems under consideration must be subject to external driving forces in order to remain in motion. We shall assume for simplicity that driving forces are stationary, i.e., that the equations of motion do not depend explicitly on time. Response to periodic driving forces presumably could be analyzed by similar methods, but it does not seem likely that much of interest can be said along the lines to be developed about driving forces without any time regularity at all.

5) We shall adopt the point of view that the chaotic behavior of turbulent fluids ought to be understandable without the necessity for introducing external random noise. It is important to distinguish here between the effects of weak and strong external noise. It would seem most unsatisfying to account for turbulent behavior by assuming the presence of random external forces

comparable in strength with the accelerations actually undergone by the fluid particles. Such an approach would require detailed assumptions about the probabilistic character of the external forces. Hence, instead of the relatively clear-cut – if complicated – problem of understanding the Navier-Stokes equation, one would be faced with the much less well-defined problem of constructing a realistic physical model for the origin of the fluctuating external forces and justifying the probabilistic hypotheses used on the basis of this model. Besides, in most situations of interest in fluid dynamics, the fluid simply is not subject to strong fluctuating external forces.

It is, on the other hand, clear that no fluid system is ever entirely isolated from its environment, so a complete theory must be able to accommodate fairly general weak external noise. We take the point of view, however, that such external noise is to be regarded as a *complication*, to be analyzed after the behavior of the system in isolation has been understood, and not as the fundamental origin of the large-scale chaotic behavior of the fluid. Whatever one feels about this issue philosophically, one of the main contributions of the set of ideas we are developing is to offer a mathematically precise explanation for chaotic behavior which does not require any external source of randomness at all.

6) The approach taken is geometrical, but the geometrical analysis is carried out in the state space of the fluid system and not in physical space. One seeks a more or less detailed geometrical view of how the system evolves in time, as a flow on the state space. For this geometrical analysis to be a useful heuristic tool, it seems necessary for the essential phenomena to be not only finite dimensional (as suggested above) but of fairly small dimension – geometric intuition falls off rapidly as the dimension increases (above three!). While the arguments for the essential finite dimensionality of fluid phenomena may well hold for arbitrarily large (but finite) Reynolds numbers, the number of relevant modes presumably increases with Reynolds number, and geometric analysis is not likely to provide useful detailed insights into "fully developed turbulence" – its usefulness is likely to be limited to the study of the onset of turbulence.

7) An important element in the approach we are describing – and one which tends to evoke suspicion – is the use of considerations of *genericity*. Put very generally, the idea is that the qualitative properties of the Navier-Stokes equation ought, in the absence of specific arguments to the contrary, to be assumed to be *unexceptional*. To take a simple example, if the equation admits two periodic solutions which are not related in any straightforward way (e.g., by a symmetry) we normally expect the two periods to be incommensurable; a rational ratio of periods would be exceptional. There are other kinds of behavior, for example, multiperiodic motion, which are exceptional for fundamentally similar but less obvious reasons; they are destroyed by most small perturbations of the equations of motion. One important area of investigation in the qualitative theory of differential equations is the determination of which properties are or are not exceptional in this sense; exceptional behavior is

referred to as *nongeneric*. A proof that some kind of behavior is nongeneric is generally taken as a strong indication that the behavior will not occur in turbulent fluids. This is an argument of desperation, useful only in the absence of detailed results about the particular equations under investigation. Nevertheless, since we are in fact extremely ignorant about even such rudimentary questions as the existence of solutions of the Navier-Stokes equations, these considerations have played a large role.

It should be noted that genericity considerations have been used before in theoretical physics. The von Neuman-Wigner theorem on noncrossing of energy levels is a familiar example. It is an empirical fact from spectroscopy that for an atom or molecule in a magnetic field energy levels corresponding to eigenstates with the same symmetry do not cross each other as the magnetic field is varied. *Von Neuman* and *Wigner* [2.2] accounted for this phenomenon, not by showing that crossing does not occur for a particular Hamiltonian and coupling term, but by showing that a crossing of eigenvalues implies a nontrivial relation between Hamiltonian and coupling term which "usually" does not hold and which has no good reason to hold in the particular case of atomic Hamiltonians interacting with external magnetic fields.

This example is instructive in another respect. If the Hamiltonian admitted no symmetry group, it would be expected that no energy levels at all could cross. The fact that both atomic Hamiltonian and coupling term admit as symmetries rotations about an axis parallel to the magnetic field means that the operators are already special and this specialness has the effect of removing the unlikeliness of the crossing of energy levels with different symmetry. Similarly, many of the standard problems in fluid mechanics admit symmetries – for example, Couette flow is axially symmetric about the common axis of the two cylinders – and for such problems one needs to classify behavior as exceptional or unexceptional among equations respecting the symmetry. This subject has not been investigated as extensively as genericity in the absence of symmetry (see, however, [2.3]).

Aside from symmetries, the Navier-Stokes equations have a number of special properties, such as locality in space, which might in principle give rise to nongeneric behavior. It is, however, hard to see how such properties could influence geometric properties of the solution flow in state space, so we shall proceed, cautiously, on the assumption that the solution flow is not exceptional except as required by the physical symmetries of the system.

2.2 Some Elements of the Qualitative Theory of Differential Equations

Having outlined the basic presuppositions of the approach we are describing, we have next to sketch the mathematical tools at our disposal. We have to begin by noting a crucial distinction between the dissipative systems we are interested in here and the more familiar conservative systems of Hamiltonian mechanics. By Liouville's theorem, the solution flow for a Hamiltonian system preserves volume in phase space. The Poincaré recurrence theorem then says (roughly) that most solution curves in phase space come back infinitely often arbitrarily near their initial points. This theorem is perfectly general, and does not depend on ergodicity of the motion; all that is required is that the solution flow preserve volume.

Dissipative systems, i.e., systems with some sort of frictional mechanism, by contrast, usually give rise to flows which *contract* volumes in phase space. This statement, unlike Liouville's theorem, is intended to be descriptive rather than a general statement of fact. Consider, as one example, a system of point particles subject to position-dependent forces together with linear frictional forces. Then

$$\frac{dq_i}{dt} = \frac{p_i}{m_i} \; ; \quad \frac{dp_i}{dt} = F_i + \sum_j A_{ij} p_j \; , \tag{2.1}$$

where the F_i depend only on the q_j's and A_{ij} is a matrix of constants. The instantaneous expansion rate of volume in state space is

$$\sum_i \left(\frac{\partial \dot{q}_i}{\partial q_i} + \frac{\partial \dot{p}_i}{\partial p_i} \right) = \mathrm{Tr}\,\{A\} \; . \tag{2.2}$$

On the other hand, the natural physical requirement that, in the absence of the F_i's, the frictional forces decrease the kinetic energy $\sum_i (p_i^2/2m_i)$ leads to

$$\sum (p_i/m_i) A_{ij} p_j \leq 0 \quad \text{for all values of} \quad p_1, ..., p_n \; , \tag{2.3}$$

which implies by a simple argument that $\mathrm{Tr}\,\{A\} \leq 0$ [with strict inequality unless equality always holds in (2.3)], so volumes contract as asserted.

Volume contraction by the solution flow is a local property in state space, i.e., it can be verified one point at a time. In examples it is frequently accompanied by a related global property, the existence of a bounded set in the state space which each solution curve eventually enters and remains in. To describe what we mean more explicitly, we need to introduce some notation. We let T^t denote the solution flow for the equation of motion, i.e., if x is a point of the state space and $x(t)$ is the solution curve with $x(0) = x$, we write

$$T^t x \quad \text{for} \quad x(t) \; . \tag{2.4}$$

The global property is then the existence of a set V of finite volume such that
  a) if the initial state x is in V, then $T^t x$ is in V for all positive t,
  b) for every initial state x, $T^t x$ is in V for all sufficiently large t.
  If the solution flow is volume-decreasing, as for the above system of particles with friction, then the sets $T^t V$ decrease as $t \to \infty$ to a set Y of zero volume.

Thus, every solution curve asymptotically approaches Y as $t \to \infty$. In other words: Any state x not in Y is transient, i.e., the solution curve $T^t x$ may (perhaps) return to the vicinity of x a finite number of times but eventually departs permanently from x and converges to Y. Since Y occupies a negligible fraction of the state space, this behavior is in sharp contrast to that of Hamiltonian systems where almost every solution curve returns infinitely often to the vicinity of its initial point.

In general, when the above situation holds, the set Y still contains some transient states. The analysis of typical orbits is simplified by concentrating on special kinds of subsets of Y, called *attractors*. There does not yet exist a universally accepted general definition of the term attractor, although some special kinds of attractors have been defined precisely. Roughly speaking, however, an attractor for a flow T^t is a subset X (generally assumed compact) of the state space with the following properties:

- X is invariant: $T^t X = X$.
- X has a shrinking neighborhood U, i.e., there is an open set $U \supset X$ with $T^t U \subset U$ for $t > 0$ and such that $T^t U$ shrinks down to X as $t \to \infty$.
- The flow T^t on X is *recurrent* and *indecomposable*. Recurrent here means that no part of X is transient, i.e., if V is any open set in the ambient space such that $V \cap X$ is nonempty, then there are arbitrarily large values of t and points x of $V \cap X$ such that $T^t x \in V \cap X$, Indecomposable means that X cannot be split into two closed nonoverlapping invariant pieces. One reasonably common circumstance which guarantees that both of these conditions hold is that T^t has an orbit which is dense in X, i.e., there is a point x_0 of X such that every x in X can be approximated arbitrarily closely by points of the form $T^t x_0$ with $t \geq 0$.

If X is an attractor, the *basin of attraction* of X is the set of initial states x such that $T^t x$ approaches X as $t \to \infty$. The basin of attraction is an open set. One pictures a well-behaved solution flow T^t as follows:

- T^t admits a finite number $X_1, \ldots, X_n$ of attractors, each of which is closed, bounded, and of zero volume.
- The set of initial states x which are not in the basin of attraction of any one of the X_i's has volume zero

In other words, there are finitely many attractors whose basins fill up essentially all of the state space.

This mathematical picture provides a general – and so far rather empty – framework for understanding the results of experiments on steadily driven dissipative systems. Ignoring external noise, each experimental run examines a particular solution curve for the differential equation governing the system's motion. It seems highly unlikely that the experiment will be started in such a way that the initial state is, either by accident or by design, in the zero-volume set of points not in the basin of any attractor. Thus, every time the experiment is done, the solution curve being examined will converge, for large t, to one of the attractors. Usually, one wants to ignore the initial period in which the solution curve is settling down into the immediate vicinity of the attractor; in any case, to understand the long-run behavior, it is necessary only to analyze the solution flow in the immediate vicinity of the attractors. It is not to be expected, in general, that the system will have only one attractor, and it may not be experimentally feasible to arrange to start out repeatedly in the same basin of attraction. Thus, different experimental runs on the same system may give rise to different long-term behaviors. For example, in experiments on Couette flow, *Coles* [2.4] found that the fluid in his apparatus, for Reynolds numbers in a certain range, could settle down into states with 20, 22, 24, 26, 28, 30, or 32 axial vortices. Each of these "states" would apparently persist indefinitely. *Fenstermacher* et al. [2.5] found that the statistical behaviors of different states were qualitatively similar but quantitatively distinguishable. These observations can be explained by identifying each of the states with a distinct attractor in the state space of the fluid system.

The next step in adding content to the above abstract framework is to try to understand what attractors look like. There are, first of all, two very simple kinds: attracting stationary solutions and limit cycles. A stationary solution is simply a point $x^{(0)}$ such that

$$T^t x^{(0)} = x^{(0)} \quad \text{for all} \quad t . \tag{2.5}$$

A stationary solution is *attracting* (or a *point sink*) if

$$\lim_{t \to \infty} T^t x = x^{(0)} \quad \text{for all } x \text{ sufficiently near to } x^{(0)} . \tag{2.6}$$

Similarly, a periodic solution $x^{(0)}(t)$ with period τ is *attracting*, or a *limit cycle*, if $T^t x$ converges to the set $X = \{x^{(0)}(t) : 0 \leq t \leq \tau\}$ for any initial x sufficiently near to X.

Linear stability analysis is a powerful technique for determining whether stationary solutions are attracting. If the equations of motion are

$$dx_i/dt = X_i(x) \quad i = 1, 2, \ldots, n \tag{2.7}$$

and if $x^{(0)}$ is a stationary solution, then $x^{(0)}$ is attracting if all eigenvalues of the matrix

$$A_{ij} = \frac{\partial X_i(x^{(0)})}{\partial x_j} \tag{2.8}$$

have strictly negative real parts[1]. We shall refer to such a stationary solution as *linearly attracting*. If $x^{(0)}$ is a linearly attracting stationary solution for (2.7),

---

1 If any eigenvalue of A lies to the right of the imaginary axis, then $x^{(0)}$ cannot be attracting. If some of the eigenvalues of A lie on the imaginary axis, and if the remainder lie in the left half-plane, then $x^{(0)}$ may or may not be attracting. Determining whether $x^{(0)}$ is or is not generally involves a delicate nonlinear analysis.

then it can be shown that any "nearby" differential equation has a linearly attracting solution near $x^{(0)}$. In other words, linearly attracting stationary solutions cannot be made to disappear by small changes in the differential equation. Bifurcation theory analyzes what happens when the differential equation is perturbed strongly enough so that part of the spectrum of the Jacobian matrix A moves into the right half-plane (and, consequently, the stationary solution ceases to be attracting). If a single complex-conjugate pair of nonreal eigenvalues crosses the imaginary axis, then one of two generic possibilities is a *normal Hopf bifurcation* in which the stationary solution, at the point it ceases to be attracting, "sheds" a limit cycle which is initially very small but which may become larger as the equation is further perturbed (see Sect. 3.5).

There is thus a deep connection between loss of attractivity of a stationary solution and appearance of an attracting periodic solution. Furthermore, there is a linearized stability analysis for periodic solutions analogous to that for stationary solutions and a corresponding bifurcation theory, due to *Neumark, Sacker, Ruelle*, and *Takens* (see the discussion in [Ref. 2.6, Sect. 6]). When a periodic solution ceases to be attracting, one of the generic possibilities – analogous to the normal Hopf bifurcation – is that it sheds an attracting two-dimensional torus. These results suggest – misleadingly, as we shall see – a simple, appealing view of the process of transition from an attracting fixed point to chaotic behavior. The picture is that the transition occurs through a sequence of higher and higher dimensional analogs of the normal Hopf bifurcation. After n of these bifurcations the motion is supposed to be quasi-periodic with n independent frequencies; for large n the motion will look chaotic. This is Landau's view of the transition to turbulence. (For a further discussion of this approach, see Sect. 3.3.) At best it is only one of a number of possibilities. We are next going to argue, moreover, on grounds of genericity, that it is a very improbable possibility and that one should expect much more complicated behavior even after the second bifurcation.

To make this argument we need to introduce a geometric way of looking at periodic and multiply periodic motion. We begin with ordinary periodic motion. A periodic solution of the equations of motion may be thought of as follows:

- It is a circle imbedded in the instantaneous state space, the image of a differentiable mapping, $\theta \mapsto x(\theta)$, from the real axis $\mathscr{R}$ into the state space; this mapping is periodic of period 2π and otherwise one to one. (We may then think of θ as a periodic coordinate for this circle.)
- The imbedded circle is invariant for the solution flow T^t.
- With a proper choice of periodic coordinate θ, the solution flow on the invariant circle takes the simple form

$$\theta(t) = \omega t + \theta_0 , \tag{2.9}$$

where ω is constant.

Actually, the existence of a coordinate θ in which the angular velocity is constant is nearly automatic; such a coordinate will always exist for an invariant imbedded circle containing no stationary solution. Note that the above discussion reflects the behavior of the solution flow only actually on the periodic solution and so does not distinguish between limit cycles and more general periodic solutions. In order that the periodic solution be a limit cycle, it is further necessary that small displacements from it be damped out by the solution flow.

Generalizing, we shall define an imbedded n *torus* (in the state space) to mean a subset of the state space which can be realized as the image of a differentiable mapping, $(\theta_1, ..., \theta_n) \mapsto x(\theta_1, ..., \theta_n)$ (from $\mathbb{R}^n$ into the state space) which is periodic with period 2π in each of the θ_i's separately, but otherwise one to one. Such a mapping establishes a set of *periodic coordinates* $(\theta_1, ..., \theta_n)$ for the torus. An n *periodic flow* will mean an imbedded n torus,

- which is mapped into itself by the solution flow T^t
- which admits a set of periodic coordinates in terms of which the solution flow on the torus itself takes the simple form

$$\theta_i(t) = \omega_i t + \theta_i(0) , \tag{2.10}$$

where the ω_i's are constants.

We shall refer to motion on the torus given by (2.10) as *constant velocity flow*.

If n is smaller than the dimensionality of the state space, an imbedded n torus occupies a negligible fraction of the volume of the state space, so an n periodic flow is unlikely to be noticed unless the torus is attracting, i.e., unless small displacements from it are damped out by the solution flow. In summary, for a system with motion governed by a differential equation (without explicit time dependence) to display *observed* n periodic motion, three separate conditions must be met:

1) The state space must contain an imbedded n torus invariant under the solution flow.
2) The motion on the torus in appropriate periodic coordinates must be constant velocity flow, i.e., must be given by (2.10).
3) The n torus must attract nearby solution curves.

We have already noted that, for $n=1$ (ordinary periodic motion), condition 2 follows almost automatically from condition 1. *There is no analogous result for* $n>1$, for $n>1$, constant velocity flow is very special and can be destroyed by small perturbations.

If a differential equation admits an invariant n torus satisfying a slightly strengthened version of condition 3 [analogous to the condition that all the eigenvalues of the matrix A in (2.8) lie strictly to the left of the imaginary axis], then making a sufficiently small perturbation on this equation gives a new equation which still has an attracting invariant torus. If the unperturbed equation also satisfies condition 2, however, it is not difficult to show that arbitrarily small perturbations can be found such that the invariant attracting torus contains a limit cycle and thus no longer satisfies condition 2. (That is, condition 2 is incompatible with the presence, on the torus, of a limit cycle or of any other kind of attractor as it rules out the possibility of having two distinct solution curves approach each other as time goes to infinity.) *Newhouse* et al. [2.7] have shown further that if n is at least three it is possible to find other arbitrarily small perturbations such that the torus contains a nontrivial attractor.

What is known to happen in the Neumark–Sacker–Ruelle–Takens bifurcation (from a limit cycle to an attracting torus) is simply the creation of an invariant 2 torus satisfying conditions 1 and 3. What the motion on the torus looks like is a delicate question which has, so far, eluded complete analysis. The Landau picture assumes that it is constant velocity motion, not just for some values of the Reynolds number but for all values slightly above the bifurcation value. The instability of constant velocity flow under small perturbations makes this assumption seem quite unlikely.

Despite the absence of definitive results, it is possible to construct a picture of how the motion on the attracting torus varies with Reynolds number just above the critical value. This picture, which is based on a combination of courageous extrapolation from what we think happens in some much simpler examples with deep and suggestive – but not directly applicable – mathematical results, is the following: Let R_0 denote the Reynolds number at which a limit cycle bifurcates into an attracting torus. Pick an R' slightly larger than R_0; let $\mathscr{R}$ denote the set of all Reynolds numbers between R_0 and R' for which the attracting torus contains a limit cycle; and let $\tilde{\mathscr{R}}$ denote the set of all Reynolds numbers in this same range for which the motion on the torus is constant velocity flow. The hypothetical picture is then

- $\mathscr{R}$ splits into infinitely many nonintersecting intervals.
- $\mathscr{R}$ is dense in the set of Reynolds numbers between R_0 and R', i.e., given any Reynolds number in this range, there are Reynolds numbers arbitrarily near to it for which the torus contains a limit cycle.
- Therefore, since $\mathscr{R}$ and $\tilde{\mathscr{R}}$ are disjoint (as they are defined by mutually exclusive conditions), $\tilde{\mathscr{R}}$ contains no interval.
- $\tilde{\mathscr{R}}$ nevertheless occupies a nonzero fraction of the total length of the range from R_0 to R', i.e., if a Reynolds number is chosen "at random" in that range, it has nonzero probability of lying in $\tilde{\mathscr{R}}$. Moreover, if R' is chosen very close to R_0, then $\tilde{\mathscr{R}}$ occupies nearly all of the range.

This picture, whether correct in the present situation or not, is a good example of the degree of complexity that it often encountered in studying systems of differential equations even in spaces of relatively low dimensionality.

The obvious question now is: What do typical attractors – aside from attracting stationary solutions, limit cycles, and attracting multiply periodic flows – look like? Here there is something of a surprise: Higher dimensional

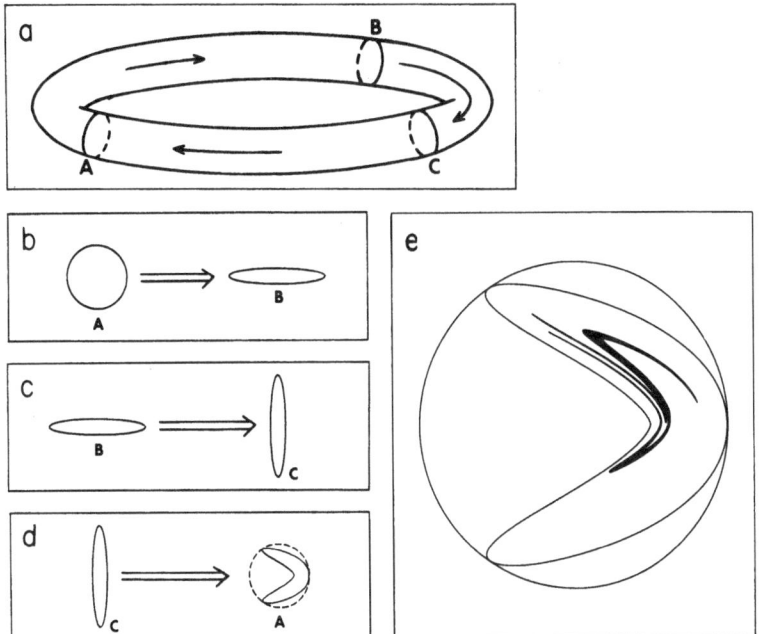

Fig. 2.1a–e. The qualitative behavior of a model strange attractor. The toroidal region in which the flow occurs is shown in (**a**). Parts (**b**), (**c**), and (**d**) show successive intersections of a set of solution curves with planes transverse to the basic flow. When the curves return to Section A after going once around in the direction perpendicular to the paper, they occupy the folded region shown in (**d**). Four iterations of this "return mapping" produce the complex layered structure shown in (**e**)

attractors tend not to be smooth surfaces like spheres and tori but rather to have a complicated, infinitely many-layered structure. For this reason, they have come to be called strange attractors. There is not yet available general structure theory for attractors; it simply is not known what all the possibilities are. We shall describe here a simple example which gives some feeling for why layered structures occur. An attractor like the one to be described was discovered by Hénon and Pomeau in the Lorenz system for fairly large Rayleigh numbers (see the discussion in [2.8]). It is not, however, the same thing as the Lorenz attractor, which occurs for smaller Rayleigh numbers and which has a more intricate structure. (For a discussion of the Lorenz attractor, see Sect. 4.3, as well as [2.8].)

We shall describe the solution flow giving rise to the attractor qualitatively rather than by writing down a specific set of differential equations. The attractor will be contained in a solid torus in three-dimensional Euclidean space, and the flow will move primarily around the torus in the longitudinal direction with a relatively small amount of transverse motion; see Fig. 2.1a. Thus, crudely, the motion looks noisily periodic. We consider a bundle of solution curves in section A, and describe how it is deformed by the transverse components of the flow as we follow it around successively to B, C, and back to A.

Between A and B, vertical separations are contracted and horizontal ones are expanded. We take the contraction to be stronger than the expansion, as shown in Fig. 2.1 b. Between B and C, the resulting flattened bundle is rotated ninety degrees; see Fig. 2.1 c. Between C and A, the vertical flattened bundle is bent into a U shape which fits back into the initial circular cross section of the bundle at A, as shown in Fig. 2.1 d. Thus, one circuit around the loop (i.e., one approximate period) carries an initially circular bundle of solutions into a subset of itself with U shaped cross section. Pairs of solution curves whose initial separation was vertical get closer together; pairs whose initial separation was horizontal get further apart. Making a second circuit around the loop maps the U shaped bundle into a subset of itself whose cross section at A has four arms; iterating gives a sequence of smaller and smaller bundles of solutions whose cross sections have more and more arms, with the width of the individual arms going rapidly to zero. Fig. 2.1 e shows the effect of four iterations.

The decreasing sequence of bundles of solutions shrinks down in the limit to an invariant set with zero volume but with infinitely many sheets. Keep in mind that the figures have represented cross sections of the bundles of solution curves. The actual flow is in each case roughly perpendicular to the plane of the paper, so to each of the infinitely many one-dimensional arms in the limiting cross section there corresponds a two-dimensional sheet in the limiting bundle of solutions.

Although we described the flow by dividing the motion around the torus into a squeezing and stretching phase, a rotation phase, and a bending phase, these phases occur simultaneously in real examples. Actually, the outcome of the construction depends only on the nature of the mapping which takes a point in the cross section A into the next place where the solution curve through that point recrosses A. This mapping, called the *return mapping* or *Poincaré mapping* is a very useful tool in analyzing the geometry of flows since it permits reduction of the number of dimensions by one.

Is the limit of this decreasing sequence of bundles of solution curves an attractor as we defined the term? It is certainly invariant, and it certainly attracts nearby solution curves. The problem is in showing that motion on the limiting set is recurrent. This is a very delicate matter. As the construction has been described, for example, it is not at all apparent that the limiting set does not contain a limit cycle, possibly with a very large period. In fact, it is a simple matter to construct flows matching the above description which *do* have limit cycles; the question is whether it is possible to carry out the construction in such a way as to avoid them. The general belief is that

– It is possible to make the construction so that there is a single solution curve which comes arbitrarily near to each point on the limiting set. This implies that the limiting set meets our definition of attractor.

– Any differential equation whose solution flow meets these criteria can be converted by an arbitrarily small perturbation into one which has a limit cycle. In other words, an attractor constructed in this way, like the attracting doubly periodic flow described above, is *unstable* under small perturbations of the differential equations.

Although these statements are quite plausible, neither of them has, to my knowledge, been proved. A partial result along the lines of the second statement has been proved by *Mañe* [Ref. 2.9, Corollary 2].

Thus far, then, we have not seen an example of an attractor other than a fixed point or limit cycle which is stable with respect to small perturbations of the differential equation from which it arises. Such attractors do exist. The Lorenz attractor, mentioned above, is stable in the sense that a sufficiently small perturbation of the differential equation gives a new differential equation which still has an attractor of the same general character. It is, however, unstable in the sense that the topological fine structure of the attractor is changed by arbitrarily small perturbations of the equations. Axiom A attractors, to be discussed in the next section, are stable in a stronger sense; all topological details are unchanged by small perturbations.

2.3 Statistical Theory

We turn now to the application of probabilistic considerations to the study of turbulence. Recall the point of view we have adopted: At least for a first approximation, we want to study turbulence as a phenomenon governed by a system of differential equations without external noise. We are thus *not* going to apply the theory of stochastic differential equations.

Two distinct statistical elements in a theory of deterministic phenomena can be distinguished.

1) *Random initial states*. Repeated runs of an experiment – e.g., to measure the instantaneous drag on the fixed cylinder in the Couette system as a function of time – do not yield the same detailed results. It is expected, however, that statistical regularities will begin to appear if the experiment is repeated a large enough number of times. These facts can be accounted for by assuming that the experimental apparatus and start-up procedure determine a probability distribution on the space of possible initial instantaneous states. Making an experimental run then means choosing a sample initial point with respect to this distribution and studying its solution curve.

2) *Time averages*. In many instances, quantities of physical interest are averages of dynamical variables over long periods of time and can be idealized as averages over all time.

There does not seem to be much of interest to be said about the first of these points. It is not common practice to make a really large number of experimental runs at fixed parameter values; what is usually done is to make one or a few long experimental runs. Moreover, there is no reason to think that the initial distribution has a universal form; it presumably depends on details of the design of the experimental apparatus, etc. There is, however, one feature of the initial distribution which is generally agreed to be universal: Any specific set of initial states of zero state space volume (i.e., zero Lebesgue measure) will also have probability zero with respect to the initial distribution. In other words, a phenomenon which occurs only for a set of initial states of zero volume will *never* be seen.

By contrast, the second point, the importance of the study of time averages, is the starting point for a deep and far-reaching analysis. Consider, for example, the experiment mentioned above, measurement of the instantaneous drag on the fixed cylinder in the Couette system while the other cylinder is rotated at a sufficient speed to generate and sustain a turbulent flow. The instantaneous drag is a function, say $D(x)$, of the instantaneous state x. For an experimental run with initial state x_0, the drag as a function of time is then $D(T^t x_0)$. For a typical x_0, this quantity will fluctuate in an irregular fashion with time, and changing x_0 even slightly will generally give quite a different function of time. Nevertheless, it is expected that, if we define

$$\bar{D}(x_0) = \lim_{\tau \to \infty} \frac{1}{\tau} \int_0^\tau dt D(T^t x_0) , \qquad (2.11)$$

then

1) $\bar{D}(x_0)$ exists.
2) $\bar{D}(x_0)$ is essentially independent of x_0 on each basin of attraction. Here, "essentially" means "except on a set of initial states with zero state-space volume". Independence of initial state is only expected to hold within a given basin of attraction; the time-average drag will normally vary from one basin to another.

Quantities like $\bar{D}$ are evidently of practical importance. The question we want to address here is: What mathematical properties of the underlying solution flow reflect the existence and essential constancy of time averages? In exploring this question, we shall adopt the point of view that there is nothing special about the instantaneous drag function and that all reasonable functions on the instantaneous state space should admit time averages.

When we say that some phenomenon admits a statistical description, we generally have in mind that, although individual events cannot be predicted, appropriate frequency ratios ("probabilities") have well-defined limits as the number of events examined becomes very large. A statistical phenomenon is not one without any predictable characteristics at all, but rather one which is predictable only in the long run. Accordingly, we shall say that a solution curve $T^t x_0$ is *statistically regular* if, for each reasonable subset $\mathscr{S}$ of the instantaneous state space, the solution curve spends a well-defined fraction of its time in $\mathscr{S}$, i.e., if the limit as τ goes to infinity of the fraction of time up to τ that $T^t x_0$ is in $\mathscr{S}$ exists. (This is not quite a complete definition since we have not explained what is meant by a "reasonable" subset of the instantaneous state space. There is in fact a completely irrelevant mathematical pathology which occurs here if we require that the limit exist for excessively general subsets.) We shall denote the limiting fraction of time spent in $\mathscr{S}$ by $\mu(\mathscr{S})$; from some points of view, $\mu(\mathscr{S})$ can be interpreted as the probability of finding the instantaneous state in $\mathscr{S}$. The set function μ defines a probability measure on the state space; we shall refer to it as the *asymptotic distribution* of the solution curve $T^t x_0$. It is a relatively simple matter to show that if $T^t x_0$ is statistically regular with asymptotic distribution μ, then limiting time averages

$$\bar{f}_{x_0} = \lim_{\tau \to \infty} \frac{1}{\tau} \int_0^\tau dt f(T^t x_0) \tag{2.12}$$

exist for all continuous functions f on the instantaneous state space, and $\bar{f}_{x_0}$ is given by the mean value of f relative to μ.

With this definition, we can ask an important question concisely: For what sorts of differential equations are all but a negligible set of solution curves statistically regular? One answer, provided by the Birkhoff pointwise ergodic theorem[2], is that Hamiltonian equations of motion have this property (provided that surface area – more precisely, the microcanonical measure – is finite on each energy surface). Perhaps surprisingly, the same is not true for general autonomous[3] differential equations; for a counterexample, see [2.11]. Indeed, although the construction of counterexamples is not obvious, and although examples arising in practice, when analyzed in detail, always turn out to have this property, there seems to be no general, directly verifiable, criterion for it to hold.

The situation of greatest physical interest is not simply one in which almost all solution curves are statistically regular; one also wants the asymptotic distribution to be essentially independent of initial state. A prototype for this behavior is provided by ergodic Hamiltonian systems. Consider a Hamiltonian solution flow restricted to an energy surface of finite area. We say that this flow is *ergodic* if there is no way of decomposing the energy surface into two pieces each invariant under the solution flow and each of nonzero area. In this case, it is an easy consequence of the Birkhoff pointwise ergodic theorem that almost every solution is statistically regular with asymptotic distribution equal to the normalized surface area (microcanonical ensemble) on the energy surface under consideration. (By "almost every" solution curve we mean that all solution curves which do not have the indicated property, taken together, occupy zero area.) Thus, every solution curve outside of a statistically negligible set spreads itself out, asymptotically, uniformly over the energy surface. The allowance for exclusion of a set of solution curves of total area zero is not simply a defect in the proof or provision for pathological situations; in most cases where it can be

---

2 See, for example, the discussion in [Ref. 2.10, Theorem 6.21].
3 That is, without explicit time dependence.

proved that a Hamiltonian system is ergodic – e.g., a particle moving in a periodic array of scatterers – it can also be shown that there are a great many solution curves which are not statistically regular, as well as a great many which are statistically regular but which have nonstandard asymptotic distributions.

For dissipative systems there is no simple criterion known for the desired kind of statistical behavior. We shall proceed by *defining* an attractor to be ergodic if it has good statistical behavior, and we shall then have to investigate what can be said about it when an attractor is ergodic. Consider an attractor X with basin $\mathscr{B}$. We shall say that X is *ergodic* with *equilibrium distribution* μ if all initial states x in $\mathscr{B}$, with the exception of a set of state-space volume zero, are statistically regular with asymptotic distribution μ. In other words, not only do time averages exist for almost all initial states, but they are essentially independent of the initial state. The equilibrium distribution μ may easily be seen to be concentrated on the attractor X itself, which normally has volume zero. This is in sharp contrast to the Hamiltonian case where the equilibrium distribution is spread uniformly over the whole energy surface. In particular, the accepted notion of a statistically negligible set of initial states – a set of zero volume in the state space – no longer coincides with the notion of a set of equilibrium probability zero.

The main piece of mathematical technology available for proving that a Hamiltonian system, or an attractor for a dissipative system, is ergodic is a sharp analysis of the notion of *sensitive dependence on initial condition*. Intuitively this is very natural – if the state at time τ (τ large) depends in a very sensitive way on the state at time zero, then the sequence of states at times 0, τ, 2τ, 3τ,... becomes effectively an independent sequence, and the law of large numbers then at least suggests that time averages should exist.

The concept we want to develop here is that the solution flow for the differential equation amplifies small perturbations of the initial state[4]. To formulate this notion cleanly, we need to look at how the flow acts on "infinitesimal" perturbations of the initial state, i.e., at the derivative of the flow with respect to the initial state. Concretely: if instantaneous states are n vectors, then the state at time τ is given by an n tuple of functions $(T^t)_1(x_1, ..., x_n), ..., (T^t)_n(x_1, ..., x_n)$. We denote by $DT^t(x_1, ..., x_n)$ the matrix of partial derivatives

$$(DT^t)_{ij}(x_1, ..., x_n) = \frac{\partial T_i^t}{\partial x_j}(x_1, ..., x_n) \ . \tag{2.13}$$

Heuristically, if the initial state x is displaced infinitesimally by δx, then the state at time t is displaced from $T^t x$ by the still infinitesimal quantity $DT^t(x) \cdot \delta x$. (Although the flow is nonlinear, infinitesimal displacements evolve linearly.)

---

4 The term *instability* is frequently used as a synonym for sensitive dependence on initial condition as well as to refer to what might be called sensitive dependence on the differential equation. These are very different concepts, and in this chapter we have used the term only with its second meaning.

A situation which lends itself to mathematical analysis is the one in which infinitesimal displacements grow exponentially with time. Exponential growth comes in several varieties. A relatively restrictive variety is exemplified by Anosov Hamiltonian systems generalized by *Smale* [2.12] to the notion of Axiom A attractor. Very roughly speaking, an attractor satisfies Axiom A provided that, for each point x of the attractor, the vector space of infinitesimal displacements δx at x may be decomposed as a direct sum of

- a one-dimensional space tangent to the solution flow
- a space of displacements which damp out exponentially as t goes to $+\infty$
- a space of displacements which are purely growing exponentials in the sense that they damp out exponentially as t goes to $-\infty$.

Although the existence of this decomposition is the central element in Axiom A, it is hedged about by a number of technical conditions – the rate of exponential damping forward and backward must be uniform over the attractor, the above decomposition must vary continuously as one moves from one point of the attractor to another, etc. – as well as the not-so-technical condition that, given any point of the attractor, there are (nonattracting) periodic solutions passing arbitrarily near to that point. A reasonably simple Axiom A attractor in four dimensions, the solenoid, was described in [2.1]. There is a rather less simple example, called the *Plykin* [2.13] attractor, in three dimensions.

I think it is fair to say that we do not yet have adequate basis to form an educated guess as to whether Axiom A attractors are likely to occur in hydrodynamics. There is some reason to think that Axiom A may be too restrictive a condition. The Lorenz attractor, for example, which arises in applications and which appears to display quite a reasonable version of sensitive dependence on initial conditions, definitely does *not* satisfy Axiom A. The same is true for the example described in the previous section. These are, however, very low-dimensional examples, and Axiom A attractors may turn out to be more common in higher dimensions.

Be that as it may, it has been shown by *Bowen* and *Ruelle* [2.14] that Axiom A attractors are ergodic in the sense defined above. The equilibrium distribution μ is an important object associated with the attractor, and it would be of great practical importance to be able to write it down explicitly; we would then be able to calculate time averages without solving the equations of motion. Unfortunately, this does not seem to be possible. In the ergodic Hamiltonian case, the equilibrium distribution (microcanonical ensemble) can in principle be written down directly in terms of the Hamiltonian. No such formula is known to exist for Axiom A attractors; available procedures for constructing the equilibrium distribution require a detailed knowledge of the *solutions* to the equations of motion as opposed to the equations themselves. Some elegant characterizations of the equilibrium distribution have been obtained, including a variational principle analogous to the free energy minimization principle which characterizes equilibrium states in equilibrium statistical mechanics, but these characterizations single out the equilibrium distribution from among all the probability distributions on the attractor invariant under the solution flow. In the absence of an effective procedure for writing down all invariant probability distributions, these characterizations do not appear to be of much practical importance.

In view of the uncertainty about whether Axiom A is general enough to apply to most situations of interest, it is worth looking for more flexible ways to exploit exponential growth of infinitesimal displacements. An approach to doing this has been developed by *Pesin* [2.15–17], *Katok* [2.18], and *Ruelle* [2.19] using the notion of Lyapunov characteristic exponent. To explain what these quantities are, we consider first an $n \times n$ matrix B with n eigenvalues which are not only distinct but which have distinct real parts: $\lambda_1 + i\mu_1, ..., \lambda_n + i\mu_n$; $\lambda_1 < \lambda_2 < ... < \lambda_n$. Let $e_1, ..., e_n$ be the corresponding eigenvectors and, for each j between 1 and n, let E_j be the j dimensional subspace spanned by $e_1, ..., e_j$. Then E_j may be characterized as

$$\left\{ \xi : \limsup_{t \to \infty} \frac{1}{t} \log(\|e^{tB}\xi\|) \leq \lambda_j \right\}. \tag{2.14}$$

Moreover, if ξ is in E_j but not in E_{j-1}, then

$$\limsup_{t \to \infty} \frac{1}{t} \log(\|e^{tB}\xi\|) = \lambda_j. \tag{2.15}$$

(Indeed, lim sup can be replaced here by the ordinary limit.) We now try to imitate as much of this as we can with $\exp(tB)$ replaced by $DT^t(x)$, where x is some point in the state space. The motivation is that, whereas $\exp(tB)$ is obtained by solving

$$\frac{d}{dt} e^{tB} = Be^{tB}, \tag{2.16}$$

$DT^t(x)$ can be obtained by solving the first variational equation which has the form

$$\frac{d}{dt} [DT^t(x)] = B(x, t)DT^t(x). \tag{2.17}$$

Since $B(x, t)$ depends (in general) on t, the behavior of $DT^t(x)$ for large t cannot readily be analyzed in terms of eigenvalues and eigenvectors. Nevertheless, we shall say that the point x has *characteristic exponents* $\lambda_1 < \lambda_2 < \lambda_3 < ... < \lambda_n$ provided that there exist subspaces $E_1 \subset E_2 \subset ... \subset E_n$ of the space $\mathbb{R}^n$ of all n vectors, with each E_j of dimension j, such that

$$E_j = \left\{ \xi : \limsup_{t \to \infty} \frac{1}{t} \log[\|DT^t(x)\xi\|] \leq \lambda_j \right\} \tag{2.18}$$

and, if ξ is in E_j but not in E_{j-1},

$$\limsup_{t \to \infty} \frac{1}{t} \log \left[\| DT^t(x)\xi \| \right] = \lambda_j . \tag{2.19}$$

(A straightforward extension of this definition, which we do not give here, covers the case where the characteristic exponents are not all distinct.) Although the definition leaves open the logical possibility that characteristic exponents do not exist anywhere, it can in fact be shown that they exist for "most" initial states x. If characteristic exponents exist at a point x, one of them is normally equal to zero (corresponding to a ξ which is parallel to the flow at that point). A plausible weakening of Axiom A is the condition that, for almost every x, no more than one characteristic exponent is equal to zero. This preserves the crucial conditions of exponential growth and decay of small displacements, but drops all uniformity and continuity conditions. The work of *Pesin, Katok*, and *Ruelle* cited above is a start on the problem of extending the theory of Axiom A attractors to this situation.

In summary: We have seen how probabilistic considerations play a natural role in the description of long-term behavior of solutions of *deterministic* differential equations. In the case of Hamiltonian systems, these probabilistic considerations are familiar from classical statistical mechanics, and the theory extends, with appropriate modifications, to at least some kinds of attractors. One crucial difference between the Hamiltonian and dissipative situations is that the equilibrium distribution is not at all easy to write down in the latter case. There is a class of particularly nice attractors – those which satisfy Smale's Axiom A – for which the statistical theory has been worked out completely; extension to more general attractors, via the notion of Lyapunov characteristic exponents, is currently a subject of active mathematical research.

References

2.1 D. Ruelle, F. Takens: On the nature of turbulence. Commun. Math. Phys. **20**, 167–192 (1971); **23**, 343–344 (1971)
2.2 J. von Neuman, E. P. Wigner: Über das Verhalten von Eigenwerten bei adiabatischen Prozessen. Phys. Z. **30**, 467–470 (1929)
2.3 D. Ruelle: Bifurcation in the presence of a symmetry group. Archiv. Ration. Mech. Anal. **51**, 136–152 (1973)
2.4 D. Coles: Transition in circular Couette flow. J. Fluid Mech. **75**, 1–15 (1965)
2.5 P. R. Fenstermacher, H. L. Swinney, J. P. Gollub. Dynamical instabilities and the transition to chaotic Taylor vortex flow. J. Fluid Mech. **94**, 103–128 (1979)
2.6 J. E. Marsden, M. McCracken: *The Hopf Bifurcation and its Applications*. Applied Math. Sci., Vol. **19** (Springer, Berlin, Heidelberg, New York 1976)
2.7 S. Newhouse, D. Ruelle, F. Takens: Occurrence of strange Axiom A attractors near quasi-periodic flows on T^m, $m \geq 3$. Commun. Math. Phys. **64**, 35–40 (1978)
2.8 O. E. Lanford III: "An Introduction to the Lorenz System", *1976 Duke Turbulence Conference*, Duke Univ. Math. Series **3** (1977)

2.9   R. Mañe: Contributions to the stability conjecture. Topology **17**, 383–396 (1978)

2.10 L. Breiman: *Probability* (Addison-Wesley, Reading 1968)

2.11 D. Ruelle: On the measures which describe turbulence, IHES preprint (1978); D. Ruelle: What are the measures that describe turbulence? Prog. Theor. Phys., Supplement No. **64**, 339–345 (1978)

2.12 S. Smale: Differentiable dynamical systems. Bull. Am. Math. Soc. **73**, 748–817 (1967)

2.13 R. V. Plykin: Sources and currents of *A*-diffeomorphisms of surfaces. Mat. Sb. **94**, No. 2 (6), 243–264 (1974)

2.14 R. Bowen, D. Ruelle: The ergodic theory of Axiom A flows. Invent. Math. **29**, 181–202 (1975)

2.15 Ya. B. Pesin: Lyapunov characteristic exponents and ergodic properties of smooth dynamical systems with an invariant measure. Dokl. Akad. Nauk SSSR **226**, No. 4, 774–777 (1976). [English translation: Sov. Mat. Dokl. **17**, No. 1, 196–199 (1976)]

2.16 Ya. B. Pesin: Invariant manifold families which correspond to the non-vanishing characteristic exponents. Izv. Akad. Nauk SSSR, Ser. Mat. **40**, No. 6, 1332–1379 (1976). [English translation: Math. USSR Izv. **10**, No. 6, 1261–1305 (1976)]

2.17 Ya. B. Pesin: Lyapunov characteristic exponents and smooth ergodic theory. Uspekhi Mat. Nauk **32**, No. 4, 55–112 (1977). [English translation: Russ. Math. Surv. **32**, No. 4, 55–114 (1977)]

2.18 S. Katok: The estimation from above for the topological entropy of a diffeomorphism; in *Global Theory of Dynamical Systems*, ed. by Z. Netecki, C. Robinson, Lecture Notes in Mathematics, Vol. 819 (Springer, Berlin, Heidelberg, New York 1980) pp. 259–264

2.19 D. Ruelle: Ergodic theory of differentiable dynamical systems. Publ. Math. IHES **50**, 27–58 (1979)

3. Hydrodynamic Stability and Bifurcation

D. D. Joseph

With 14 Figures

The goal of hydrodynamics is to describe and predict the motions of fluids under applied forces. For incompressible Navier-Stokes fluids, in many circumstances, these forces scale with the Reynolds number. When the Reynolds number is small, hydrodynamics is not so difficult because there is a unique correspondence between the given boundary and internal forcing data and the predicted motions. But when the Reynolds number is larger, hydrodynamics is complicated; there are many solutions; nonuniqueness is the rule; sets of solutions must be described, and stable and observable subsets must be separated from the others.

A mathematical basis for the study of these hard problems is the theory of stability and the theory of bifurcation. In this review I will discuss some basic features of these two theories, their relation to one another, and their hydrodynamic applications[1].

3.1 The Navier-Stokes Equations and the Prescribed Data

The starting point for the study of laminar and turbulent motion of a viscous, incompressible, Newtonian fluid is the initial-value problem for the Navier-Stokes equations. We assume that the fluid occupies a confined region $\mathscr{V}$ of space and that the fluid velocity $U(x, t)$ is prescribed in $\mathscr{V}$ at instant $t=0$. The mass of the incompressible fluid with density ϱ will be conserved in each small part of $\mathscr{V}$ if div $U = 0$, and its momentum will be balanced if

$$\frac{\partial U}{\partial t} + U \cdot \nabla U = -\nabla P + \nu \nabla^2 U + G(x, t) . \tag{3.1}$$

Here $\nu = \mu/\varrho$ is the kinematic viscosity and P is a scalar field, called the "reaction pressure". P may be regarded as one of four unknown fields (U, P) governed by the four equations (3.1) and div $U = 0$. $G(x, t)$ is a prescribed body force field. We shall also assume that the motion of the boundary $\partial \mathscr{V}$ of $\mathscr{V}$ is

---

1 Readers who wish to pursue the study of stability and bifurcation will find a relatively complete list of books, monographs, and review papers in the references. In the course of this review I will make frequent reference to materials in my two-volume treatment *Stability of Fluid Motions* [3.1, 2].

prescribed and that the fluid adheres to this boundary,

$$U(x, t) = U_B(x, t) \quad \text{for} \quad x \in \partial \mathscr{V}. \tag{3.2}$$

When other fields, like temperature or magnetic fields, enter into the dynamics, it is necessary to modify (3.1) and to supplement (3.1) and (3.2) with additional equations.

We may write the Navier-Stokes equations in dimensionless form, with velocity scale $\tilde{U}$ and length scale l. The dimensionless and dimensional equations are identical except that the kinematic viscosity v is replaced with the reciprocal of the Reynolds number $R = \tilde{U}l/v$. Then all conclusions about the Navier-Stokes equations hold for all flows with different velocities $\tilde{U}$, in different domains $\mathscr{V}$ (proportional to l^3) and for fluids with different v provided only that they share a common R. It will be convenient to think of v^{-1} in (3.1) as the Reynolds number.

I call the boundary velocities $U = U_B$ and the body forces G which drive the motion, the *data*. We can think of $\mathscr{V}$ as a black box of fluid which makes contact with the outside world through the data. To a limited extent the fluid does what the data do, but in most cases the fluid has a mind of its own; it does what it wants. The data could be steady and flow inside $\mathscr{V}$ time periodic or even turbulent.

The flow through a round pipe is a good example of this anthropomorphic character of fluid motions. Suppose that $\mathscr{V}$ is an infinitely long pipe with axis x and radius l and that the data are given by

$$U_B = 0 \quad \text{on} \quad r = l, \quad G(x, t) = -e_x \hat{P}, \tag{3.3}$$

where $\hat{P}$ = constant is the pressure drop per unit length. Then (3.1) has a simple solution with one nonzero component of velocity

$$U(r) = \tilde{U}(1 - r^2/l^2), \tag{3.4}$$

where $\tilde{U} = \hat{P}l^2/4v$ is the maximum velocity. The motion (3.4) is called laminar for obvious reasons. When $R = \tilde{U}l/v$ is small, this flow is always realized in experiments. But when R is larger, the flow is time dependent and erratic even though the data are steady and do not vary from place to place. The erratic flow is called turbulent and for a fixed value of $\hat{P}$ it has a smaller mass flux than the laminar flow (3.4). (It would be good if it were possible to run oil through pipes in laminar flow at large values of R; we could then reduce the cost of moving oil from Alaska to Minnesota. But such is not nature's design.)

The first serious scientific investigation of transition to turbulence in pipes was that of *Reynolds* [3.3] in 1883. His experiments revealed that the breakdown of laminar flow to "sinuous" motion occurred at a critical R which is the same for different diameter pipes, different velocities, and fluids with different viscosities. In the same classic paper *Reynolds* posed the question of why and how transition from laminar flow to turbulent flow takes place and he proposed an answer, which he attributed to Stokes, in terms of the breakdown of stability of the laminar flow. "The general cause of the change from steady to eddying motion was in 1843 pointed out by Professor Stokes, as being that under certain circumstances the steady motion becomes unstable, so that an indefinitely small disturbance may lead to a change to sinuous motion". *Reynolds* first postulated that the circumstances referred to are that R, gradually increased, reaches a critical value at which the laminar flow becomes unstable to infinitesimal disturbances. But investigation based on infinitesimal disturbances did not give critical values like those observed in experiments. In *Reynolds'* [3.3, 4] address "On two manners of the motion of water" he noted pessimistically that "it has long been a matter of very general regret to those who are interested in Natural Philosophy, that in spite of the most strenuous efforts of the ablest mathematicians, the theory of fluid motion fits very ill with the actual behaviour of fluids, and this for unexplained reasons. The theory itself appears to be tolerably complete, and affords the means of calculating the results to be expected in almost every case of fluid motion, but while in many cases the theoretical results agree with those actually obtained, in other cases they are altogether different".

Pipe flow is evidently stable to infinitesimal disturbances at all values of the Reynolds number [Ref. 3.1, p. 120]. The instability observed by *Reynolds* seems to depend on nonlinearity in an important way. *Reynolds* [3.5] himself noted that the abruptness of the transition from laminar to turbulent flow "at once suggested the idea that the condition might be one of instability for disturbances of a certain magnitude, and stability for smaller disturbances".

The hydrodynamics of flow through round pipes is not very different from other types of shearing flow. For most of these, linearized theories of stability give a critical R but it is much larger than experimentally observed values. In still other types of hydrodynamical situations linearized theories give results which are close to actual behavior. The superficially anomalous relation of linearized theories to experiments has been clarified to a considerable extent by concepts which arise from bifurcation theory.

Bifurcation theory, in its broadest sense, attempts to classify and characterize the properties of all of the solutions which the initial-value problem can support when the transients have died away, the initial values have been forgotten, and the interior motions are driven by the data. In its more usual, less ambitious form, bifurcation theory classifies and characterizes all solutions which can arise from the instability of a given solution. I do not want to give a too cryptic description of the application of bifurcation theory to hydrodynamical problems at this point. For now it will suffice to remark that the theory allows one to make useful statements about the behavior of classes of problems even when explicit computations are not possible. By treating the problem from a more general point of view we learn how to distinguish the forest from the trees.

3.2 Uniqueness and Stability of Solutions when the Reynolds Number is Small

We start our discussion of stability and bifurcation with the comforting observation that when the Reynolds number is small, all solutions of the Navier-Stokes equations tend to a single one, determined by the data after the initial conditions have died away. So if the data are steady, the solution is steady; if the data are time periodic, so is the solution. It is of interest, and not too hard to show this. A perturbed solution (U^*, P^*) satisfies (3.1) and (3.2) but differs from (U, P) initially. The difference $(U - U^*, P - P^*) = (u, p)$ satisfies

$$\frac{\partial u}{\partial t} + (U \cdot \nabla)u + (u \cdot \nabla)U + (u \cdot \nabla)u = -\nabla p + \nu \nabla^2 u , \tag{3.5}$$

where

$$u \in H = \{u : \text{div } u = 0, u|_{\partial \mathcal{V}} = 0, \langle |\nabla u|^2 \rangle < \infty \} , \tag{3.6}$$

and

$$\langle |\nabla u|^2 \rangle = \int_{\mathcal{V}} |\nabla u|^2 d\mathcal{V} \equiv \int_{\mathcal{V}} \frac{\partial u_i}{\partial x_j} \frac{\partial u_i}{\partial x_j} d\mathcal{V} . \tag{3.7}$$

Of course, $u(x, 0) = u_0$ is prescribed.

To get the stability and uniqueness result, we work with an energy identity derived from (3.5–7). In preparation for the derivation of this identity we note that

$$u \cdot (U \cdot \nabla u + u \cdot \nabla u + \nabla p) = \nabla \cdot (U|u|^2/2 + u|u|^2/2 + up) . \tag{3.8}$$

Then multiplying (3.5–7) by u and integrating we get

$$\frac{1}{2} \frac{d \langle |u|^2 \rangle}{dt} = -\langle u \cdot \nabla U \cdot u \rangle - \nu \langle |\nabla u|^2 \rangle \leq \langle |\nabla u|^2 \rangle (\nu_E - \nu) , \tag{3.9}$$

where the existence of

$$\nu_E = \max_H \left(\frac{-\langle u \cdot \nabla U \cdot u \rangle}{\langle |\nabla u|^2 \rangle} \right) > 0 \tag{3.10}$$

is guaranteed by the calculus of variations (provided that $\mathcal{V}$ can be confined between two parallel planes). We get monotonic decay of $\langle |u|^2 \rangle$ when the dissipation $\nu \langle |\nabla u|^2 \rangle$ is larger than the "production" $|\langle u \cdot \nabla U \cdot u \rangle|$ of energy. The Navier-Stokes equations are special in that the only nonlinear term $(u \cdot \nabla)u$ is inertial and it vanishes after integration. This feature, which is not typical of nonlinear problems, allows one to get (3.9) in a form independent of the amplitude u; we get (3.9) for v if u is replaced by av, even when $a \to 0$. So (3.9) applies to (3.5–7) equally when $u \cdot \nabla u$ is finite or, as in the linear theory, when $(u \cdot \nabla)u$ is set to zero. Since the ratio of quadratic forms in u mentioned in (3.10) is standard in the calculus of variations, the existence of v_E when u is a member of the set H, which contains at least all solutions of (3.5–7), is guaranteed, and $\langle |u|^2 \rangle$ certainly decays when $v > v_E$, that is, for small Reynolds numbers. In fact, the same theory guarantees the existence of $\Lambda > 0$ such that

$$\langle |\nabla u|^2 \rangle > \frac{\Lambda}{2} \langle |u|^2 \rangle . \tag{3.11}$$

Combining (3.9) with (3.11) we find, after integration, that

$$\langle |u(t)|^2 \rangle \leqq \langle |u_0|^2 \rangle \exp \left[\Lambda \int_0^t (v - v_E) dt \right], \tag{3.12}$$

where v_E depends on v and t because $U(x, t, v)$ does. If $u(t)$ is steady, then $u(t) = u_0$ and (3.12) holds if and only if $\langle |u_0|^2 \rangle = \langle |u|^2 \rangle = 0$. So steady flows are unique when $v > v_E$ [3.6]. Similarly, we can show that almost periodic motions are unique [Ref. 3.1, Chap. 1]. More generally, (3.12) shows that $\langle |u|^2 \rangle = \langle |U - U^*|^2 \rangle \to 0$ as $t \to \infty$ so that $U^* \to U$ in the mean when v is large or $R = 1/v$ is small. So for small Reynolds numbers there is just one solution U of (3.1) and (3.2) determined uniquely by the data U_B and $G(x, t)$ independent of the initial values $U(x, 0)$.

There is another interesting way to state the theorem of global stability just proved: *Flows which perturb rigid motions are globally stable*. We may define a rigid motion as a motion $U(x, t)$ for which the stretching (rate of strain) tensor $D = (\nabla U + \nabla U^T)/2 = 0$ $(2D_{ij} = \partial U_i / \partial x_j + \partial U_j / \partial x_i)$ in $\mathscr{V}$. The velocity gradient ∇U may be decomposed into symmetric and antisymmetric parts, $\nabla U = D + \Omega$, where $\Omega = (\nabla U - \nabla U^T)/2$ and $2\Omega_{ij} = \partial U_i / \partial x_j - \partial U_j / \partial x_i$, is the vorticity tensor with components $\Omega_{ij} = -\Omega_{ji}$. Since $u \cdot \Omega \cdot u = u_i \Omega_{ij} u_j \equiv 0$, we find that $\langle u \cdot \nabla U \cdot u \rangle = \langle u \cdot D \cdot u \rangle$ always, and $\langle u \cdot D \cdot u \rangle = 0$ for rigid motions. The energy of any disturbance of motions with $D = 0$ decays monotonically since, by (3.9),

$$\frac{1}{2} \frac{d \langle |u|^2 \rangle}{dt} = -v \langle |\nabla u|^2 \rangle . \tag{3.13}$$

If D is small, the second term on the right of (3.9) will outweigh the first, leading again to the decay of the energy of arbitrary disturbances.

Further discussion of energy methods in the theory of hydrodynamic stability can be found in [3.1, 2]. The results proved here are due to *Orr* [3.7], *Thomas* [3.8], *Hopf* [3.9], and *Serrin* [3.6].

3.3 Instability and Transition into Turbulence

We have seen that when the Reynolds number is small, the flow which evolves after a time is uniquely determined by the data independent of initial conditions. This unique flow has the maximum symmetry consistent with the data. At higher values of the Reynolds number we lose uniqueness. Other flows with different, usually more complicated patterns of symmetry are then observed after transients have decayed away. For example, motions which are spatially uniform in certain directions can be replaced by motions which are spatially periodic, quasi-periodic, or aperiodic. And motions which are steady can be replaced by motions which are time periodic, quasi-periodic, or aperiodic. So at larger values of the Reynolds numbers we may observe flows which do not follow the symmetry of the data (boundary conditions).

This breakdown in the symmetry of solutions is especially dramatic in the flows we call turbulent. In such flows the connection between the data and the flow is very elusive; even with steady data we can observe flows whose behavior after a long time is aperiodic with no (as yet identifiable) regularity. The spatial structure of turbulent flow is also very complicated with many little eddies and fluctuations which are sometimes called random because nobody knows how to characterize them in a precise way. Turbulent flows are extremely sensitive to initial conditions. Two flows with the same data but slightly different initial conditions evolve into two very different flows (see the discussions in Chaps. 2 and 4). There is surely a sense in which the data make themselves known in turbulent flow, but the connections between the data and the flow are subtle and elusive.

The following casual observations about transition to turbulence may be of value. When the data are steady and the Reynolds number is increased beyond the point at which turbulence first appears, the structure of the apparently chaotic flow does not seem to exhibit further qualitative changes. The most noticeable changes are that the parts of the flow which are turbulent gradually consume the whole flow. And, of course, the intensity of the fluctuating motions increases. It may be useful to think of turbulence as the least symmetric state of motion consistent with the given data. This way of thinking may or may not have intrinsic merit, but I use it to say that when we get turbulence you should read another book because I have arrived at the end of this story.

It is the very good and important observation of *Ruelle* and *Takens* [3.10] that there is actually an end to this story. They have given good theoretical support to the idea that we arrive at turbulence after a few bifurcations. This idea seems to be verified in the experiments known to me but it contradicts the appealing ideas about transition to turbulence which are associated with the names of *Landau* [3.11; see also 3.12] and *Hopf* [3.13, 14].

Landau and *Hopf* base their conjectures on ideas which derive from successive loss of stability of solutions to small disturbances. (The Landau idea is also discussed in Chap. 2 and Sect. 4.3.) At small Reynolds number $R = 1/v$ there is just one steady flow, $U = U(x, R)$, and the scalar field $p = p[U]$

belonging to this U^2. The evolution of a disturbance u of U is governed by (3.5–7)

$$\frac{\partial u}{\partial t} = F(u, R) - \nabla p, \quad u \in H, \tag{3.14}$$

where

$$F(u, R) = F_u(0, R|u) - u \cdot \nabla u, \tag{3.15}$$

and

$$F_u(0, R|u) = \frac{1}{R} \nabla^2 u - U(R) \cdot \nabla u - u \cdot \nabla U(R) \tag{3.16}$$

is a linear operator, the derivative of the operator $F(u, R)$, evaluated on the function $u = 0$. Since $F(0, R) = 0$, $u = 0$ and $p[0] = 0$ solve (3.14–16) for all R. When u is sufficiently small, we may seek conditions for the stability of U from the linearized problem,

$$\frac{\partial u}{\partial t} = F_u(0, R|u) - \nabla p, \quad u \in H, \tag{3.17}$$

which arises from (3.14–16) when $u \cdot \nabla u$ is set to zero. $u = e^{\sigma t} \zeta$, and $p[u] = \exp(\sigma t) p[\zeta]$, $\zeta \in H$ solve (3.17) if ζ and $\sigma(R) = \xi(R) + i\eta(R)$ solve the spectral problem,

$$\sigma \zeta = F_u(0, R|\zeta) - \nabla p[\zeta], \quad \zeta \in H. \tag{3.18}$$

When $\mathscr{V}$ is a bounded domain[3] there are a countably infinite number of isolated eigenvalues $\sigma(R)$ and all of these eigenvalues lie inside a parabola opening out to the left in the complex $\sigma(R)$ plane [3.15], as in Fig. 3.1.

The stability of U may be determined from the spectral problem: U is conditionally stable if for all eigenvalues $\xi(R) < 0$ and is unstable if $\xi(R) > 0$ for some eigenvalue. Conditionally stable means that U is stable to small disturbances; a conditionally stable flow may be unstable to large disturbances whose evolution is not governed by the linearized theory. As we increase R, there is a first critical value R_0 for which $\xi(R_0) = 0$ for some eigenvalues $\sigma(R_0) = i\eta(R_0)$. If $\eta(R_0) \equiv \omega_0 \neq 0$, then $\pm i\omega_0$ are equally eigenvalues of F_u with

---

[2] Velocities and pressures are fields which depend on the position x at which they are evaluated. This dependence on x is suppressed in our notations.

[3] The spectrum of F_u on unbounded domains is not fully understood. It is believed that there is a point spectrum (eigenvalues) controlling stability and a continuous spectrum which is confined to the left-hand side of the complex plane [$\xi(R) < 0$ for all R].

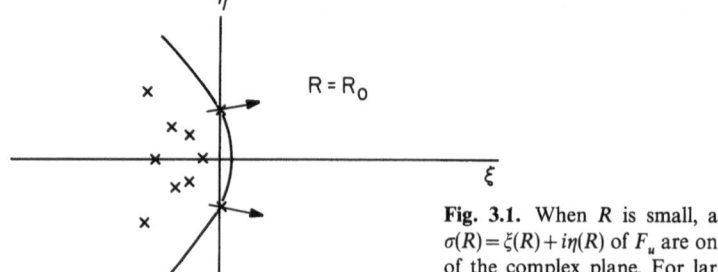

Fig. 3.1. When R is small, all the eigenvalues $\sigma(R) = \xi(R) + i\eta(R)$ of F_u are on the left-hand side of the complex plane. For larger values of R a finite number of eigenvalues have $\xi(R) > 0$

eigenfunctions ζ and $\bar{\zeta}$, where $\bar{\zeta}$ is the complex conjugate of ζ. In general, a complex conjugate pair of eigenvalues crosses to the positive side of the complex σ plane as R is increased past R_0. If $\omega_0 = 0$, a single eigenvalue crosses at the origin.

In the usual situation the loss of stability of U implies the existence of a new solution $U + u$, which bifurcates from U. The usual situation is as follows: ζ solving (3.18) is determined uniquely up to a multiplicative constant when $\sigma(R_0) = i\omega_0$ [where $\sigma(R_0)$ is a simple eigenvalue of F_u], $\pm i\omega_0$ are the only eigenvalues of F_u at criticality and the loss of stability of U is strict, $d\xi(R_0)/dR > 0$. In this case, the solution which bifurcates is a steady solution if $\omega_0 = 0$ (see Sects. 3.5 and 3.8) and is a time-periodic solution if $\omega_0 \neq 0$ (see Sects. 3.7, 8).

Returning now to the conjecture of *Landau-Hopf*, we suppose that $\omega_0 \neq 0$, and a stable time-periodic flow with a characteristic frequency $\omega(\varepsilon)$, where ε is the amplitude of u, exists for $R = R(\varepsilon) \geq R_0$. [$\omega(\varepsilon)$ and $R(\varepsilon)$ are even functions (see Sect. 3.7).] So $u(t, \varepsilon) = u[t + 2\pi/\omega(\varepsilon), \varepsilon]$ exchanges stability with U and is unique up to a choice of phase (a choice of the origin of t).

We need now to study the stability of the new periodic flow as R is increased. It is assumed that $R(\varepsilon)$ is an increasing function. We consider the evolution of a small disturbance v of $U + u$ and derive the linearized problem for v,

$$\frac{\partial v}{\partial t} = F_u(u(t, \varepsilon), R(\varepsilon)|v) - \nabla p , \quad v \in H . \tag{3.19}$$

Equation (3.19) has time-periodic coefficients of period $2\pi/\omega(\varepsilon)$. A spectral problem for such equations may be obtained using the method of *Floquet* (see [Ref. 3.51, Chap. VII]). According to this method we set

$$v = e^{\tilde{\sigma}t} \tilde{\zeta}(t) , \quad \tilde{\sigma}(\varepsilon) = \tilde{\xi}(\varepsilon) + i\tilde{\eta}(\varepsilon) , \tag{3.20}$$

where $\tilde{\zeta}(t) = \tilde{\zeta}[t + 2\pi/\omega(\varepsilon)]$ has the same period as the coefficients of (3.19) and is governed by the spectral problem

$$\tilde{\sigma}\tilde{\zeta} + \frac{\partial \tilde{\zeta}}{\partial t} = F_u(u(t, \varepsilon), R(\varepsilon)|\zeta) - \nabla p , \quad \tilde{\zeta} \in H . \tag{3.21}$$

The eigenvalues $\tilde{\sigma}(\varepsilon)$ of (3.21) are called Floquet exponents. The solution $U + u$ is stable to small disturbances when $\tilde{\xi}(\varepsilon) < 0$ and is unstable when $\tilde{\xi}(\varepsilon) > 0$. The value ε_0 for which $\tilde{\xi}(\varepsilon_0) = 0$ as ε is increased past ε_0 is the critical value for the loss of stability of $u + U$ and at criticality $\tilde{\eta}(\varepsilon_0) = \tilde{\omega}_0$.

If we assume that $\tilde{\sigma}(\varepsilon_0) = i\tilde{\omega}_0$ is a simple eigenvalue of (3.21) and $\pm i\tilde{\omega}_0$ are the only eigenvalues of (3.21) and $d\tilde{\xi}(\varepsilon_0)/d\varepsilon > 0$, then a new solution bifurcates. The properties of this solution depend on the ratio of the frequencies $\tilde{\omega}_0/\omega(\varepsilon_0)$, $0 \leq \tilde{\omega}_0/\omega(\varepsilon_0) < 1$. If $\tilde{\omega}_0/\omega(\varepsilon_0) = r/n$ is a fraction with $n = 1, 2, 3,$ or 4 the new solution has a period which is about n times the period of the bifurcating solution at criticality and the new periodic solution has a new frequency which depends on the amplitude. In the case of forced periodic motion the frequency $\omega(\varepsilon)$ is independent of amplitude and the bifurcating solution is strictly subharmonic with a fixed period $\tilde{T} = 2\pi/\tilde{\omega}_0 = 2\pi n/\omega r = nT/r$ independent of amplitude. In both cases, supercritical solutions $[R > R(\varepsilon_0)]$ are stable and subcritical $[R < R(\varepsilon_0)]$ solutions are unstable when $n = 1$ and 2. The $3T$-bifurcating solution is unstable on both sides of criticality when its amplitude is small. The stability properties of the $4T$-periodic bifurcating solution are slightly more complicated because there are several possibilities [3.16, 17]. If $n \neq 1, 2, 3,$ or 4, the solutions which bifurcate are, in general, not periodic. They may be visualized as living on a torus which encircles the limit cycle representing the periodic solution. If the torus bifurcates subcritically $[R < R(\varepsilon_0)]$, all the bifurcating solutions are unstable; if it bifurcates supercritically $[R > R(\varepsilon_0)]$, the torus is stable but the solutions on the torus need not be stable. The analytical properties of the solution on the torus are not fully understood [3.16–19].

Landau [3.11, 12] conjectured that the solution which bifurcates when the periodic solution loses stability and ε [and $R(\varepsilon)$] is increased past ε_0 is a quasi-periodic solution with two frequencies. A quasi-periodic function of n variables $f(\omega_1 t, \omega_2 t, \ldots, \omega_n t)$ is a function containing a finite number of rationally independent frequencies $\omega_1, \omega_2, \ldots, \omega_n$ which is periodic with period 2π in each of its variables. For example, the function $f(\omega_1 t, \omega_2 t) = \cos t \cos \pi t$ is a quasi-periodic function with frequencies $\omega_1 = 2\pi$ and $\omega_2 = 2$. This function has the value $f(0, 0) = 1$ when $t = 0$. $f(\omega_1 t, \omega_2 t) < 1$ for any $t \neq 0$ but there is always a $t(\mu) > 0$ such that $|f(\omega_1 t, \omega_2 t) - 1| < \mu$ for preassigned $\mu > 0$. The function

$$e^{i\tilde{\omega}_0 t}\tilde{\zeta}(t), \quad \tilde{\zeta}(t) = \zeta[t + 2\pi/\omega(\varepsilon_0)] \tag{3.22}$$

is a quasi-periodic function with two frequencies when $\omega(\varepsilon_0)/\tilde{\omega}_0$ is irrational.

So in the Landau-Hopf conjecture that transition to turbulence occurs by repeated quasi-periodic branching of solutions as R is increased, we get a two-frequency (ω_1, ω_2) solution

$$v(t) = \tilde{v}(\omega_1 t + \alpha_1, \omega_2 t + \alpha_2) \tag{3.23}$$

with two arbitrarily independent phases (α_1, α_2) (and an increasingly complicated spatial structure) when the periodic solution loses stability. A sequence of bifurcations like that just described does appear to occur in certain of the Bénard and Couette flow experiments described in Sect. 3.4 and elsewhere in this volume[4]. But this sequence is by no means universal because in many problems, like those described in the beginning of Sect. 3.4, we get a direct transition to turbulence even when U is stable according to the criteria of the linearized theory of stability. These cases, which defy *Landau* and *Hopf*, are examples of subcritical (or inverted) bifurcation which is a theoretical idea which I am going to explain in Sects. 3.5 and 3.7.

In the next step, *Landau* [3.11, 12] made a mistake and suggested a generalized Floquet theory which lacks a theoretical foundation and leads to results which contradict experience. He speculated that a spectral problem for the stability of the two-frequency solution $U + u + v$ where v is given by (3.23) can be obtained by linearizing with disturbances of the form $\zeta(\omega_1 t, \omega_2 t)\exp(\tilde{\sigma} t)$, where $\zeta(\omega_1 t, \omega_2 t)$ has the same two periods as the coefficients of the linearized equation for the stability of $U + u + v$. At criticality $\tilde{\sigma} = i\tilde{\omega}$, introducing a third frequency, and a three-frequency solution with three arbitrary phases is said to bifurcate. So we get manifolds of quasi-periodic solutions of increasing dimension as R is increased and the process of transition to almost periodic turbulence with a discrete set of countably infinite frequencies occurs only in the limit $R \to \infty$.

The conjectures of *Landau* and *Hopf* can be criticized in several ways. *Bass* [3.21] called attention to the fact that, unlike turbulent solutions, quasi-periodic solutions do not phase mix. The values of quasi-periodic functions at two distant times are correlated as strongly as at close times. If $f(t)$ is defined on a turbulent field, then experiments show that the autocorrelation function

$$g(\tau) = \lim_{T \to \infty} \frac{1}{T} \int_0^T f(t+\tau)\bar{f}(t)dt \to 0 \quad \text{as} \quad \tau \to \infty . \tag{3.24}$$

Solutions with the property (3.24) are called *phase mixing*. If $f(t)$ is almost periodic (or quasi-periodic; see [3.1, p. 220]), then $f(t)$ has a Fourier series

$$f(t) = \sum_{-\infty}^{\infty} f_n e^{-i\lambda_n t} , \tag{3.25}$$

and

$$g(\tau) = \sum_{-\infty}^{\infty} |f_n|^2 e^{-i\lambda_n \tau} \tag{3.26}$$

does not tend to zero as $\tau \to \infty$.

---

4 Quasi-periodic flows with two frequencies usually lose stability to nonperiodic flow. An interesting experiment on forced periodic convection in bounded domains which undergoes transition from a flow with two frequencies to nonperiodic flow was reported by *Gollub* and *Benson* [3.20].

A second criticism of the Landau-Hopf conjecture raised by *Ruelle* and *Takens* [3.10] is that turbulent (phase mixing) solutions should appear after just a few bifurcations. In their early work they showed that even simple dynamical systems in four dimensions could be dominated by phase mixing solutions. A later analysis, leading to similar results in a simpler context, was given by *McLaughlin* and *Martin* [3.22]. Even earlier, *Lorenz* [3.23] and *Baker* et al. [3.24] had exhibited simple nonlinear ordinary differential equations in three dimensions with complicated turbulentlike dynamics. In all cases known to me turbulence appears after a finite number of bifurcations. In many flows, like the pipe flow discussed in the introduction, we get a direct transition to turbulence apparently without intervening bifurcations. In other flows, like the flow between rotating cylinders or rotating spheres, there are some symmetry-breaking bifurcations intervening between the unique basic flow which exists at small values of R and turbulence which exists at larger values.

3.4 Examples of Hydrodynamic Stability and Bifurcation

According to what I have already said, it is impossible to understand the hydrodynamics of flows which arise as the Reynolds number is increased without understanding instability and bifurcation of flows. To fix this important idea more firmly, it is useful to give a descriptive account of instability and bifurcation in some hydrodynamic examples. These examples are restricted to problems with steady data enjoying a high degree of spatial symmetry and are selected so as to represent certain general principles whose statement requires prior specification of some ideas from bifurcation theory. (Some of these ideas are developed in Sects. 3.5–7.)

Our first examples are the Poiseuille flows through the annulus between two long concentric cylinders. When there is no inner cylinder, we have the flow through a round pipe (Hagen-Poiseuille flow) whose most symmetric laminar state is given by (3.4). We have already discussed Reynolds experiments on transition to turbulence in pipe flow. The main features of the observations are also characteristic of flow through the annulus between concentric cylinders. The main features are: 1) There are no stable symmetry-breaking bifurcations of simple type; instead we go directly from laminar flow to turbulent flow. 2) In the low Reynolds turbulent regime only some parts of the flow are turbulent [3.25, 26]; the flow is spatially segregated into distinct packets of laminar and turbulent flow (turbulent "puffs" when R is slightly above the critical R, and "slugs" at higher values of R). The transition from laminar to turbulent flow at a fixed place occurs suddenly as a puff or slug sweeps over the place, and the reverse transition occurs just as suddenly when it leaves the place. These observations suggest a sort of cycling in phase space between two weakly attracting solutions, one of which is laminar (see Fig. 3.2).

The second example is Couette flow between rotating cylinders when the inner cylinder is at rest and the outer one rotates with a steady angular velocity

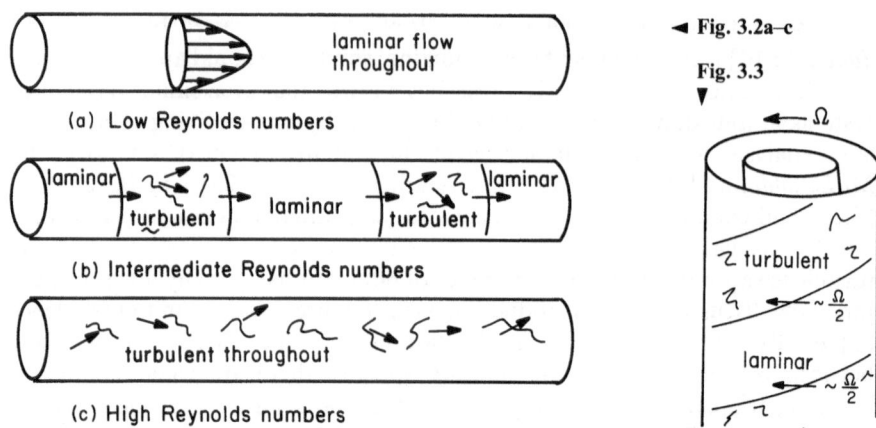

◄ **Fig. 3.2a–c**

Fig. 3.3
▼

(a) **Low Reynolds numbers**

(b) **Intermediate Reynolds numbers**

(c) **High Reynolds numbers**

Fig. 3.2a–c. Sketch of the observational events in the transition to turbulence in pipes. The laminar flow which moves down the pipe at intermediate values of the Reynolds number is not the same as the Hagen-Poiseuille flow which exists at low Reynolds number. The alternate patches of laminar flow in (**b**) have a considerable amount of vorticity

Fig. 3.3. Spiral bands of turbulence which arise in the direct transition from Couette flow to turbulence. The outer cylinder rotates and the inner cylinder is stationary (or rotates slowly in the opposite direction). The spiral bands rotate at about one-half the angular speed of the outer cylinder. At higher Reynolds numbers the turbulence fills the whole annulus

Ω. The flows which evolve when the inner cylinder is at rest and the outer one rotates are much different than those which evolve when the outer cylinder is at rest and the inner one rotates. In the former case we get direct transition to turbulence, without intermediate bifurcations, as in Poiseuille flow. At low Reynolds numbers the flow is the laminar flow of maximum symmetry; in the idealized infinitely long annulus the flow is axisymmetric and does not vary along the cylinder axis, all components of velocity except the one in the azimuth vanish and it depends on the radius only. As in Poiseuille flow, Couette flow with the outer cylinder rotating undergoes direct transition to turbulence which first appears as alternate spiral bands of laminar and turbulent flow [3.27] (see Fig. 3.3), again as in Poiseuille flow.

The two examples just reviewed are examples of the direct transition to turbulence. Examples of transition to turbulence through a repeated finite number of symmetry-breaking bifurcations occurs in the Bénard problem, in the problem of Couette flow between cylinders when the outer cylinder is at rest, and in the problem of flow between concentric spheres when the outer sphere is at rest. The Bénard and Couette flow problems are reviewed in Chaps. 5 and 6, respectively. Here, I wish only to summarize features which appear to be common to all three problems and to draw attention to "abnormal" solutions which run against physical intuition but which are observed. These "abnormal" solutions appear much less abnormal when viewed in terms of general principles stemming from bifurcation theory (see remarks concluding Sect. 3.6).

The common features which appear in the three problems mentioned above are

1) The first bifurcation appears to break the spatial symmetry of the basic solution and to replace this solution of maximum symmetry with another steady flow having a different pattern of symmetry.

2) At higher values of the Reynolds number[5] new solutions, each with a different pattern of symmetry, appear to gain stability. Eventually there is a bifurcation of steady flow into a time-periodic motion with a single frequency. *Coles'* [3.27] study of stability and bifurcation of flow between rotating cylinders with the outer cylinder at rest draws attention to the marked degree of nonuniqueness and hysteresis which characterize these flows. For supercritical speed of a rotating inner cylinder up to about ten times critical, *Coles* finds that in one and the same apparatus the number of vortices and the number of waves travelling around these vortices are not uniquely determined by the speed. The number of Taylor cells in his apparatus ranges from 18 to 32 and the number of waves which travel around the axis of the cells ranges from 3 to 7. Moreover, "as many as 20 or 25 different states (each state being defined by the number of Taylor cells and the number of tangential waves) have been observed at a given speed".

3) At still higher values of Reynolds number the periodic solution with one frequency appears to give up its stability to a quasi-periodic solution with two frequencies both of which appear to vary smoothly with Reynolds number [3.28–30] (see Chaps. 5 and 6). As the Reynolds number is increased further, there is a continuous amplification of dynamical noise (turbulence) which eventually wipes out the sharp spectral peaks associated with discrete frequencies.

Now I will describe the "abnormal" solutions which are observed in the process of repeated bifurcation. In the problem of Couette flow between cylinders in which the outer cylinder is at rest, such abnormal solutions seem to have been observed first by *Benjamin* [3.31] and a possibly correct theoretical explanation of them has been given by him. Benjamin's apparatus is a short annulus which accommodates from two to four cells. (See the discussion of finite annulus height effects in Sect. 6.6.) In such experiments, cell cross sections never deviate very much from squares. He found that there are certain states, characterized by the number of cells and the sense of circulation in the cells, which can be reached only through sudden acceleration from rest. Such cells seem to be isolated from the basic cellular solution for most values of the height l of the rotating cylinders. In Benjamin's experiments and in the experiments on flow between spheres which are described below, there are an integral number of cells and an integral number of time-periodic undulations around the cells.

---

5 In the Bénard problem the relevant parameter is the Rayleigh number, measuring the temperature difference driving the motion, rather than the Reynolds number.

The different solutions correspond to different integers and the change from one to another set of integers appears always to be discontinuous. This suggests that we are dealing here with the problem of isolated solutions. (See Sect. 3.6 for a theoretical discussion of such possibilities.)

The abnormal cell observed by *Benjamin* appears to be one of the hard-to-reach solutions which can be reached only through special initial conditions like sudden acceleration from rest. To understand the abnormal cells we specify what is meant by a normal cell on the top or bottom of the finite cylinder. The top and bottom, like the outer cylinder, do not rotate. The cell is normal because the circulation in it is such that the fluid near the top or the bottom moves from the outer cylinder toward the inner cylinder. The reason is that in the center of the apparatus, away from the ends, the flow is like Couette flow (there are also weak secondary motions) in which an inward-pointing pressure gradient (pointing to the inner cylinder) is balanced against centripetal accelerations which in the absence of the opposing pressure gradient would throw the fluid outward. At the end plates this balance is broken because the centripetal accelerations are nullified by the boundary condition which requires that the fluid stick to the stationary end plates. So the fluid moves toward the inner cylinder driven by the unbalanced pressure gradient. In the "abnormal" cell observed by *Benjamin* the circulation is in the opposite sense.

Turning next to the allied problem of flow between rotating spheres we may cite experiments by *Sawatzki* and *Zierep* [3.32], *Munson* and *Menguturk* [3.33], *Wimmer* [3.34], *Belyaev* et al. [3.35], and *Yavorskaya* et al. [3.36]. The most symmetric laminar flow between spheres with the outer sphere at rest is a steady axisymmetric spiral flow superposing motion in circles driven by shearing at the boundary and secondary motion between the equator and the poles driven by centripetal accelerations. This flow, unlike idealized Poiseuille and Couette flow between infinitely long cylinders, changes with the Reynolds number. The basic flow can be obtained by perturbation analysis or numerical analysis [3.37].

The characteristics of the flows which break the symmetries of the basic flow seem to depend strongly on the size of the gap between the spheres. When the gap is very small, the flow near the equator is very much like the flow between rotating cylinders and the sequence bifurcating flow is like that in the flow between cylinders. First, the basic flow loses stability and is replaced by a flow with vortices near the equator, but the flow near the poles is undisturbed. The vortices near the equator are like Taylor vortices; the cross section of each vortex is very nearly square, but the axis of the vortices spirals with a very slight pitch angle. For higher values of the Reynolds number the steady bifurcating flow with spiral vortices loses stability to a time-periodic motion in which the cell boundary becomes wavy and the waviness propagates around the sphere axis. This mode of instability is like the wavy vortex state in the Taylor problem (see Chap. 6). At higher values of the Reynolds number the wavy vortex mode gives up its stability to turbulence. The turbulent motion does not destroy the vortex structure when the gap is small; instead dynamical noise (fluctuations) of increasing (with the Reynolds number) amplitude is superposed on the equatorial vortices.

The situation is much more complicated when the gap between the spheres is large. The following observation may be helpful in interpreting the results. If we take the diameter of a vortex as the scale of length and then let the gap size tend to zero, the region near the equator looks locally like the infinitely long cylinders of the idealized Taylor problem. When the gap is large, we may expect that the vortices are influenced by the flow near the poles and the flow is more closely allied to a Taylor apparatus of finite height in which end effects are important.

Instability and bifurcation may be studied experimentally by measuring the torque the fluid exerts on the spheres, or by flow visualization. The two methods agree when the instability manifests itself as a break in the spatial symmetry of the solution. But *Munson* and *Menguturk* [3.33] found that when the gap is large (inner radius/outer radius = 0.44) there is no break in the spatial symmetry of the flow at the first value at which there is a break in the torque curve. They conjectured that the radial distribution of velocities is altered at this point of instability, but that the symmetry pattern of the new flow is basically the same as the old flow. *Munson* and *Menguturk* found three more breaks in the torque curve as the Reynolds number is increased, each of which is associated with some visual event, the last one leading to a bifurcation into turbulence.

The experimental results of *Munson* and *Menguturk* for stability of flow between spheres when the inner one rotates have been checked, but not reproduced, in the course of careful experiments of *Belyaev* et al. [3.35] and *Yavorskaya* et al. [3.36]. They did not observe the first three breaks in the torque curve, and the fourth break is observed as a three-dimensional wave, periodic in time. They said that this bifurcation leads to a new laminar flow and not to turbulence.

Similar experiments carried out by *Sawatzki* and *Zierep* [3.32] and *Wimmer* [3.34] revealed that five different types of flow can occur at the same Reynolds number when the gap between the cylinders is relatively wide. The five solutions observed in the experiments may be described as follows:

1) There is one mode which is visually like the basic laminar flow but has a different, nonlinear relation between the measured torque and the Reynolds number. This is possibly the same mode observed by *Munson* and *Menguturk* [3.33]. It is a very persistent solution which exists for a very large range of Reynolds numbers.

2, 3) There are paired steady solutions of normal and abnormal cellular motions in which there are a few vortices bordering the equator whose axes are parallel to the equator. In the normal cell the fluid is thrown out at the equator. This cell is normal because it is consistent with the physical fact that centripetal accelerations pushing the fluid out are greatest at the equator. In the abnormal flow the fluid at the equator moves inward, violating physical intuition (see Fig. 3.4). According to *Wimmer* [3.34] the abnormal flow has the lowest critical

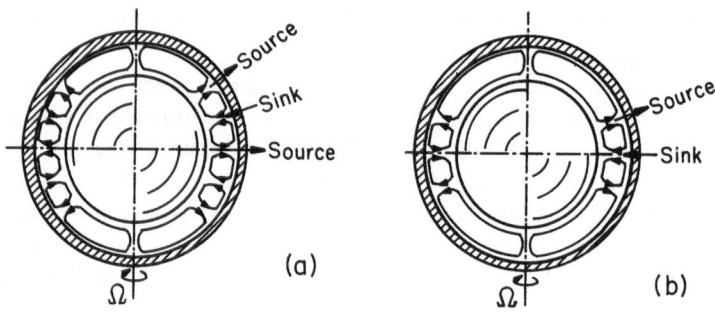

Fig. 3.4a, b. Two steady flows, (a) normal and (b) abnormal, which have been observed in the experiments of *Sawatzki* and *Zierep* [3.32] and *Wimmer* [3.34]

Reynolds number, which means that it is the first solution to replace the basic flow. It is also said to be the most stable solution, because it still occurs at very high Reynolds numbers close to the onset of turbulence.

4, 5) Corresponding to the normal and abnormal steady solutions is an analogous pair of unsteady solutions which may be time periodic. These solutions are like the ones described under 2) except that the axes of the vortices are no longer parallel but are inclined to the equator and the vortices penetrate this plane. Since the axes of the vortices now end in the midst of the flow field, single spots of vorticity are hurled off and travel to the poles. The whole process is said to be periodic.

The examples of instability, nonuniqueness and transition to turbulence which we have reviewed in this section will be discussed from a theoretical point of view in Sects. 3.6–8. The examples are merely a small sample of the types of problems which come up in trying to understand observed motions in fluids. A particularly interesting class of problems which is not well understood concerns the various common flows which occur in unbounded regions which cannot be contained between parallel planes. These include the instability of boundary layers, jets, shear layers, wakes, vortex streets, and thermal plumes. Some information about these problems can be found in Chaps. 7 and 8 and in the books, monographs, and review papers cited Sect. 3.10.

3.5 A Simplified Mathematical Discussion of some General Properties of Stability and Bifurcation

The difficulty I face in explaining the theory of hydrodynamic stability and bifurcation is that the equations are very complicated and admit very many different kinds of solutions. To my knowledge there are no good examples of realistic hydrodynamic situations which are simple enough to be understood at a glance. And in this subject physical intuition can betray you. For example, if you rely on physical intuition, you would dismiss as impossible the "abnormal" flows which appear between rotating cylinders and rotating spheres (see Sect. 3.4). I do not see any way to intuit the qualitative conditions under which a periodic flow will replace a steady one or a quasi-periodic flow a periodic one. What happens to flows depends mathematically on nonintuitive operations which in the simplest of cases involve roots of algebraic equations and the calculation of eigenvalues. It is dangerous to imagine that you can know such things without calculations, and the calculations in the hydrodynamic case are very involved.

To some extent the procedures of bifurcation theory simplify the problem by abstracting from specific problems the features which are essential in the description of the loss of stability and bifurcation. But in the process of abstraction we get to a more general problem in which the hydrodynamical equations are just one realization. The usual form for the abstracted problem is an evolution equation of the form

$$\frac{d\boldsymbol{u}}{dt} = F(\mu, \boldsymbol{u}) \, , \tag{3.27}$$

where μ is a scalar parameter and $\boldsymbol{u}$ is a vector-valued field. In fact, the Navier-Stokes equations (3.14–16) can be written as in (3.27) when (3.14–16) are projected by the method explained in Sect. 3.8 with vectors $\boldsymbol{\zeta}$ which have $\mathrm{div}\,\boldsymbol{\zeta} = 0$ in $\mathscr{V}$ and $\boldsymbol{\zeta} \cdot \boldsymbol{n} = 0$ on $\partial\mathscr{V}$. The method of projections is fundamental in theoretical studies of the Navier-Stokes equations (see, for example, [3.38–40]). I am not going to get into an involved discussion of projections here. For now it is enough to note that (3.27) is a general problem of which the Navier-Stokes equations is but one realization. There are a lot of properties of stability and bifurcation which are shared by all of the realizations of (3.27). And, in fact, we can learn a tremendous amount about the hydrodynamic case by studying the simplest realization of (3.27), the problem which arises when u is a scalar parameter and

$$\frac{du}{dt} = F(\mu, u) \tag{3.28}$$

governs its evolution. Of course (3.28) is not a hydrodynamical problem, but it is a good model and, in fact, it can be shown (for example, in [3.41, 42] that in important cases we can actually reduce the Navier-Stokes problem to the study of bifurcation and stability of steady solutions of (3.28). I do some of this showing in Sect. 3.8.

Some preliminary remarks about (3.28) are necessary for the analysis and some are useful for understanding the physical significance of the analysis. We first specify that both u and μ live on the real line, $-\infty < u < \infty$, $-\infty < \mu < \infty$. It is essential that $F(\mu, u)$ be a nonlinear function of two variables and it is necessary to assume that it and all of its first and second partial derivatives are continuous. It is good to regard (3.28) as an equation of motion. In most (nearly all) of the studies of bifurcation and stability it is conventional to imagine that $u=0$ satisfies (3.28) for all values of μ in an interval of special interest

$$F(\mu,0)=0 . \qquad (3.29)$$

I call the assumption (3.29) a reduction to local form. This reduction follows automatically when u is the difference between any solution and some special solution which exists for all values of μ (see [3.43]). In the hydrodynamic case the special solution is conventionally taken to be the unique one (the basic solution) which exists at small Reynolds numbers R and has the maximum symmetry consistent with the data [see discussion following (3.5–7) and (3.14–16)]. It is assumed that this solution continues to exist and is a sufficiently smooth function of R for all R, even for values $R>R_c$ for which the basic solution is unstable. In the hydrodynamic case we would put $\mu=(-1/R)$ $+(1/R_c)$ so that the basic solution would be stable for $-\infty<\mu<0$ and unstable for positive $\mu\leq1/R_c$.

There is no great loss of generality in reducing bifurcation problems to local form. To see the small loss, suppose that there is a steady solution $u(\mu)$ of (3.28). Then a disturbance $x(t)$ of $u(\mu)$ satisfies

$$\frac{dx}{dt} = F[\mu, u(\mu)+x] . \qquad (3.30)$$

So $\tilde{F}(\mu, x)=F[\mu, u(\mu)+x]$ obviously vanishes when $x=0$ for all μ such that $u(\mu)$ is defined. In the analysis of (3.28) it is neither necessary nor desirable to reduce $F(\mu, u)$ to local form. Instead we shall only introduce the notation $u=\varepsilon$, $d\varepsilon/dt=0$ for steady solutions of (3.28).

Now I am going to do some analysis, the simplest type of analysis I can do, which will still help in understanding broad features of stability and bifurcation in fluids. Until the end of this section, I suppress hydrodynamics and emphasize mathematics. We shall return to hydrodynamics, better prepared, in subsequent sections.

I start the mathematical discussion with a brief but thorough study of bifurcation and stability of steady solutions of (3.28). The bifurcation part of the study is just a review of the classical theory of singular points of plane curves. To this study we add results concerning the stability of bifurcating solutions (see [3.41–43]).

In our study of steady solutions of (3.28) it is desirable to introduce the following classification of points:

i) A *regular point* of $F(\mu, \varepsilon)=0$ is one for which the implicit function theorem works,

$$F_\mu \neq 0 \quad \text{or} \quad F_\varepsilon \neq 0 . \qquad (3.31)$$

If (3.31) holds, then we can find a unique curve $\mu=\mu(\varepsilon)$ or $\varepsilon=\varepsilon(\mu)$ through the point.

ii) A *regular turning point* is a point at which $\mu_\varepsilon(\varepsilon)$ changes sign and $F_\mu(\mu, \varepsilon) \neq 0$.

iii) A *singular point* of the curve $F(\mu, \varepsilon) = 0$ is a point at which

$$F_\mu = F_\varepsilon = 0 . \tag{3.32}$$

iv) A *double point* of the curve $F(\mu, \varepsilon) = 0$ is a singular point through which pass two and only two branches of $F(\mu, \varepsilon) = 0$ possessing distinct tangents.

v) A *singular turning (double) point* of the curve $F(\mu, \varepsilon)$ is a double point at which μ_ε changes sign.

vi) A *cusp point* of the curve $F(\mu, \varepsilon) = 0$ is a point of second-order contact between two branches of the curve. The two branches of the curve have the same tangent at a cusp point.

vii) A *higher order singular point* of the curve $F(\mu, \varepsilon) = 0$ is a singular point at which all three second derivatives of $F(\mu, \varepsilon)$ are null.

It is necessary to be precise about double points. Suppose that (μ_0, ε_0) is a singular point. The equilibrium curves passing through the singular point satisfy

$$2F(\mu, \varepsilon) = F_{\mu\mu}(\delta\mu)^2 + 2F_{\varepsilon\mu}\delta\varepsilon\delta\mu + F_{\varepsilon\varepsilon}(\delta\varepsilon)^2 + O((\delta\mu)^2 + \delta\varepsilon\delta\mu + (\delta\varepsilon)^2) = 0, \tag{3.33}$$

where $\delta\mu = \mu - \mu_0$, $\delta\varepsilon = \varepsilon - \varepsilon_0$, and $F_{\mu\mu} = F_{\mu\mu}(\mu_0, \varepsilon_0)$, etc. In the limit as $(\mu, \varepsilon) \to (\mu_0, \varepsilon_0)$, the equation (3.33) for the curves $F(\mu, \varepsilon) = 0$ reduces to the quadratic equation

$$F_{\mu\mu}(d\mu)^2 + 2F_{\varepsilon\mu}d\varepsilon d\mu + F_{\varepsilon\varepsilon}(d\varepsilon)^2 = 0 \tag{3.34}$$

for the tangents to the curve. We find two roots, designated with superscripts

$$\begin{pmatrix} \mu_\varepsilon^{(1)}(\varepsilon_0) \\ \mu_\varepsilon^{(2)}(\varepsilon_0) \end{pmatrix} = -\frac{F_{\varepsilon\mu}}{F_{\mu\mu}} \begin{pmatrix} 1 \\ 1 \end{pmatrix} + \left(\frac{D}{F_{\mu\mu}^2}\right)^{1/2} \begin{pmatrix} 1 \\ -1 \end{pmatrix} \tag{3.35}$$

or

$$\begin{pmatrix} \varepsilon_\mu^{(1)}(\mu_0) \\ \varepsilon_\mu^{(2)}(\mu_0) \end{pmatrix} = -\frac{F_{\varepsilon\mu}}{F_{\varepsilon\varepsilon}} \begin{pmatrix} 1 \\ 1 \end{pmatrix} - \left(\frac{D}{F_{\varepsilon\varepsilon}^2}\right)^{1/2} \begin{pmatrix} 1 \\ -1 \end{pmatrix}, \tag{3.36}$$

where

$$D = F_{\varepsilon\mu}^2 - F_{\mu\mu}F_{\varepsilon\varepsilon} . \tag{3.37}$$

If $D < 0$ there are no real tangents through (μ_0, ε_0) and the point (μ_0, ε_0) is an isolated point solution of $F(\mu, \varepsilon) = 0$.

We shall consider the case when (μ_0, ε_0) is *not* a higher order singular point. Then (μ_0, ε_0) is a double point if and only if $D > 0$. If $D = 0$, then the slope at the singular point of higher order contact is given by (3.35) or (3.36). If $D>0$ and $F_{\mu\mu}\neq0$, then there are two tangents with slopes $\mu_\varepsilon^{(1)}(\varepsilon_0)$ and $\mu_\varepsilon^{(2)}(\varepsilon_0)$ given by (3.35). If $D>0$ and $F_{\mu\mu}=0$, then $F_{\varepsilon\mu}\neq0$ and

$$d\varepsilon(2d\mu F_{\varepsilon\mu}+d\varepsilon F_{\varepsilon\varepsilon})=0\ , \tag{3.38}$$

and there are two tangents $\varepsilon_\mu(\mu_0)=0$ and $\mu_\varepsilon(\varepsilon_0)=-F_{\varepsilon\varepsilon}/2F_{\varepsilon\mu}$. If $\varepsilon_\mu(\mu_0)=0$ then $F_{\mu\mu}(\mu_0,\varepsilon_0)=0$. So all possibilities are covered in the following two cases:
  A) $D>0$, $F_{\mu\mu}\neq0$ with tangents $\mu_\varepsilon^{(1)}(\varepsilon_0)$ and $\mu_\varepsilon^{(2)}(\varepsilon_0)$.
  B) $D>0$, $F_{\mu\mu}=0$ with tangents $\varepsilon_\mu(\mu_0)=0$ and $\mu_\varepsilon(\varepsilon_0)=-F_{\varepsilon\varepsilon}/2F_{\varepsilon\mu}$.
  Now I am going to connect stability and bifurcation. To study the stability of the solution $u=\varepsilon$ we study the linearized equation

$$Z_t=F_\varepsilon(\mu,\varepsilon)Z \tag{3.39}$$

by the spectral method

$$Z=e^{\gamma t}Z'\ , \tag{3.40}$$

where

$$\gamma=F_\varepsilon(\mu,\varepsilon)\ . \tag{3.41}$$

The solution $u=\varepsilon$ is stable when $\gamma<0$ and is unstable when $\gamma>0$.

Theorem 1 (Factorization Theorem): For every equilibrium solution $F(\mu,\varepsilon)=0$ *for which* $\mu=\mu(\varepsilon)$ *we have*

$$\gamma(\varepsilon)=F_\varepsilon[\mu(\varepsilon),\varepsilon]=-\mu_\varepsilon(\varepsilon)F_\mu[\mu(\varepsilon),\varepsilon]\equiv-\mu_\varepsilon\hat{\gamma}(\varepsilon)\ . \tag{3.42}$$

The proof of Theorem 1 follows from (3.41) and the equation

$$\frac{dF}{d\varepsilon}[\mu(\varepsilon),\varepsilon]=F_\varepsilon[\mu(\varepsilon),\varepsilon]+\mu_\varepsilon(\varepsilon)F_\mu[\mu(\varepsilon),\varepsilon]=0\ . \tag{3.43}$$

One of the main implications of the factorization theorem is that $\gamma(\varepsilon)$ *must change sign as ε is varied across a regular turning point*. This implies that the solution $u=\varepsilon$, $\mu=\mu(\varepsilon)$ is stable on one side of a regular turning point and is unstable on the other side (see Fig. 3.5).

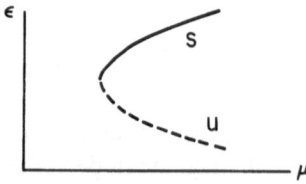

Fig. 3.5. Change of stability at a regular turning point

In the following theorem I make a connection between double-point bifurcation and the hypothesis that the loss of stability of a solution as a singular point is traversed is strict. This hypothesis seems to have been introduced by *Hopf* [3.44].

Theorem 2: Suppose that (μ_0, ε_0) *is a singular point and* (A) $\gamma_\varepsilon(\varepsilon_0) \neq 0$ *or* (B) $\gamma_\mu(\mu_0) \neq 0$. *Then* (μ_0, ε_0) *is a double point.* The proof of this theorem is elementary (see [3.41]) and will be omitted. The bifurcation picture is more complicated when the hypotheses of strict crossing in Theorem 2 are relaxed. If $\gamma_\varepsilon = 0$ when $\gamma = 0$, we may get cusp bifurcation; or if all three second derivatives vanish, then the cubic equation can give a triple point (three real roots for the slopes) or no bifurcation (two complex conjugate roots). If third derivatives also vanish, we face the problem of classifying the roots of a quartic. For example, we may get four bifurcating branches.

It is possible to make precise statements about the stability of solutions near double points of bifurcation. All of the possibilities for the stability of double-point bifurcation can be described by the cases A and B which were fully specified under (3.38). In case A, two curves $\mu^{(1)}(\varepsilon)$ and $\mu^{(2)}(\varepsilon)$ pass through the double point (μ_0, ε_0). In case B, two curves, $\varepsilon^{(1)}(\mu)$ [with $\varepsilon_\mu^{(1)}(\mu_0) = 0$] and $\mu_\varepsilon^{(2)}$, pass through the double point. The eigenvalue $\gamma^{(1)}$ belongs to the curve with superscript (1) and $\gamma^{(2)}$ to the curve with superscript (2).

Theorem 3: Suppose that (μ_0, ε_0) *is a double point. Then, in case* A,

$$\gamma^{(1)}(\varepsilon) = -\mu_\varepsilon^{(1)}(\varepsilon)[\hat{s}\sqrt{D}(\varepsilon - \varepsilon_0) + o(\varepsilon - \varepsilon_0)]. \tag{3.44}$$

and

$$\gamma^{(2)}(\varepsilon) = \mu_\varepsilon^{(2)}(\varepsilon)[\hat{s}\sqrt{D}(\varepsilon - \varepsilon_0) + o(\varepsilon - \varepsilon_0)], \tag{3.45}$$

where $\hat{s} = F_{\mu\mu}/|F_{\mu\mu}|$ *and* D *and* $F_{\mu\mu}$ *are evaluated at* $\varepsilon = \varepsilon_0$. *And in case* B,

$$\gamma^{(1)}(\mu) = s\sqrt{D}(\mu - \mu_0) + o(\mu - \mu_0), \tag{3.46}$$

and

$$\gamma^{(2)}(\varepsilon) = -s\mu_\varepsilon^{(2)}(\varepsilon)[\sqrt{D}(\varepsilon - \varepsilon_0) + o(\varepsilon - \varepsilon_0)], \tag{3.47}$$

where $s = F_{\varepsilon\mu}/|F_{\varepsilon\mu}|$.

Proof: If $\mu = \mu(\varepsilon)$ we have (3.42) in the form,

$$\gamma(\varepsilon) = -\mu_\varepsilon(\varepsilon)F_\mu[\mu(\varepsilon), \varepsilon]$$
$$= -\mu_\varepsilon(\varepsilon)\{[F_{\mu\mu}(\mu_0, \varepsilon_0)\mu_\varepsilon(\varepsilon_0) + F_{\varepsilon\mu}(\mu_0, \varepsilon_0)](\varepsilon - \varepsilon_0) + o(\varepsilon - \varepsilon_0)\}. \tag{3.48}$$

The formulas (3.44) and (3.45) arise from (3.48) when $\mu_\varepsilon(\varepsilon_0)$ is replaced with the values given by (3.35). If $\varepsilon = \varepsilon(\mu)$ with $\varepsilon_\mu(\mu_0) = 0$, then $F_{\mu\mu}(\mu_0, \varepsilon_0) = 0$, $F_{\varepsilon\mu}^2(\mu_0, \varepsilon_0) = D$ and

$$\gamma(\mu) = F_\varepsilon[\mu, \varepsilon(\mu)] = F_{\varepsilon\mu}(\mu_0, \varepsilon_0)(\mu - \mu_0) + o(\mu - \mu_0)$$

$$= s\sqrt{D}(\mu - \mu_0) + o(\mu - \mu_0). \tag{3.49}$$

Theorem 3 gives an exhaustive classification relating the stability of solutions near a double point to the slope of the bifurcation curves near the point. The result may be summarized as follows. Suppose that $|\varepsilon - \varepsilon_0| > 0$ is small. Then (3.44) and (3.45) show that $\gamma^{(1)}(\varepsilon)$ and $\gamma^{(2)}(\varepsilon)$ have the same (different) sign if $\mu_\varepsilon^{(1)}(\varepsilon)$ and $\mu_\varepsilon^{(2)}(\varepsilon)$ have different (the same) sign. The same conclusion can be drawn from (3.46) and (3.47). The possible distributions of stability of solutions is sketched in Fig. 3.6.

The analysis of double-point bifurcation is even easier when one first makes the reduction (3.29) to local form. It may be helpful to make a few remarks about the bifurcation diagrams which follow from analysis of (3.29). Nearly all the literature, not only the hydrodynamic literature, starts from a setup in which $u = 0$ is a solution of the evolution problem. If $F(\mu, 0) = 0$, then $F_\mu(0, 0) = F_{\mu\mu}(0, 0) = 0$ and the strict loss of stability of the solution $u = 0$ as μ is increased past zero is

$$\gamma_\mu^{(1)}(0) = F_{\mu\varepsilon}(0, 0) < 0. \tag{3.50}$$

Then we get $D > 0$ and

$$\gamma^{(2)}(\varepsilon) = -\mu_\varepsilon^{(2)}(\varepsilon)\gamma_\mu^{(1)}(0)[\varepsilon + o(\varepsilon)]. \tag{3.51}$$

The bifurcation diagrams which follow from these results and the conventional statements which we make about them are given by the diagrams and caption to Fig. 3.7.

The results of our local analysis can be collected into a global theorem about double-point bifurcation.

Theorem 4: Assume that all singular points of solutions of $F(\mu, \varepsilon) = 0$ are double points. The stability of such solutions must change at each regular turning point and at each singular point (which is not a turning point) and only at such points.

A marvelous demonstration which can help to fix the ideas embodied in Theorem 4 has been found by *Benjamin* (private communication). Benjamin's demonstration is an example of the buckling of a simple structure under the action of gravity. His apparatus is a board with two holes through which a viscoelastic wire is passed. The wire forms an arch above the board whose area length is l. The wire which is actually used in Benjamin's demonstration is like a

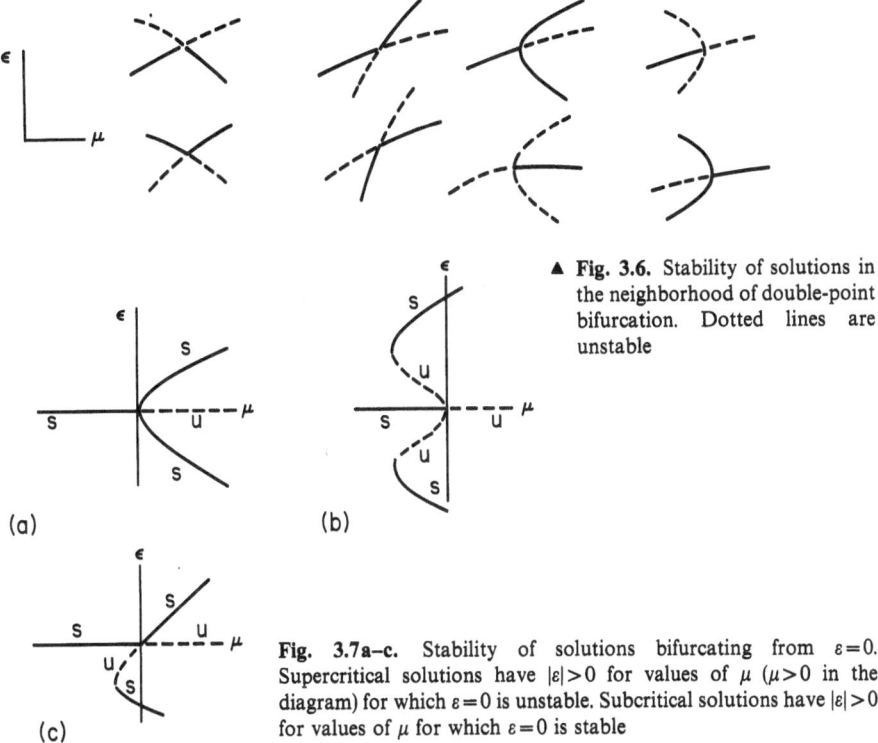

▲ **Fig. 3.6.** Stability of solutions in the neighborhood of double-point bifurcation. Dotted lines are unstable

(a)   (b)

(c)

Fig. 3.7a–c. Stability of solutions bifurcating from $\varepsilon=0$. Supercritical solutions have $|\varepsilon|>0$ for values of μ ($\mu>0$ in the diagram) for which $\varepsilon=0$ is unstable. Subcritical solutions have $|\varepsilon|>0$ for values of μ for which $\varepsilon=0$ is stable

bicycle brake cable; it is wound like a tight coil spring and covered with a plastic sheaf. The demonstration apparatus is sketched in Fig. 3.8.

We imagine that the equation of motion for the wire arch is

$$\frac{d\theta}{dt} = F(l, \theta) .$$ (3.52)

The steady solutions of (3.52) are imagined to be in the form $F[l(\theta), \theta]=0$ shown in Fig. 3.9. Here $\theta=0$ is one solution (the upright one) and $l(\theta)$ is another solution (the bent arch). In fact there is a one-to-one correspondence between Benjamin's demonstration and the bifurcation diagram in Fig. 3.9; nothing is seen in the demonstration that does not appear in the diagram and there is nothing in the diagram that is not in the demonstration. The interpretation of events in the demonstration is given in the caption for Fig. 3.9.

Double-point bifurcation is perhaps the most common form of bifurcation which can occur at a singular point. Other types of bifurcation, cusp points, triple points, etc., are less common because they require that some higher order derivatives of $F(\mu, \varepsilon)$ vanish. Such situations are sometimes called

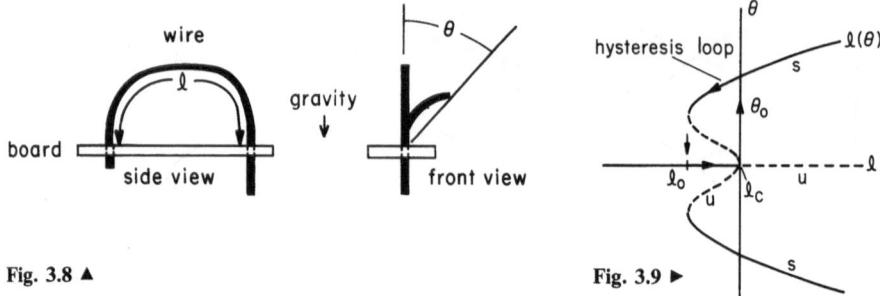

Fig. 3.8 ▲

Fig. 3.9 ►

Fig. 3.8. Benjamin's apparatus for demonstrating the buckling of a viscoelastic arch under gravity loading. The bifurcation diagram which fits this system is shown in Fig. 3.9. When l is small the only stable solution of (3.52) is the upright one ($\theta=0$). When $l>l_c$ is large the upright position is unstable and the arch falls to the left or to the right as shown in the front view. The bent position of the wire is also stable when $l<l_c$. When $l<l_c$ there are three stable steady solutions, the upright one ($\theta=0$) and the left or right bent one ($|\theta|\neq0$)

Fig. 3.9. Bifurcation diagram for the buckling of the viscoelastic arch. When l is small the only equilibrium of (3.52) is the upright one ($\theta=0$). The solution $\theta=0$ loses stability when $u=l-l_c$ is increased past zero. A new solution $\mu(\theta)=l(\theta)-l_c$ corresponding to the bent arch then undergoes double-point bifurcation at a singular turning point $(l,\theta)=(l_c,0)$. The system is symmetric in θ. When $l<l_c$ only the left and right bent equilibrium configurations are stable. The points $(l,\theta)=(l_0,\pm\theta_0)$ are regular turning points. When $l_0\leq l\leq l_c$ there are three stable solutions $\theta=0$ and the symmetric left and right bent positions. In this region the system exhibits hysteresis. If the length l of the arch of the wire above the board is decreased while the wire is bent, the bent configuration will continue to be observed until $l=l_0$. When $l=l_0$ the bifurcating bent position is a regular turning point and for $l<l_0$ only $\theta=0$ is stable. So when l is reduced below l_0 the arch snaps through to the upright solution. Now if we increase l, the arch stays in the vertical position until $l=l_c$. When $l>l_c$, the upright solutions lose stability and the arch falls back into the left or right stable bent position

nongeneric. There is a technical mathematical sense for the word generic (having to do with dense open coverings). But most of the time the word is just a fancy alternative for the plain English word "typical". Analysis of typical problems does not help you if your problem is not typical. For example, at a cusp point where $D=0$ and all second derivatives are not null we get the bifurcation diagrams like those shown in Fig. 3.10 (see [3.42]).

All the results which we have asserted so far can be shown to apply to problems of partial differential equations, like the Navier-Stokes equations, under a condition, to be explained in Sect. 3.8, called bifurcation at simple eigenvalue.

It is very important that at this point we note with emphasis that it is not necessary for equilibrium solutions of evolution equations to be connected by bifurcations. There are isolated solutions, which are as common as rain, which are not connected to other solutions through bifurcation. Such isolated solutions of $F(\mu,\varepsilon)=0$ occur even in one-dimensional problems (see Fig. 3.11 for one typical example). In the one-dimensional case it is possible to prove

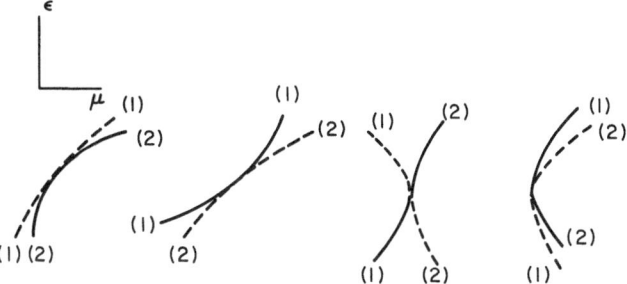

Fig. 3.10. Stability of solutions bifurcating at cusp point of second order

[3.42] that the stability of solutions which pierce the line $\mu=$ constant is of alternating sign as shown in Fig. 3.11. This result, however, is strictly one dimensional and does not apply to one-dimensional projections of higher dimensional problems in which curves of solutions which appear to intersect when projected onto the plane of the bifurcation diagram actually do not intersect in the higher dimensional space.

The possibilities for bifurcation which we have already indicated have important hydrodynamic applications. In particular we note that the super-critical form of bifurcation shown in Fig. 3.7 is typical of certain problems which have a high degree of spatial symmetry. The problem of bifurcation of Couette flow between rotating cylinders when the outer one is at rest is of this type when the cylinders are idealized to have an infinite length and the

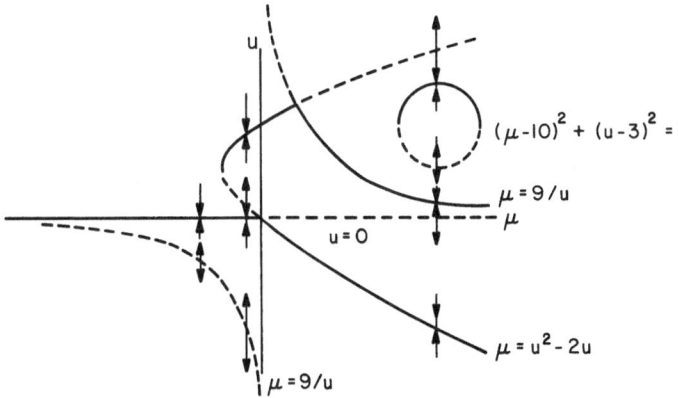

Fig. 3.11. Bifurcation, stability, and domains of attraction of equilibrium solutions of

$$du/dt = u(9-\mu u)(\mu+2u-u^2)\cdot[(\mu-10)^2+(u-3)^2-1].$$

The equilibrium solution $\mu=9/u$ in the third quadrant and the circle are isolated solutions which cannot be obtained by bifurcation analysis

disturbances may be assumed to be spatially periodic along the axis of the cylinders. In finite cylinders with closed ends the bifurcating flow is possibly of the transcritical type (see [3.31]). The first bifurcation of the idealized Bénard problem, in which the basic state is heat conduction without motion, is also of the supercritical type. Various modifications of this problem [Ref. 32, Chap. X; 3.45] lead to transcritical bifurcations. And the results of *Munson* and *Menguturk* [3.33] suggest that the first bifurcation of flow between rotating spheres may be of the transcritical type.

There are several important differences between bifurcations of the super-critical and transcritical type. Supercritical bifurcation is stable, and the stable flow is well approximated by eigenfunctions of linearized theory. In the transcritical case there is the possibility of hysteresis of the type shown in Fig. 3.9. In this case there are stable solutions which differ from the basic flow (here represented by the solution $u=0$) by a large amount ($|\varepsilon|$ in Fig. 3.7, θ in Fig. 3.9) at values of the Reynolds number (here $\mu<0$ in Fig. 3.7 and $l<l_c$ in Fig. 3.10) below those for which the basic flow is stable. So in the case of supercritical bifurcation we get agreement between what we find from studying the linear stability of the basic solution ($\varepsilon=0$) and what we observe. In the subcritical case and the transcritical case we may be observing the large amplitude solutions for which the linearized theory of stability of $\varepsilon=0$ is not relevant.

It is necessary to add that the conclusion which we have drawn from the study of problems in one dimension are only suggestive; they apply strictly in situations which may be described as bifurcation from a simple, real, isolated eigenvalue. We discuss these situations and draw closer to the actual complexities involved in the study of bifurcation and stability of solutions of the Navier-Stokes equation in Sect. 3.8.

3.6 Isolated Solutions Which Perturb Bifurcation

Isolated solutions are probably very common in hydrodynamical problems. One way to treat them is as a perturbation of problems which do bifurcate. This method of studying isolated solutions which are close to bifurcating solutions is known as imperfection theory. Some of the basic ideas involved in imperfection theory can be understood by comparing the bending of an initially imperfect, say bent, column (see Fig. 3.12). The first column will remain straight under end loadings P until a critical load P_c is reached. The column then undergoes supercritical, one-sided, double-point bifurcation (Euler buckling). In this perfect problem there is no way to decide if the column will buckle to the left or to the right. The situation is different for the initially bent column. The sidewise deflection starts as soon as the bent column is loaded and it deflects in the direction $x<0$ of the initial bending. If the initial bending is small, the deflection will resemble that of the perfect column. There will be small, nonzero

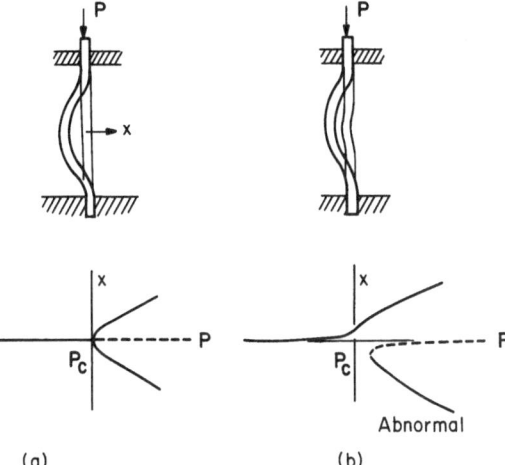

Fig. 3.12. (a) Buckling of a straight column: double-point supercritical bifurcation. (b) Bending of a bent column: isolated solutions which perturb double-point bifurcation

deflection with increasing load until a neighborhood of P_c is reached; then the deflection will increase rapidly with increasing load. When P is large it will be possible to push the deflected bent column into a stable "abnormal" position ($x > 0$) opposite to the direction of initial bending.

To understand the hydrodynamical implications of isolated solutions which perturb bifurcation, it is desirable to examine the possibilities with some generality. It is possible to do this simply, again by studying steady solutions of one-dimensional problems[6].

We first consider the problem

$$\frac{dx}{dt} = \tilde{F}(\mu, x, \delta) , \qquad (3.53)$$

where δ and μ are parameters, and F has at least three derivatives with respect to each of its three variables at the point $(0, 0, 0)$. To simplify notations we drop the overbar on $\tilde{F}$ and the partial derivatives of $\tilde{F}$ at $(0, 0, 0)$. For example

$$F = \tilde{F}(0, 0, 0) ,$$
$$F_\mu = \tilde{F}_\mu(0, 0, 0) , \quad \text{etc.} \qquad (3.54)$$

It is assumed that $(\mu, x) = (0, 0)$ is a double point of $\tilde{F}(\mu, x, 0) = 0$ and that $\tilde{F}$ is in local form, that is, $x = 0$ is a solution for all μ in a neighborhood of zero when $\delta = 0$. Since $\tilde{F}(\mu, 0) = 0$ is an identity in μ, all the partial derivatives of $\tilde{F}(\mu, x, \delta)$

---

6 The results given here have much in common with the work of *Koiter* [3.46], *Thom* [3.47], *Benjamin* [3.48], *Matkowsky* and *Reiss* [3.49], and *Golubitsky* and *Schaeffer* [3.50]. The particular formulation of this section, the iterative procedures, and the stability results are taken from the forthcoming book on stability and bifurcation theory by *Iooss* and *Joseph* [3.51].

with respect to μ vanish at $(0,0,0)$. Since $(0,0,0)$ is a double point

$$F=F_x=0 \quad \text{and} \quad D=F_{\mu x}^2>0 . \tag{3.55}$$

We are interested in the steady solutions $x=\varepsilon$, $d\varepsilon/dt=0$ of

$$\tilde{F}(\mu,\varepsilon,\delta)=0 \tag{3.56}$$

which break the solutions which bifurcate at the double point into isolated solutions. For this, it is enough that

$$F_\delta \neq 0 . \tag{3.57}$$

Let us derive the form of the isolated solutions which break the bifurcation. The implicit function theorem and (3.57) guarantee that there is a function $\delta = \varDelta(\mu,\varepsilon)$ with $\varDelta(0,0)=0$ and

$$\tilde{F}[\mu,\varepsilon,\varDelta(\mu,\varepsilon)]\equiv 0 . \tag{3.58}$$

It follows from (3.58) and the fact that $\tilde{F}(\mu,0,0)=0$ that we may take $\varDelta(\mu,0)=0$ so that all the partial derivatives of $\varDelta(\mu,\varepsilon)$ with respect to μ alone vanish when $(\mu,\varepsilon)=(0,0)$. Differentiating (3.58) with respect to ε and μ at $(\mu,\varepsilon)=(0,0)$, we find that

$$F_\varepsilon+F_\delta \varDelta_\varepsilon=0 , \tag{3.59}$$

$$F_{\varepsilon\varepsilon}+F_\delta \varDelta_{\varepsilon\varepsilon}=0 , \tag{3.60}$$

$$F_{\mu\varepsilon}+F_\delta \varDelta_{\mu\varepsilon}=0 . \tag{3.61}$$

Now $F_\mu=0$ identically, and (3.59) shows that $\varDelta_\varepsilon=0$ so that the surface $\delta = \varDelta(\mu,\varepsilon)$ is tangent to the plane $\delta=0$ in the three-dimensional space with coordinates (μ,ε,δ) at the point $(0,0,0)$. From (3.61) we learn that

$$D[\varDelta(0,0)]=\varDelta_{\mu\varepsilon}^2=F_{\mu\varepsilon}^2/F_\delta>0 . \tag{3.62}$$

so that the point $(0,0,0)$ is a saddle point.

We may find all of the derivatives of $\varDelta(\mu,\varepsilon)$ at $(0,0)$ in terms of the derivatives of $\tilde{F}$ at $(0,0,0)$ by differentiating (3.58) repeatedly with respect to μ and ε. We find that

$$\varDelta(\mu,\varepsilon)= \frac{1}{2}(\varDelta_{\varepsilon\varepsilon}\varepsilon^2+2\varDelta_{\varepsilon\mu}\varepsilon\mu)$$

$$+ \frac{1}{3!}(\varDelta_{\varepsilon\varepsilon\varepsilon}\varepsilon^3+3\varDelta_{\mu\varepsilon\varepsilon}\mu\varepsilon^2+3\varDelta_{\mu\mu\varepsilon}\mu^2\varepsilon)+ \dots \tag{3.63}$$

$$= a\varepsilon^2+2b\varepsilon\mu+d\varepsilon^3+e\mu\varepsilon^2+f\mu^2\varepsilon+ \dots ,$$

where $\Delta_{\varepsilon\varepsilon}$ and $\Delta_{\mu\varepsilon}$ are given by (3.60) and (3.61) with similar equations for third derivatives.

Our problem now is to solve (3.63) for $\mu(\varepsilon,\delta)$ for a fixed value δ. The intersection of the surface $\delta=\Delta(\mu,\varepsilon)$ and the planes $\delta=$ constant determine these curves. The plane $\delta=$ constant may be written in parametric form with ε as a parameter.

$$\delta=\varepsilon\hat{\delta}=\varepsilon\hat{\Delta}(\mu,\varepsilon)\,, \tag{3.64}$$

$$\hat{\delta}=\hat{\Delta}(\mu,\varepsilon)=a\varepsilon+2b\mu+d\varepsilon^2+e\varepsilon\mu+f\mu^2+\dots\,. \tag{3.65}$$

We can solve (3.64, 65) by successive approximations. First solve for

$$\mu=\frac{1}{2b}(\hat{\delta}-a\varepsilon-d\varepsilon^2-e\mu\varepsilon-f\mu^2+\dots)\,. \tag{3.66}$$

The first approximation is given by

$$\mu\sim\mu^{(1)}=\frac{1}{2b}(\hat{\delta}-a\varepsilon)=\frac{1}{2b}\left(\frac{\delta}{\varepsilon}-a\varepsilon\right)\,. \tag{3.67}$$

Equation (3.67) gives two isolated solutions which break double-point bifurcation. For example, if $a=0$, as in supercritical bifurcation, we get two bifurcating solutions when $\delta=0$: $\varepsilon=0$ and $\mu=0$. The isolated solutions which perturb these bifurcating solutions when $\delta\neq0$ are given by the pair of hyperbolas $\mu=\delta/2b\varepsilon$.

The second approximation is given by

$$\mu\sim\mu^{(2)}=\frac{1}{2b}(\hat{\delta}-a\varepsilon-d\varepsilon^2-e\varepsilon\mu^{(1)}-f\mu^{(1)^2})$$

$$=\frac{1}{2b}\left[\frac{\delta}{\varepsilon}-a\varepsilon-d\varepsilon^2-\frac{e}{2b}(\delta-a\varepsilon^2)-\frac{f}{4b^2}\left(\frac{\delta}{\varepsilon}-a\varepsilon\right)^2\right]\,. \tag{3.68}$$

For example, if $a=0$ we get two bifurcating solutions when $\delta=0$. These bifurcating solutions are given locally by

$$\varepsilon=0 \quad \text{and} \quad \mu=-\frac{d}{2b}\varepsilon^2\,, \tag{3.69}$$

corresponding to one-sided supercritical bifurcation if $d/2b<0$. The isolated solutions which break bifurcation when $\delta\neq0$ are given by (3.68). In the supercritical case one possible pair of isolated solutions is that shown in Fig. 3.12b.

The stability of isolated solutions on the curve $\mu(\varepsilon,\delta)$ may be obtained from the factorization theorem. Perturbing the solutions $[\mu(\varepsilon,\delta),\varepsilon]$ of (3.53) with

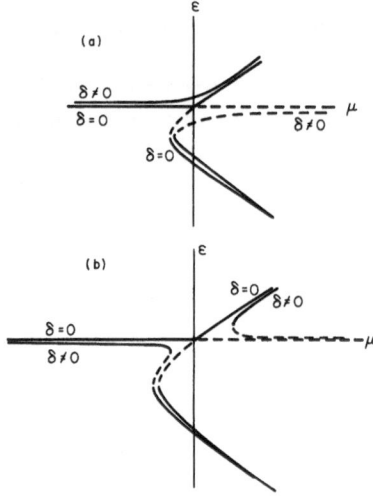

Fig. 3.13a, b. Isolated solutions which perturb transcritical bifurcation. The solution $\delta \neq 0$, $\mu < 0$ in **(b)** exhibits hysteresis as shown

small disturbances proportional to $\exp \gamma t$, we find that

$$\gamma(\varepsilon) = F_\varepsilon[\mu(\delta, \varepsilon), \varepsilon, \delta] = -\mu_\varepsilon(\varepsilon, \delta) F_\mu[\mu(\delta, \varepsilon), \varepsilon, \delta] \ . \tag{3.70}$$

Then we have:

I) Stable branches in the bifurcation diagram perturb, by continuity, into stable branches of the perturbed solutions except at

II) regular turning points where the stability index of a solution changes sign.

Typical cases of the breaking of bifurcation with δ, in the second approximation, combined with the stability principles, are shown in Fig. 3.12 and in Fig. 3.13.

At a fixed value of μ there is one stable solution when μ is small. At larger values of μ there are two stable solutions, one for $\varepsilon > 0$ and one for $\varepsilon < 0$. In the hydrodynamic context $\varepsilon > 0$ implies a cellular motion with one sense of circulation and $\varepsilon < 0$ a circulation with the other sense. So the isolated solutions, like the bifurcating solutions, come in pairs. It follows then, even from the most elementary theoretical arguments, that stable cellular motions with different circulations are to be expected in flows which perturb bifurcation. In this sense the normal and abnormal cells which were discussed in Sect. 3.4 are to be expected, despite the fact that the "abnormal" cells appear to violate physical intuition.

3.7 Bifurcation of Steady Flow into Time-Periodic Flow

The problem of bifurcation of steady flow into time-periodic flow is basically two dimensional. It is not possible for a time-periodic solution to bifurcate from a steady one in one dimension. Time-periodic bifurcations are very important in hydrodynamics. In fact, a time-periodic bifurcation seems always to play a role in the transition to turbulence. It appears as one of the bifurcations in the finite sequence of supercritical bifurcations which lead to turbulence in the Bénard problem and in the Taylor problem. And a subcritical, unstable, time-periodic bifurcation may be characteristic of flows which undergo a direct, snap-through transition from steady laminar flow to turbulent flow.

The problem of bifurcation of steady solutions of the Navier-Stokes equations into time-periodic solutions can be reduced, after analysis involving projections (see Sect. 3.8), to a two-dimensional problem associated with the following pair of nonlinear ordinary differential equations:

$$\frac{dx_i}{dt} = F_i(\mu, x_1, x_2) = A_{ij}(\mu)x_j + k_{ijk}(\mu)x_j x_k$$

$$+ \text{ higher order terms ,} \tag{3.71}$$

where the summation convention holds, $i = 1, 2$, $k_{ijk} = k_{ikj}$, and $A_{ij}(\mu)$ are components of

$$A(\mu) = \begin{pmatrix} a(\mu) & b(\mu) \\ c(\mu) & d(\mu) \end{pmatrix} . \tag{3.72}$$

We suppose that $(a - d)^2 + 4bc > 0$ in a neighborhood of $\mu = 0$. Then the eigenvalues $\sigma(\mu) = \xi(\mu) + i\eta(\mu)$ and eigenvectors $\zeta(\mu)$ of $A(\mu)$ are complex conjugates, and

$$\sigma(\mu)\zeta = A\zeta \quad (\sigma\zeta_i = A_{ij}\zeta_j) , \tag{3.73}$$

and

$$\sigma(\mu)\bar{\zeta}^* = A^T\bar{\zeta}^* , \tag{3.74}$$

where ζ^* is the adjoint eigenvector with eigenvalue $\bar{\sigma}(\mu)$ in the scalar product, $\langle x, y \rangle = x \cdot \bar{y}$. We may normalize so that

$$\langle \zeta, \zeta^* \rangle = \zeta \cdot \bar{\zeta}^* = \zeta_k \bar{\zeta}_k^* = 1 ,$$

$$\langle \zeta, \bar{\zeta}^* \rangle = \zeta_k \zeta_k^* = 0 . \tag{3.75}$$

The eigenvalues $\sigma(\mu) = \xi(\mu) + i\eta(\mu)$ arise in the spectral problem for the stability of the solution $x_i = 0$ of (3.71). We suppose that this loss of stability occurs at $\mu = 0$ so that $\xi(0) = 0$. We will get bifurcation into periodic solutions if

$$\eta(0) = \omega_0 \neq 0 \quad \text{and} \quad \xi_\mu(0) \neq 0 \, [\text{say } \xi_\mu(0) > 0] . \tag{3.76}$$

To prove bifurcation into periodic solutions under conditions (3.76), we note that ζ and $\bar{\zeta}$ are independent so that any real-valued two-dimensional vector $x=(x_1, x_2)$ may be represented as

$$x_i = a(t)\zeta_i + \bar{a}(t)\bar{\zeta}_i \ . \tag{3.77}$$

Substitute (3.77) in (3.71) and use (3.73) to find

$$\mathring{a}\zeta_i + \mathring{\bar{a}}\bar{\zeta}_i = \sigma(\mu)\zeta_i + \bar{\sigma}(\mu)\bar{\zeta}_i + a^2 k_{ijk}\zeta_j\zeta_k$$
$$+ 2|a|^2 k_{ijk}\zeta_i\bar{\zeta}_k + \bar{a}^2 k_{ijk}\bar{\zeta}_i\bar{\zeta}_k$$
$$+ O(|a|^3) \ . \tag{3.78}$$

The orthogonality properties (3.75) are now employed to reduce (3.78) to a single, complex-valued, amplitude equation

$$\mathring{a} = f(\mu, a) = \sigma(\mu)a + \alpha(\mu)a^2 + 2\beta(\mu)|a|^2 + \gamma(\mu)\bar{a}^2 + O(|a|^3), \tag{3.79}$$

where, for example, $\alpha(\mu) = k_{ijk}(\mu)\zeta_j\zeta_k\zeta_i^*$. The linearized stability of the solution $a=0$ of (3.79) is determined by $\mathring{a} = \sigma(\mu)a$, $a = (\text{constant}) \times \exp[\sigma(\mu)t]$. At criticality ($\mu=0$), $a = (\text{constant}) \times \exp(i\omega_0 t)$ is 2π periodic in $s = \omega_0 t$.

A bifurcating time-periodic solution may be constructed from the solution of the linearized problem at criticality. This bifurcating solution is in the form

$$a(t) = b(s, \varepsilon), \quad s = \omega(\varepsilon)t, \quad \omega(0) = \omega_0, \quad \mu = \mu(\varepsilon) \ , \tag{3.80}$$

where ε is the amplitude of a defined by

$$\varepsilon = \frac{1}{2\pi} \int_0^{2\pi} \exp(-is)b(s, \varepsilon)ds = [b] \ . \tag{3.81}$$

The solution (3.80) of (3.79) is unique to within an arbitrary translation of the origin of the time. This means that under translation $t \to t+c$ the solution $b(s, \varepsilon) \to b[s+c\omega(\varepsilon), \varepsilon]$ shifts its phase. This unique solution is analytic in ε when $f(\mu, a)$ is analytic in both variables and it may be expressed as a series

$$\begin{pmatrix} b(s, \varepsilon) \\ \omega(\varepsilon) - \omega_0 \\ \mu(\varepsilon) \end{pmatrix} = \sum_{n=1}^{\infty} \varepsilon^n \begin{pmatrix} b_n(s) \\ \omega_n \\ \mu_n \end{pmatrix} . \tag{3.82}$$

The perturbation problems which govern $b_n(s)$, ω_n, and μ_n can be obtained by identifying the coefficients of ε^n which arise when (3.82) is substituted into the two equations: $\omega\overset{\circ}{b} = f(\mu, b)$ and $\varepsilon = [b]$. We find that at order one

$$\omega_0\overset{\circ}{b}_1 - i\omega_0 b_1 = 0, \quad [b_1] = 1, \quad b_1(s) = e^{is}. \tag{3.83}$$

At order two, we find that $[b_2] = 0$ and

$$\omega_0[\overset{\circ}{b}_2 - ib_2] + \omega_1\overset{\circ}{b}_1 = \mu_1\sigma_\mu b_1 + \alpha_0 b_1^2 + 2\beta_0|b_1|^2 + \gamma_0\bar{b}_1^{\,2}, \tag{3.84}$$

where $\sigma_\mu = d\sigma(0)/d\mu$ and, for example, $\alpha_0 = \alpha(0)$.

Equations of the form $\overset{\circ}{b}(s) - ib(s) = f(s) = f(s + 2\pi)$ are solvable for $b(s) = b(s + 2\pi)$ if and only if $f(s)$ has no term proportional to $\exp(is)$. Hence

$$\mu_1 = \omega_1 = 0 \quad \text{in (3.84)} \tag{3.85}$$

and

$$\overset{\circ}{b}_2 - ib_2 = [\alpha_0\exp(2is) + 2\beta_0 + \gamma_0\exp(-2is)]/\omega_0. \tag{3.86}$$

We find that

$$b_2(s) = (\alpha_0 e^{2is} - 2\beta_0 - \gamma_0 e^{-2is}/3)/i\omega_0. \tag{3.87}$$

The problem which governs at order 3 is $[b_3] = 0$ and

$$\overset{\circ}{b}_3 - ib_3 = [-\omega_2\overset{\circ}{b}_1 + \mu_2\sigma_\mu b_1 + 2\alpha_0 b_1 b_2$$
$$+ 2\beta_0(b_1\bar{b}_2 + \bar{b}_1 b_2) + 2\gamma_0\bar{b}_1\bar{b}_2]/\omega_0. \tag{3.88}$$

To solve (3.88) we must eliminate terms proportional to $\exp(is)$ from the right side of (3.88). This is done if

$$i\omega_2 - \mu_2\sigma_\mu = (4\alpha_0\beta_0 - 4|\beta_0|^2 - 2\alpha_0\beta_0 - 2|\gamma_0|^2/3)/i\omega_0. \tag{3.89}$$

The real part of (3.89) is solvable for μ_2 provided that $\xi' \neq 0$. The imaginary part of (3.89) is always solvable for ω_2.

Proceeding to higher orders, it is easy to verify that all of the perturbation problems are solvable when (3.76) holds and, in fact, $\omega(\varepsilon) = \omega(-\varepsilon)$, $\mu(\varepsilon) = \mu(-\varepsilon)$ are even functions. It follows that periodic solutions which bifurcate from steady solutions bifurcate to one or the other side of criticality and never to both sides; periodic bifurcating solutions cannot undergo two-sided or transcritical bifurcation (cf. Figs. 3.14 and 3.7c).

We now search for the conditions under which the bifurcating periodic solutions are stable. We consider a small disturbance $z(t)$ of $b(s, t)$. Setting $a(t) = b(s, \varepsilon) + z(t)$ in (3.79) we find the linearized equation $\dot{z}(t) = f_a[\mu(\varepsilon), b(s, \varepsilon)]$ $z(t)$ where $f_a = \partial f/\partial a$ and $s = \omega(\varepsilon)t$. Then, using Floquet theory (see [3.51]), we

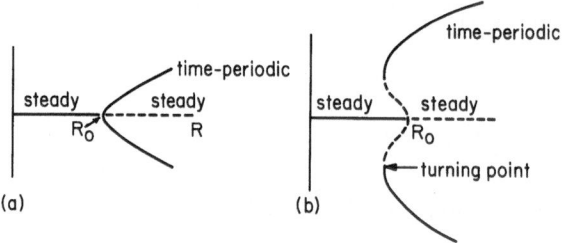

Fig. 3.14a, b. Supercritical (**a**) and subcritical (**b**) Hopf bifurcation. The part of the bifurcation diagram giving the values $R(\varepsilon)$ for which periodic solutions exist are symmetric, $R(\varepsilon)=R(-\varepsilon)$. For Navier-Stokes problems we always expect turning points at which the sign of the slope of $R(\varepsilon)$ changes, as in (**b**). It is not possible to have a periodic solution of a problem with steady data when R is small (see Sect. 3.2). One example of supercritical Hopf bifurcation (**a**) in hydrodynamics occurs in the problem of loss of stability of the steady Ekman layer [3.52]. The problem of loss of stability of Poiseuille flow is associated with subcritical Hopf bifurcation of type (**b**) (see Chap. 7)

set $z(t)=\exp(\gamma t)y(s)$ where $y(s)=y(s+2\pi)$ and find that

$$\gamma y = -\mathring{y} + f_a(\mu,b)y \equiv J[\mu(\varepsilon),b(s,\varepsilon)]y . \tag{3.90}$$

The stability result we need may be stated as a factorization theorem. To prove this theorem we use the fact that $\gamma=0$ is always an eigenvalue of J with eigenfunction $\mathring{b}(s,\varepsilon)$,

$$J\mathring{b}=0 , \tag{3.91}$$

and the relation

$$\omega_\varepsilon(\varepsilon)\mathring{b}(s,\varepsilon)=\mu_\varepsilon(\varepsilon)f_\mu[\mu(\varepsilon),b(s,\varepsilon)]+Jb_\varepsilon , \tag{3.92}$$

which arises from differentiating $\omega\mathring{b}=f(\mu,b)$ with respect to ε at any ε.

Factorization Theorem:

$$\left\{ \begin{array}{l} y(s,\varepsilon)=c(\varepsilon)\left[\dfrac{\tau}{\gamma}\mathring{b}(s,\varepsilon)+b_\varepsilon(s,\varepsilon)+\mu_\varepsilon(\varepsilon)q(s,\varepsilon)\right] , \\[2mm] \tau(\varepsilon)=\omega_\varepsilon(\varepsilon)+\mu_\varepsilon(\varepsilon)\hat{\tau}(\varepsilon) , \\[2mm] \gamma(\varepsilon)=\mu_\varepsilon(\varepsilon)\hat{\gamma}(\varepsilon) , \end{array} \right\} \tag{3.93}$$

where $c(\varepsilon)$ is an arbitrary constant and $q(s,\varepsilon)=q(s+2\pi,\varepsilon)$, $\hat{\tau}(\varepsilon)$, and $\hat{\gamma}(\varepsilon)$ satisfy the equation

$$\hat{\tau}\mathring{b}+\hat{\gamma}b_\varepsilon+f_\mu(\mu,b)+\varepsilon(\gamma q-Jq)=0 \tag{3.94}$$

and are real analytic functions in a neighborhood of $\varepsilon = 0$. *Moreover,* $\hat{\tau}(\varepsilon)$ *and* $\hat{\gamma}(\varepsilon)/\varepsilon$ *are even functions and such that*

$$\hat{\gamma}_\varepsilon(0) = -\xi_\mu(0), \qquad \hat{\tau}(0) = -\eta_\mu(0) . \tag{3.95}$$

Proof: Substitute the representations (3.93) into (3.90) utilizing (3.91) to eliminate $J\overset{\circ}{b}$ and (3.92) to eliminate Jb_ε. This leads to (3.94), and (3.94) may be solved by series

$$\begin{pmatrix} q(s, \varepsilon) \\ \hat{\gamma}(\varepsilon)/\varepsilon \\ \hat{\tau}(\varepsilon) \end{pmatrix} = \sum_{l=0} \begin{pmatrix} q_l(s) \\ \hat{\gamma}_l \\ \hat{\tau}_l \end{pmatrix} \varepsilon^l , \tag{3.96}$$

where $\hat{\gamma}_0 = \hat{\gamma}_\varepsilon(0)$ and $\hat{\tau}_0 = \hat{\tau}(0)$. Using the fact that to the lowest order, $b = \varepsilon \exp(is)$, $\gamma = 0(\varepsilon^2)$ and [from (3.79)] $f_\mu(\mu, b) = \sigma_\mu(0) \exp(is)\varepsilon$, we find that

$$\exp(is)[\hat{\tau}(0) - \hat{\gamma}_\varepsilon(0) - \sigma_\mu] + J_0 q_0 = 0 . \tag{3.97}$$

Equation (3.97) is solvable for $q_0(s) = q_0(s + 2\pi)$ if and only if the term in the bracket vanishes; that is, (3.95) holds. The remaining properties asserted in the theorem may be obtained by mathematical induction using the power series (3.96) (see [Ref. 3.1, Chap. 2]).

The linearized stability of the periodic solution for small values of ε can now be obtained from the spectral problem: $x(s, \varepsilon) = x(s + 2\pi, \varepsilon)$ is stable when $\xi(\varepsilon) < 0$ and is unstable when $\xi(\varepsilon) > 0$ where

$$\begin{aligned} \gamma(\varepsilon) &= \xi(\varepsilon) + i\eta(\varepsilon) = \mu_\varepsilon(\varepsilon)\hat{\gamma}(\varepsilon) \\ &= -\mu_\varepsilon(\varepsilon)\{[\xi_\mu(0) + i\eta_\mu(0)]\varepsilon + O(\varepsilon^3)\} . \end{aligned} \tag{3.98}$$

We have already assumed that the basic flow loses stability strictly when μ is increased past zero, $\xi_\mu(0) > 0$. So the branches for which $\mu_\varepsilon(\varepsilon) > 0$ are stable and the ones for which $\mu_\varepsilon(\varepsilon)\varepsilon < 0$ are unstable. There are two possibilities when ε is small: supercritical bifurcation (Fig. 3.14a) [3.52] or subcritical bifurcation (Fig. 3.14b). It is not possible to have transcritical periodic bifurcations as in Fig. 3.7c because $\mu(\varepsilon) = \mu(-\varepsilon)$.

3.8 Finite Dimensional Projections

You may think that the analysis of stability and bifurcation in one and two dimensions is merely suggestive of some typical hydrodynamical situations. In fact, the analysis applies strictly to general problems of bifurcation of solutions of the Navier-Stokes equations when certain typical conditions are satisfied. The typical conditions I have in mind are those associated with bifurcation at a simple eigenvalue; they are precisely described in the paragraph two up from (3.19) and in the paragraph one down from (3.21). The one- and two-dimensional problems arise as finite-dimensional projections of the evolution equation (3.14–16) by a procedure which I will now describe.

First we define a scalar product

$$\langle a, b \rangle \equiv \int a \cdot \bar{b} \, dr = \langle \overline{b, a} \rangle . \tag{3.99}$$

Associated with this scalar product is a linear operator $F_u^*(0, R|\cdot)$ which is called the adjoint of $F_u(0, R|\cdot)$ and is defined by the relation

$$\langle F_u(0, R|a), b \rangle = \langle a, F_u^*(0, R|b) \rangle \tag{3.100}$$

for all fields a and b in H. It is readily verified that

$$\langle \nabla \phi, b \rangle = \langle \nabla \cdot (b\phi) - \phi \nabla \cdot b \rangle = 0 \tag{3.101}$$

for all $b \in H$ and any scalar field ϕ. It follows now from (3.101) and (3.18) that

$$\langle F_u(0, R|\zeta), b \rangle = \sigma \langle \zeta, b \rangle \tag{3.102}$$

for all $b \in H$. An adjoint eigenvector ζ^* may be associated with the second member of (3.100)

$$\langle a, F_u^*(0, R|\zeta^*) \rangle = \sigma \langle a, \zeta^* \rangle = \langle a \bar{\sigma} \zeta^* \rangle \tag{3.103}$$

for all $a \in H$, so that

$$\bar{\sigma} \zeta^* = F_u^*(0, R|\zeta^*) - \nabla p[\zeta^*] , \quad \zeta^* \in H \tag{3.104}$$

is the eigenvector problem adjoint to (3.18). It has exactly the same set of eigenvalues as (3.18) [cf. (3.74)].

There is an infinite number of eigenvalues $\sigma(R)$ which may be arranged in a sequence corresponding to the size of their real parts

$$\xi_1 \geq \xi_2 \geq \xi_3 \geq \cdots \geq \xi_n \geq \cdots \tag{3.105}$$

clustering at $-\infty$ (see Fig. 3.1). To each eigenvalue there corresponds at most a finite number of eigenvectors ζ_n and adjoint eigenvectors ζ_n^*.

An eigenprojection is the scalar product of a field in H with one of adjoint eigenvectors. So we may form eigenprojections of the evolution equation (3.14–16)

$$\frac{d\langle u, \zeta_n^* \rangle}{dt} = \langle F(u, R), \zeta_n^* \rangle$$

$$= \langle F_u(0, R|u), \zeta_n^* \rangle - \langle u \cdot \nabla u, \zeta_n^* \rangle ,$$

$$= \langle u, F_u^*(0, R|\zeta_n^*) \rangle - \langle u \cdot \nabla u, \zeta_n^* \rangle ,$$

$$= \sigma_n \langle u, \zeta_n^* \rangle - \langle u \cdot \nabla u, \zeta_n^* \rangle . \tag{3.106}$$

There are as many projections as there are adjoint eigenvectors. It is evident that when u is small

$$\langle u(t), \zeta_n^* \rangle \simeq \langle u(0), \zeta_n^* \rangle e^{\xi_n(R)t} e^{i\eta_n(R)t} \tag{3.107}$$

so that all the projections with $\xi(R) < 0$ decay to zero. It follows that the important parts of the evolution problem (3.18) are the eigenprojections associated with eigenvalue $\sigma_n(R)$ for which $\xi_n(R) > 0$.

In the problem of bifurcation at a simple eigenvalue we suppose that $\xi_n(R) < 0$ when $n > 1$. The important eigenprojection is then the one associated with $\zeta_1^* \equiv \zeta^*$ and $\sigma_1 \equiv \sigma$,

$$\frac{d\langle u, \zeta^* \rangle}{dt} = \langle F(u, R), \zeta^* \rangle = \sigma(R)\langle u, \zeta^* \rangle - \langle u \cdot \nabla u, \zeta^* \rangle . \tag{3.108}$$

If $\sigma(R)$ is real valued, (3.108) is a one-dimensional problem. If $\sigma(R)$ is complex values, there are two equations, the real and imaginary parts of (3.108), and the problem is essentially two-dimensional

I now want to delineate the sense in which the essentially two-dimensional problem is strictly two dimensional. We first decompose the bifurcating solution u into a real-valued orthogonal sum

$$u = a(t)\zeta + \bar{a}(t)\bar{\zeta} + w , \tag{3.109}$$

where

$$\langle \zeta, \zeta^* \rangle - 1 = \langle \zeta, \bar{\zeta}^* \rangle = \langle w, \zeta^* \rangle = 0 . \tag{3.110}$$

Substitute (3.108) into (3.14–16) and use (3.18) to derive

$$[\mathring{a} - \sigma(R)a]\zeta + [\mathring{\bar{a}} - \bar{\sigma}(R)\bar{a}]\bar{\zeta} + \frac{\partial w}{\partial t} = F_u(0, R|w) - u \cdot \nabla u - \nabla p[w] , \tag{3.111}$$

where $\nabla p[w] = \nabla(p[u] - ap[\zeta] - \bar{a}p[\bar{\zeta}])$. Now project (3.111) with ζ^*, use (3.109),

$$\left\langle \frac{\partial w}{\partial t}, \zeta^* \right\rangle = \frac{d\langle w, \zeta^* \rangle}{dt} = 0 ,$$

$$\langle F_u(0, R|w), \zeta^* \rangle = \langle w, F_u^*(0, R|\zeta^*) \rangle = \sigma \langle w, \zeta^* \rangle = 0 , \tag{3.112}$$

and show that

$$\mathring{a} - \sigma(R)a = -\langle u \cdot \nabla u, \zeta^* \rangle = \alpha(R)a^2 + 2\beta(R)|a|^2 + \gamma(R)\bar{a}^2$$
$$- [a\langle (\zeta \cdot \nabla w + w \cdot \nabla \zeta), \zeta^* \rangle + \bar{a}\langle (\bar{\zeta} \cdot \nabla w + w \cdot \nabla \bar{\zeta}), \zeta^* \rangle]$$
$$- \langle w \cdot \nabla w, \zeta^* \rangle , \tag{3.113}$$

where

$$\alpha(R) = -\langle \zeta \cdot \nabla \zeta, \zeta^* \rangle ,$$
$$2\beta(R) = -\langle (\bar{\zeta} \cdot \nabla \zeta + \zeta \cdot \nabla \zeta), \zeta^* \rangle , \tag{3.114}$$
$$\gamma(R) = -\langle \bar{\zeta} \cdot \nabla \bar{\zeta}, \zeta^* \rangle .$$

Returning now to (3.111) with (3.113), we find that

$$\frac{\partial w}{\partial t} = F_u(0, R|w) - (u \cdot \nabla u - \langle u \cdot \nabla u, \zeta^* \rangle \zeta - \langle u \cdot \nabla u, \bar{\zeta}^* \rangle \bar{\zeta}) - \nabla p[w] , \tag{3.115}$$

and, using (3.108), we conclude from (3.115) that $w = O(|a|^2)$. Equation (3.113) governs the evolution of the projection of the solution u into the eigensubspace belonging to the eigenvalue $\sigma(R) = \sigma(R)$, and (3.115) governs the evolution of the part of the solution orthogonal to ζ^*.

In bifurcation problems the complementary projection w plays a minor role; it arises only as a response generated by nonlinear coupling to the component of the solution spanned by ζ and $\bar{\zeta}$. Since $w = O(|a|^2)$, we may dramatize the two-dimensional structure by comparing (3.113), written as

$$\mathring{a} = \sigma(R)a + \alpha(R)a^2 + 2\beta(R)|a|^2 + \gamma(R)\bar{a}^2 + O(|a|^3) , \tag{3.116}$$

with the equation (3.79) which governs in the strictly two-dimensional problem.

The nature of the relatively unimportant variation of the present problem with $w \neq 0$ from the strictly two-dimensional problem studied in Sect. 3.7 can be best appreciated by carrying out an explicit computation of Hopf bifurcation starting from (3.113) and (3.115). We seek a 2π periodic solution $u(s, \varepsilon) = u(s + 2\pi, \varepsilon)$, where $s = \omega(\varepsilon)t$ and $R = R(\varepsilon)$, $z^* = \exp(is)\zeta^*$, and

$$\varepsilon = [u(s, \varepsilon), z^*(s)] \equiv \frac{1}{2\pi} \int_0^{2\pi} \langle u(s, \varepsilon), z^* \rangle ds . \tag{3.117}$$

Since the solution is to bifurcate from $u(s,0)=0$ when $R(0)=R_0$ and $\omega(0)=\omega_0$, we may set $a(t)=b(s,\varepsilon)=\varepsilon B(s,\varepsilon)$, where $B(s,0)=b_1(s)$ is bounded. Equation (3.115) then shows that

$$w(s,\varepsilon)\Rightarrow\varepsilon^2 w(s,\varepsilon) , \tag{3.118}$$

so that (3.108) becomes

$$u=\varepsilon[B(s,\varepsilon)\zeta+\bar{B}(s,\varepsilon)\bar{\zeta}]+\varepsilon^2 w(s,\varepsilon) , \tag{3.119}$$

where $[w,z^*]=0$. It follows from (3.119), (3.117), and (3.109) that

$$1=\frac{1}{2\pi}\int_0^{2\pi}e^{-is}B(S,\varepsilon)ds, \equiv[B] , \tag{3.120}$$

which is exactly the normalizing condition (3.81). Moreover, (3.113) becomes

$$\omega(\varepsilon)\mathring{B}-\sigma[R(s)]B=\varepsilon[\alpha(R)B^2+2\beta(R)|B|^2+\gamma(R)\bar{B}^2]$$
$$-\varepsilon^3[B\langle(\zeta\cdot\nabla w+w\cdot\nabla\zeta),\zeta^*\rangle$$
$$+\bar{B}\langle(\bar{\zeta}\cdot\nabla w+w\cdot\nabla\bar{\zeta}),\zeta^*\rangle]$$
$$-\varepsilon^5\langle w\cdot\nabla w,\zeta^*\rangle \tag{3.121}$$

and (3.115) becomes

$$\omega(\varepsilon)\frac{\partial w}{\partial s}=F_u[0,R(\varepsilon)|w]-M(B\zeta+\bar{B}\bar{\zeta}+\varepsilon w,B\zeta+\bar{B}\bar{\zeta}+\varepsilon w)$$
$$-\nabla p[w] , \tag{3.122}$$

where

$$M(u,u)=u\cdot\nabla u-\langle u\cdot\nabla u,\zeta^*\rangle\zeta-\langle u\cdot\nabla u,\bar{\zeta}^*\rangle\bar{\zeta} . \tag{3.123}$$

We may construct the solution of (3.121) and (3.122) in a series equivalent to (3.82)

$$\begin{aligned}
B(s,\varepsilon)&=b_1(s)+\varepsilon b_2(s)+\varepsilon^2 b_3(s)+\dots\\
\omega(\varepsilon)&=\omega_0+\omega_1\varepsilon+\omega_2\varepsilon^2+\dots\\
R(\varepsilon)&=R_0+R_1\varepsilon+R_2\varepsilon^2+\dots\\
w(s,\varepsilon)&=w_0(s)+\varepsilon w_1(s)+\dots .
\end{aligned} \tag{3.124}$$

Substituting these representations into (3.121) and (3.122), we find that $\omega_1=R_1=0$ and $b_1(s), b_2(s), b_3(s), \omega_2$, and R_2 are exactly as given in Sect. 3.7 (with μ replaced by R). The function $w(s,\varepsilon)$ first enters into the analysis of $O(\varepsilon^3)$.

The equations governing the coefficients $w_n(s) = w_n(s + 2\pi)$ are in the form

$$\omega_0 \frac{\partial w_n}{\partial s} - F_u(0, R_0 | w_n) = g_n - \nabla p[w_n] , \qquad (3.125)$$

where g_n depends on lower order terms; for example,

$$g_0 = - M(b_1 \zeta + \bar{b}_1 \bar{\zeta}, b_1 \zeta + \bar{b}_1 \bar{\zeta}) . \qquad (3.126)$$

The necessary and sufficient condition for the solvability (3.125) is that g_n should be orthogonal to 2π periodic eigenvectors $z^* = \exp(is)\zeta^*$ and $\bar{z}^*$ of the operator

$$\omega_0 \frac{\partial}{\partial s} + F_u^*(0, R_0 | \cdot) , \qquad (3.127)$$

which is adjoint to the operator on the left of (3.125) relative to the scalar product $[\cdot, \cdot]$ defined by (3.117). It is easy to verify that the problems (3.125) are uniquely solvable among vectors for which $[w_n, z^*] = 0$.

All of the results of bifurcation theory, hydrodynamic or otherwise, are based on the method of projections. In mathematically strict discussions of projections it is necessary to introduce concepts from functional analysis. The end result of the application of functional analysis to partial differential equations is a reduction to problems of finite dimension. The existence and properties of bifurcation are associated with the finite dimensional projections and require classical analysis of functions and ordinary differential equations rather than functional analysis for their elucidation.

We have studied problems which could be reduced to one or two dimensions. It is also possible to consider multiple eigenvalue problems of multiplicity $N > 2$. Such problems are N dimensional and in the case of symmetry-breaking bifurcations of steady uniform solutions into steady solutions they may be reduced to the study of nonlinear algebraic equations. Such breaking of symmetry is well known in the Bénard problem between infinite planes. *Sattinger* [3.53] has shown how to use group theory to solve such problems.

The problems which arise when $N > 2$ in genuinely dynamic problems are much more difficult than in the symmetry-breaking bifurcations at eigenvalues of higher multiplicity. For example, the nonlinear problem of fluid convection which *Lorenz* [3.54] approximated with $N = 3$ gives rise to very complicated dynamics, characterized by nonperiodic attractors with an exotic structure. It is possible to get interesting models of complicated dynamics, including turbulent-like behavior, with finite systems, of N nonlinear ordinary differential equations with $N \geq 3$. For example, *Curry's* [3.55] $N = 14$ mode truncation of the system of convection equations which Lorenz truncated at $N = 3$ gives a sequence of bifurcations which is in remarkable agreement with recent experiments of *Gollub* and *Benson* [3.20]. The same sequence of bifurcations is predicted by Curry's model and observed in the following experiments: bifurcation of a stationary solution into a time-periodic solution; bifurcation of the T periodic solution into another $2T$ periodic solution (a subharmonic solution); bifurcation of the $2T$ periodic solution into a quasi-periodic solution with two frequencies, and bifurcation of the quasi-periodic solution into a nonperiodic, turbulent like solution. The investigation of nonperiodic attractors generated by systems of nonlinear ordinary differential equations in $\mathbb{R}^N$, with N not too large, may have a big potential for understanding turbulence. (For further discussion see Chap 4 and [3.20].)

3.9 Bifurcation, Stability, and Transition in Poiseuille and Couette Flows

The bifurcation diagram for spatially periodic, time-periodic, axisymmetric solutions of the problem of Poiseuille flow through the annulus between two concentric cylinders is probably like those shown in Fig. 3.14b. This problem was studied and numerical calculations were carried out for small values of ε by *Joseph* and *Chen* [3.56] under the assumption that the bifurcating solution was axisymmetric and spatially periodic along the axis of the annulus. The Reynolds number $R(\varepsilon, \eta) = R(-\varepsilon, \eta)$ and the frequency $\omega(\varepsilon, \eta) = \omega(-\varepsilon, \eta)$ depend on the amplitude ε and radius ratio $\eta = a/b \leqq 1$, where a is the inner and b is the outer radius of the cylinders. The critical Reynolds numbers $R_0(\eta) = R(0, \eta)$ of the linearized theory of stability [3.57] depend strongly on η and $R_0(\eta) \to \infty$ as $\eta \to 0$. It is not known if the limiting flow $\eta = a/b \to 0$ is the same as Hagen-Poiseuille flow $(a = 0)$, but in both cases $R_0 \to \infty$ and the time-periodic bifurcating solution is isolated from the laminar solution $(\varepsilon = 0)$.

When the gap between the cylinders is very small, $\eta \to 1$, the effects of curvature of the walls decreases to zero and the flow reduces to plane Poiseuille flow between two, infinitely extended, parallel plates. The problem of stability and bifurcation of plane Poiseuille flow has been studied by many authors (see [3.1, 58, 59] and Chap. 7). For plane flows a well-known theorem of Squires (see [3.60]) asserts that among all the critical spatially quasi-periodic disturbances a two-dimensional disturbance, in the plane of the flow, has the smallest critical value of R. This means that the most unstable disturbance in the annulus is axisymmetric when $\eta \to 1$. There is no special reason to believe that axisymmetric disturbances are most destabilizing when $\eta \neq 1$.

If we define $R = U_{max}(b - a)/2v$, then $R(0, \eta)$ varies monotonically between $R(0, 1) = 5800$ and $R(0, 0) = \infty$ when $R(0, \eta)$ is computed for axisymmetric disturbance. Axisymmetric disturbances are the most unstable when $\eta \to 1$ and $R(0, 0)$ is probably infinite even when computed on nonaxisymmetric disturbances. Unlike the critical values of the linear theory of stability, the observed limit of stability to natural disturbances is not sensitive to η and seems to be fixed somewhere between 1000 and 1500 (see [Ref. 3.1, Fig. 35.1]).

It is of interest to discuss the two limiting cases $\eta \to 1$ and $\eta \to 0$ separately. We shall take the case $\eta \to 0$ as representative of the class of flows in which there is no finite critical value of $R(0, 0)$. In all such flows there is a direct transition to turbulence sharing common properties which I will describe later. The flows which appear to be in this class are the Poiseuille flows through an annular pipe in the limiting case $\eta \to 0$; Hagen-Poiseuille flow, $\eta = 0$; the flow between infinitely long rotating cylinders when the inner one is at rest (see Sect. 3.4); and plane Couette flow which is a limiting case of the shearing flow between cylinders when $\eta \to 1$ and the difference in the speed of the two cylinders Ω is reduced so that $\Omega/(1 - \eta)$ is finite. In such flows the loss of stability is necessarily due to a disturbance of finite amplitude since infinitesimal disturbances may be presumed to decay at all finite Reynolds numbers. There do not appear to be strong differences between flows which have no finite linear stability limit and the flows close to them which have very large critical Reynolds numbers, for example, in pipe flow with small nonzero values of η or in the flow between rotating cylinders of finite height when the inner cylinder rotates and the outside, top, and bottom of the cylinders are stationary. The flow between rotating spheres when the inner one is at rest probably falls into the class of flows with large finite critical Reynolds numbers that undergo a direct transition to turbulence.

The subcritical finite-amplitude motions which occur in the cases when there is no finite critical Reynolds number seem to be characterized by the spatial segregation of the flow into laminar and turbulent patches as in the Poiseuille flow shown in Fig. 3.2 or the Couette flow shown in Fig. 3.3.

In the motions close to these, with finite large critical values of the Reynolds number, we also get subcritical direct transition to turbulence and, in addition, we may compute a subcritical, unstable, time-periodic bifurcation of the type described in Chaps. 7 and 8. We may regard Poiseuille flow between cylinders as representative of shearing flows which are linearly stable at high Reynolds numbers, and even the most studied case of plane Poiseuille flow with $\eta = 1$ is in this class. For such flows the unstable time-periodic solutions on the small ε part of the bifurcation curve shown in Fig. 3.14b)the dotted lines with $\varepsilon \neq 0$ near R_0) have been observed in the experiments of *Nishioka* et al. [3.61]. Their experiments were for the case $\eta \to 1$ and they found the unstable time-periodic bifurcating solution as a "metastable" or slowly changing transient state.

We could claim perfect agreement between bifurcation theory and experiments if the observed solutions on the stable large amplitude branch of the bifurcation curve in Fig. 3.14b were time periodic with the predicted values of $R(\varepsilon, \eta)$ and the frequencies $\delta(\varepsilon, \eta)$. In fact stable time-periodic solutions are not observed; instead, we see direct snap-through instability to turbulence. The reason for this discrepancy is that the stability analysis which leads to the conclusion that the upper branch of time-periodic bifurcating Poiseuille flow is stable is insufficiently general because it is carried out in a too-restricted set of disturbances in which three-dimensional disturbances and spatially aperiodic disturbances are arbitrarily excluded.

The difficulties involved in arriving at a correct mathematical interpretation of the mechanisms involved in the observed instability, bifurcation, and transition of shearing flows are enormous. Even the linear stability problem is difficult and proofs of various important points, like the linear stability of laminar flow in pipes at all finite values of the Reynolds number, have yet to be established in a mathematically secure way. Of course, the nonlinear problem is even more difficult and nearly all analytical results are restricted to small amplitudes. This type of restriction is especially serious for problems like the ones under discussion where, in the already cited words of Reynolds, "the condition might be one of instability for disturbances of a certain magnitude, and stable for smaller disturbances".

So it is nearly if not strictly true to say that the ordinary analytical methods of nonlinear stability theory and bifurcation theory have been unavailing for the problems of stability and transition in shear flows. Numerical methods have been only slightly more successful until recently. The recent numerical study of *Orszag* and *Kells* [3.59] of the fully nonlinear initial value problem for the stability of plane Poiseuille flow and plane Couette flow appears to represent a real breakthrough. The study of *Orszag* and *Kells* utilizes spectral, Galerkin-type methods of analysis combined with an efficient numerical method of computation of nonlinear terms. The methods used have been developed by *Orszag* and associates, and the mathematical foundations, justifications, and guides for use can be found in the recently published monograph of *Gottleib* and *Orszag* [3.62].

The results of *Orszag* and *Kells* [3.59] are summarized below.

1) The bifurcation picture for two-dimensional solutions of the equations governing disturbances of plane Poiseuille flow is like that shown in Fig. 3.14b. This confirms the qualitative picture given by the theory of bifurcation at a simple eigenvalue. The minimum value of $R(\varepsilon, \eta)$ for which a bifurcating solution of the classical Hopf type exists is about 2800. Earlier, *Zahn* et al. [3.63] found the minimum value of $R(\varepsilon, \eta) = 2707$ for the same type of disturbances, but in a severely truncated approximation. These maximum values are more than twice those observed in experiments. The numerical methods of *Orszag* and *Kells* enable one to compute solutions with large amplitudes and to separate the study of two- and three-dimensional disturbances. It is impossible to suppress three-dimensional disturbances in experiments and these evidently lead to the lower values of Reynolds number for which turbulent solutions are observed in experiments.

2) *Orszag* and *Kells* found that all spatially periodic two-dimensional disturbance of plane Couette flow decay to zero. It is generally agreed that plane Couette flow is stable against infinitesimal disturbances at all finite Reynolds numbers. But the result suggested by the computation of *Orszag* and *Kells* is new because it implies that Couette flow is also stable against finite amplitude two-dimensional disturbances. This suggests that a similar global

stability result holds for axisymmetric disturbances of Poiseuille flow through pipes when $\eta \to 0$ and when $\eta = 0$ and for axisymmetric disturbances of Couette flow between infinitely long cylinders when the inner cylinder is at rest and the outer one rotates.

3) *Orszag* and *Kells* found that small three-dimensional disturbances of plane Couette flow decay but that large amplitude disturbances persist and appear to evolve into turbulent flow at Reynolds numbers as low as 1000. And turbulent solutions will persist at slightly lower Reynolds numbers. This result is in rather good agreement with the experiments of *Reichardt* [3.64] and extends, but does not contradict, previous results on the linear stability of plane Couette flow. Three-dimensional disturbances of plane Poiseuille flow are also very destabilizing and lead to turbulentlike solutions at Reynolds numbers of about 1000, again in agreement with experiments. We may say that this shows that the stable two-dimensional solutions of the Hopf type, those in Fig. 3.14b, are unstable to three-dimensional disturbances and lead to turbulence. All disturbances of Couette and Poiseuille flow decay at Reynolds numbers of about 500.

3.10 Bibliographical Notes and Comments on Methods of Analysis

An extensive review of the analytical methods used in studying stability and bifurcation in fluids can be found in the Notes for Chap. II of my books [3.1, 2][7].

Linear stability theory is applied to interesting special problems in the following books and general reviews: *Chandrasekhar* [3.65] treated many kinds of special problems; *Lin* [3.60] also considered different problems but he emphasized the Orr-Sommerfeld equation; *Synge* [3.66], *Stuart* [3.67], *Shen*

---

7 A corrigendum for errors and misprints which I have found in these books so far is as follows:

Vol. I [3.1]:

p.15 $v_g(\infty)$ is bounded.
p.29   Replace line 11 with "Then $y(t) = \Phi(t) \cdot \psi$ and".
        Replace (7.18) with "$y(t+T) = \Phi(t+T) \cdot \psi = \Phi(t)\Phi(T) \cdot \psi = e^{-\gamma T}\Phi(T) \cdot \psi = e^{-\gamma T}y(t)$".
        Replace lines 17 through 27 with "and $w(t+T) = e^{\gamma t}e^{\gamma T}y(t+T) = e^{\gamma t}y(t) = w(t)$ is a T-periodic function".
p.167   Correct Eq. (44.20).
p.223   Change "$\phi \in C^3$" on line 11 to "$\phi \in C^2$"; change "$C^1(\mathscr{V})$" to "$C^3(\mathscr{V})$" on line 16.
p.224   Change lines 5 and 6 to "Since $\phi \in C^2(\mathscr{V})$ and $p^+ - p^-$ is continuous on S", we find, using Lemma 1, that $p^+ = p^-$ and p is single-valued.

Vol. II [3.2]:

p.59   Replace "$\mathscr{R}^{-1}$" with "$\mathscr{R}$" in (70.23–25).
p.60   Replace "$n \cdot D \cdot e_z$" with "$2n \cdot D \cdot e_z$" in (70.26) and in the first equality below (70.26).

[3.68], and *Monin* and *Yaglom* [3.69] gave general reviews; *Reid* [3.70] and the computer-oriented book of *Betchov* and *Criminale* [3.71] confined their analysis to the Orr-Sommerfeld theory; *Drazin* and *Howard* [3.72] considered the stability of inviscid flow; *Greenspan* [3.73] gave results for the stability of rotating flow; and *Yih* [3.74] dealt with the stability of stratified flows. *Davis* [3.75] is the only one on this list who treated the stability of time-dependent flows. Nonlinear problems of thermal convection are discussed in the review paper of *Segel* [3.76]. Other nonlinear problems of hydrodynamics were reviewed by *Stuart* [3.58]. A wider range of problems of interest can be found in the book of *Denn* [3.77]. The books of *Monin* and *Yaglom*, and *Denn* also treat nonlinear problems. The general problem of transition to turbulence has been reviewed in the paper of *Swinney* and *Gollub* [3.78].

The works listed in the last paragraph are concerned with linear stability theory and some of them with nonlinear stability theory. Nonlinear stability theory is different than bifurcation theory. The main differences are in the way you get hold of a problem and in what you aim to achieve. In nonlinear stability theory it is necessary to have explicit representations of some flow. Then one can do some explicit, if approximate, analysis of disturbances of this flow. Most of the time, studies of nonlinear stability theory are concerned with what happens when the unique flow which exists when the Reynolds number is small gives up its stability. Problems of secondary and higher bifurcation are hard to treat by the methods of nonlinear stability theory because explicit representations are known, usually in series form, only in asymptotic limits and the resulting analysis is, at the very best, difficult and approximate. When the explicit methods work they yield very good quantitative results suitable for comparison with experiments. But explicit methods lack generality because we cannot hope to have explicit representations for most flows and if we did, they would usually be too complicated to analyze by the ordinary methods of applied analysis.

---

p.70   Replace "θ_0" in (73.4a) with "θ_0^*".

p.71   Replace "θ" in (73.9) with "$\bar\theta^*$".

p.82   Replace "$M(a^2)\neq0$" in Exercise 73.1 with "$M(a^2)=0$".

p.99   Replace "$-\sigma M\cdot q$" in (75.11) with "$-\sigma M\cdot q^*$".

p.196   In line 4, replace "…multilinear…" with "…multilinear and symmetric…".

p.197   Replace "$3\mathscr{F}_2$," in (93.6) with "$2\mathscr{F}_2$,".

p.201   Replace "$d/d\tau^m$" on the last line with "$d^n/d\tau^m$".

p.203   Change "$A_m[u]x,t)]$" to "$A_m[u(x,t)]$" at the end of (94.8)

p.211   Change "$\mathscr{F}^{\langle-\rangle}$" to "$\mathscr{F}^{\langle2\rangle}$" in line 12.

p.219   Replace "(94.52)" in the last line with "(94.51)".

p.225   Replace "…average of (94.67)…" with "…average of (94.67b)…".

p.226   On line 10, change "$\mu\langle u^{\langle1\rangle}\cdot(\nabla\cdot A_1^{(2)})-u^{\langle2\rangle}\cdot A_1^{(1)}\rangle+…$" to "$\mu\langle(u^{\langle1\rangle}\cdot\nabla)A_1^{(2)}$
        $-(u^{\langle2\rangle}\cdot\nabla)A_1^{(1)}\rangle+…$".

p.230   Replace the right-hand term with "$\varepsilon\partial U_j^{\langle1\rangle}(X,\tau)/\partial X_i$".

p.256   Change the term on the right-hand side of (99.4b) from "$1-G\varepsilon^2/2+…$" to "$-1-G\varepsilon^2/2+…$".
        Change the last term in the inequality below (99.4b) from "$\min_r$" to "$\min_{\bar r}$".

Bifurcation theory does not require detailed knowledge of the flow which loses stability. Instead, analysis attaches itself to the problem through a classification of possibilities based on the spectrum, the eigenvalues of the linearized problem. For example, we already discussed the nature of bifurcation at an algebraically simple isolated eigenvalue. And in a longer work we would discuss bifurcation at multiple eigenvalues, bifurcation in the presence of continuous spectra, etc. So theory leads to a classification of the possible forms of bifurcation, and makes qualitative statements about their properties without computations based on explicit representations of flow. But, of course, many interesting details about the flow are lost in qualitative analysis. And to be certain, if a specific problem falls in one or the other classification, it is frequently necessary to verify the assumptions about the eigenvalues by explicit computations.

An introduction to the basic features of bifurcation theory, its principal applications, and an extensive bibliography was given by *Stakgold* [3.79]. The monographs of *Sattinger* [3.40] and *Iooss* [3.80] treat bifurcation problems in a Banach space. Excellent collections of papers on applications of bifurcation theory are found in the volumes edited by *Keller* and *Antman* [3.81], *Haken* [3.82], and *Rabinowitz* [3.83]. Many problems of bifurcation in hydrodynamical problems are studied in [3.1, 2]. Finally, an introductory treatment of bifurcation theory has been written as a textbook on "Elementary Bifurcation Theory" by *Iooss* and *Joseph* [3.51].

In my discussion of problems of stability and bifurcation in fluids I have tried to emphasize the main methods and I relied mostly on analytical methods. Most types of pure analysis are possible only when the amplitude of the disturbance is small. Explicit quantitative results can usually be obtained only when the amplitude is small and, in addition, when the domain occupied by the fluid and the data prescribed on the boundary of the domain have considerable symmetry. Many of the results we need in the theory of stability, bifurcation, and transition in fluids cannot be obtained easily, and it is even possible that the only way to obtain them is by numerical methods.

The most successful numerical methods used so far are Galerkin-type methods, and the most far-reaching consequences of these methods have been reached by *Busse* and his associates (see his article in this volume, Chap. 5, and [Ref. 3.2, Sect. 81]) and, more recently, by the introduction of powerful spectral-numerical methods by *Orszag* and his associates (see, e.g., [3.62]). The various methods which I call Galerkin type involve expanding the solution in some set of eigenfunctions, truncating and examining the convergence of the method relative to some "residual" which gives the difference between the true solution and the approximating one. The application of this method to stability problems is explained in the book by *Finlayson* [3.84], and the newer and apparently more powerful spectral-numerical methods in the book by *Gottleib* and *Orszag* [3.62]. The explanation of numerical methods is beyond the scope of this chapter.

It seems likely to me that the theory of hydrodynamic stability, bifurcation, and transition will, in the future, come to rely increasingly on abstract methods for qualitative analysis and on numerical analysis for explicit results. I would expect an important role, but a decreasingly important one, to be played by the traditional methods of applied analysis.

Acknowledgement. This work was supported by the Fluid Mechanics Program of the National Science Foundation.

References

3.1 D.D.Joseph: *Stability of Fluid Motions I.* Springer Tracts in Natural Philosophy, Vol. 27 (Springer, Berlin, Heidelberg, New York 1976)

3.2 D.D.Joseph: *Stability of Fluid Motions II.* Springer Tracts in Natural Philosophy, Vol. 28 (Springer, Berlin, Heidelberg, New York 1976)

3.3 O.Reynolds: An experimental investigation of the circumstances which determine whether the motion of water shall be direct or sinuous, and of the law of resistance in parallel channels. Phil. Trans. R. Soc. **174**, 935 (1883)

3.4 O.Reynolds: In *Papers on Mechanical and Physical Subjects*, Vol. II (Cambridge University Press, Cambridge 1901) p. 51

3.5 O.Reynolds: On the dynamical theory of incompressible viscous fluids and the determination of the criterion. Phil. Trans. R. Soc. A **186**, 123 (1895)

3.6 J.Serrin: On the stability of viscous fluid motions. Arch. Ration. Mech. Anal. **3**, 1 (1959)

3.7 W.McF.Orr: The stability or instability of steady motions of a liquid. Part II. A viscous liquid. Proc. R. Irish. Acad. A **27**, 69 (1907)

3.8 T.Y.Thomas: Qualitative analysis of the flow of fluid in pipes. Am. J. Math. **64**, 754 (1942)

3.9 E.Hopf: Ein allgemeiner Endlichkeitssatz der Hydrodynamik. Math. Ann. **117**, 764 (1941)

3.10 D.Ruelle, F.Takens: On the nature of turbulence. Commun. Math. Phys. **20**, 167 (1971)

3.11 L.D.Landau: On the problem of turbulence. C.R. Acad. Sci. USSR **44**, 311 (1944)

3.12 L.D.Landau, E.M.Lifshitz: *Fluid Mechanics* (Pergamon Press, Oxford 1959)

3.13 E.Hopf: A mathematical example displaying features of turbulence. Commun. Pure Appl. Math. **1**, 303 (1948)

3.14 E.Hopf: Repeated branching through loss of stability, an example. Proc. Conf. Diff. Eqs., Univ. of Maryland (1956) p. 49

3.15 G.Prodi: Teoremi di tipo locale per il sistema di Navier-Stokes l'stabilita delle soluzioni stazionarie. Rend. Sem. Univ. Padova **32**, 374 (1962)

3.16 G.Iooss, D.D.Joseph: Bifurcation and stability of nT-periodic solutions branching from T-periodic solutions at points of resonance. Arch. Ration. Mech. Anal. **66**, 135 (1977)

3.17 G.Iooss: "Topics in Bifurcation of Maps and Applications", University of Minnesota, Lecture Notes (1978)

3.18 G.Iooss: Bifurcation of a periodic solution of the Navier-Stokes equations into an invariant torus. Arch. Ration. Mech. Anal. **58**, 35 (1975)

3.19 G.Iooss: Sur la deuxième bifurcation d'une solution stationnaire de systèmes du type Navier-Stokes. Arch. Ration. Mech. Anal. **64**, 339 (1977)

3.20 J.P.Gollub, S.V.Benson: Chaotic response to periodic perturbation of convecting fluid. Phys. Rev. Lett. **41**, 948 (1978)

3.21 J.Bass: *Les Fonctions Pseudo-Aléatoires* (Mémorial Des Sciences Mathématiques) (Gauthier-Villars, Paris 1962)

3.22 J.B.McLaughlin, P.C.Martin: Transition to turbulence of a statically stressed fluid system. Phys. Rev. A **12**, 186 (1975)

3.23 E.N.Lorenz: Deterministic nonperiodic flow. J. Atmos. Sci. **20**, 130 (1963)

3.24 N.H.Baker, D.W.Moore, E.A.Spiegel: Aperiodic behaviour of a non-linear oscillator. Quart. J. Mech. Appl. Math. **XXIV**, 391 (1971)

3.25 I.J.Wygnanski, F.H.Champagne: On transition in a pipe. Part 1. The origin of puffs and slugs and the flow in a turbulent slug. J. Fluid Mech. **59**, 281 (1973)

3.26 I.J.Wygnanski, M.Sokolov, D.Friedman: On transition in a pipe. Part 2. The equilibrium puff. J. Fluid Mech. **69**, 283 (1975)

3.27 D.Coles: Transition in circular Couette flow. J. Fluid Mech. **21**, 385 (1965)

3.28 H.Swinney, P.R.Fenstermacher, J.P.Gollub: "Transition to turbulence in a fluid flow", in *Synergetics, a Workshop*, ed. by H. Haken (Springer, Berlin, Heidelberg, New York 1977) p. 60

3.29 P.R.Fenstermacher, H.L.Swinney, J.P.Gollub: Transition to chaotic Taylor vortex flow. J. Fluid Mech. **94**, 103 (1979)

3.30 G.Ahlers, R.P.Behringer: Evolution of turbulence from the Rayleigh-Bénard Instability. Phys. Rev. Lett. **40**, 712 (1978)

3.31 T.B.Benjamin: Applications of Leray-Schauder degree theory to problems of hydrodynamic stability. Math. Proc. Camb. Phil. Soc. **79**, 373 (1976)

3.32 O.Sawatzki, J.Zierep: Flow between a fixed outer sphere and a concentric rotating inner sphere (in German). Acta Mech. **9**, 13 (1970)

3.33 B.R.Munson, M.Menguturk: Viscous incompressible flow between concentric rotating spheres. Part 3. Linear stability and experiment. J. Fluid Mech. **69**, 705 (1975)

3.34 M.Wimmer: Experiments on a viscous fluid flow between concentric rotating spheres. J. Fluid Mech. **78**, 317 (1976)

3.35 J.N.Belyaev, A.A.Monaxov, I.M.Yavorskaya: Stability of a spherical Couette flow in thick layers with an interior rotating sphere. Mech. Fluid Gas No. **2** (1978)

3.36 I.M.Yavorskaya, J.N.Belyaev, A.A.Monaxov: Research on stability and secondary flows within rotating spherical layers under arbitrary Rossby numbers. Rep. Acad. Sci. USSR **237**, 801 (1977)

3.37 B.R.Munson, D.D.Joseph: Viscous incompressible flow between concentric rotating spheres. Part 1. Basic flow. Part 2. Hydrodynamic stability. J. Fluid Mech. **49**, 289 (1971)

3.38 O.A.Ladyzhenskaya: *The Mathematical Theory of Viscous Incompressible Flow*, 2nd ed. (Gordon and Breach, New York 1963)

3.39 O.A.Ladyzhenskaya: Mathematical analysis of the Navier-Stokes equations for incompressible liquid. Annu. Rev. Fluid Mech. **7**, 249 (1975)

3.40 D.H.Sattinger: *Topics in Stability and Bifurcation Theory*, Lecture Notes in Mathematics, Vol. 309 (Springer, Heidelberg, New York 1973)

3.41 D.D.Joseph: Factorization theorems, stability and repeated bifurcation. Arch Ration. Mech. Anal. **66**, 99 (1977)

3.42 D.D.Joseph: Factorization theorems and repeated branching of solutions at a simple eigenvalue. Ann. N.Y. Acad. Sci. **316**, 150 (1979)

3.43 S.Rosenblat: Global aspects of Hopf bifurcation and stability. Arch. Ration. Mech. Anal. **66**, 2 (1977)

3.44 E.Hopf: "Abzweigung einer Periodischen Lösung eines Differentialsystems", Berichte der Mathematisch-Physikalischen Klasse der Sächsischen Akademie der Wissenschaften zu Leipzig **XCIV** (1942)

3.45 D.D.Joseph: Stability of convection in containers of arbitrary shape. J. Fluid Mech. **47**, 257 (1971)

3.46 W.T.Koiter: On the Stability of Elastic Equilibrium, Delft, 1945 (in Dutch); translated into English as NASA TTF-10, 833 (1967)

3.47 R.Thom: Topological methods in biology. Topology **8**, 313 (1968)

3.48 T.B.Benjamin: Bifurcation phenomena in steady flows of a viscous fluid. Part 1, Theory and Part 2, Experiments. Proc. R. Soc. London A **359**, 1 (1978)

3.49 B.J.Matkowsky, E.Reiss: Singular perturbation of bifurcations. Soc. Ind. Appl. Math. J. Appl. Math. **33**, 230 (1977)

3.50 M.Golubitsky, D.Schaeffer: A theory for imperfect bifurcation via singularity theory. Commun. Pure Appl. Math. **32**, 21 (1979)

3.51 G.Iooss, D.D.Joseph: *Elementary Stability and Bifurcation Theory* (Springer, Berlin, Heidelberg, New York, 1980)

3.52 G.Iooss, H.Nielsen, H.True: Bifurcation of the stationary Ekman flow in a stable periodic flow. Arch. Ration. Mech. Anal. **68**, 227 (1978)

3.53 D.H.Sattinger: Selection mechanisms for pattern formation. Arch. Ration. Mech. Anal. **66**, 31 (1977)

3.54 E.Lorenz: Deterministic nonperiodic flow. J. Atmos. Sci. **20**, 130 (1963)

3.55 J.H.Curry: A generalized Lorenz system. Commun. Math. Phys. **60**, 193 (1978)

3.56 D.D.Joseph, T.S.Chen: Friction factors in the theory of bifurcating flow through annular ducts. J. Fluid Mech. **66**, 189 (1974)

3.57 T.Mott, D.D.Joseph: Stability of parallel flow between concentric cylinders. Phys. Fluids **11**, 2065 (1968)

3.58 J.T.Stuart: Nonlinear stability theory. Annu. Rev. Fluid Mech. **3**, 347 (1971)

3.59 S.A.Orszag, L.C.Kells: Transition to turbulence in plane Poiseuille and plane Couette flow. J. Fluid Mech. **96**, 159–207 (1980)

3.60 C.C.Lin: *The Theory of Hydrodynamic Stability* (Cambridge University Press, Cambridge 1955)

3.61 M.Nishioka, S.Iida, Y.Ichikawa: An experimental investigation of the stability of plane Poiseuille flow. J. Fluid Mech. **72**, 731 (1975)

3.62 D.Gottleib, S.A.Orszag: *Numerical Analysis of Spectral Methods: Theory and Applications*, NSF-CBMS Monograph **26**, Soc. Ind, App. Math. Philadelphia (1978)

3.63 J.Zahn, J.Toomre, E.Spiegel, D.Gough: Nonlinear cellular motions in Poiseuille channel flow. J. Fluid Mech. **64**, 319 (1974)

3.64 H.Reichardt: Über die Geschwindigkeitsverteilung in einer geradlinigen turbulenten Couette-Strömung. Z. Angew. Math. Mech. **36**, S26 (1956)

3.65 S.Chandrasekhar: *Hydrodynamic and Hydromagnetic Stability* (Oxford University Press, London 1961)

3.66 J.L.Synge: "Hydrodynamic Stability", in Semicentennial Publications of the American Mathematical Society, Vol. 2 (Amer. Math. Soc. 1938) p. 227

3.67 J.T.Stuart: "Hydrodynamic Stability", in *Laminar Boundary Layers*, ed. by L. Rosenhead (Oxford University Press, London 1963)

3.68 S.F.Shen: *Stability of Laminar Flows: Theory of Laminar Flows* (High Speed Aerodynamics and Jet Propulsion, Vol. 4) Sect. G, ed. by F. K. Moore (Princeton University Press, 1964)

3.69 A.S.Monin, A.M.Yaglom: *Statistical Fluid Mechanics*, Vol. 1. *Mechanics of Turbulence* (The MIT Press, Cambridge 1971)

3.70 W.H.Reid: "The Stability of Parallel Flows", in *Basic Developments in Fluid Dynamics*, Vol. 1, ed. by M. Holt (Academic Press, New York 1965) p. 249

3.71 R.Betchov, W.O.Criminale: *Stability of Parallel Flow* (Academic Press, New York 1967)

3.72 P.Drazin, L.N.Howard: Hydrodynamic stability of parallel flow of inviscid fluid. Adv. Appl. Mech. **9**, 1 (1966)

3.73 H.P.Greenspan: *The Theory of Rotating Fluids* (Cambridge University Press, Cambridge 1969)

3.74 C.S.Yih: *Dynamics of Nonhomogeneous Fluids* (Macmillan, New York 1965)

3.75 S.Davis: The stability of time-periodic flow. Annu. Rev. Fluid Mech. **8**, 57 (1976)

3.76 L.Segel: "Nonlinear Hydrodynamic Stability Theory and Its Application to Thermal Convection and Curved Flow", in *Non-Equilibrium Thermodynamics: Variational Techniques and Stability*, ed. by R. J. Donnelly, I. Prigogine, R. Herman (University of Chicago Press, Chicago 1966)

3.77 M.Denn: *Stability of Reaction and Transport Processes* (Prentice-Hall, Englewood Cliffs, N.J. 1975)

3.78 H. Swinney, J. Gollub: The transition to turbulence. Phys. Today **31**, No. 8, 41 (August, 1978)

3.79 I. Stakgold: Branching of solutions of nonlinear equations, Soc. Ind. Appl. Math. Rev. **13**, 289 (1971)

3.80 G. Iooss: *Bifurcation et stabilité*, Pub. Math. d'Orsay No. 31 (1974)

3.81 J. Keller, S. Antman (eds.): *Bifurcation Theory and Nonlinear Eigenvalue Problems* (Benjamin, New York 1969)

3.82 H. Haken (ed.): *Synergetics, a Workshop* (Springer, Berlin, Heidelberg, New York 1977)

3.83 P. Rabinowitz (ed.): *Applications of Bifurcation Theory* (Academic Press, New York 1977)

3.84 B. A. Finlayson: *The Method of Weighted Residuals and Variational Principles* (Academic Press, New York 1972)

4. Chaotic Behavior and Fluid Dynamics

J. A. Yorke and E. D. Yorke

With 4 Figures

In the early phases of many branches of physics, model systems are studied in which most "real world" complications are thrown away. If, nonetheless, key properties may be found which are interesting and which mimic reality with a minimum of computational effort, then the model is useful. Examples of such systems include the ideal gas and the one- or two-dimensional Ising model.

Fluid turbulence is characterized by unpredictable motions of the fluid. While fluids are very complicated systems having many degrees of freedom, it is possible that the basic mechanism underlying the observed chaotic behavior is one which does not require consideration of many degrees of freedom (see Chap. 2). Certain simple dynamical systems (that is, systems which evolve in time) display erratic and chaotic behavior which is reminiscent of turbulence. The key to this behavior is "sensitivity to initial conditions", a term which was introduced by *Ruelle* [4.1] to describe such systems. In this chapter, we shall describe a few rather simple models which are sensitive to initial data and so have allowed workers to study this strange yet widespread mode of behavior in some detail.

4.1 Background

To examine further what "sensitivity to initial conditions" means, let us first consider an *in*sensitive system of the sort studied in elementary physics courses – a simple pendulum. If two pendulums are started out with slightly different initial positions and velocities and the two trajectories in phase space [that is, $(x, \dot{x})$ space] are followed, it is found that the trajectories remain close to each other for all time. This insensitivity has consequences on both computational and experimental levels. Experimentally, it means that sufficiently fine control of the conditions under which the system is investigated will result in a similar description of the detailed behavior of the system each time the experiment is performed. Neither small, uncontrollable errors in initial conditions nor the equally uncontrollable presence of noise during the experiment will produce major deviations in measurements. Computationally it means that if an analytic expression for the time evolution of the system cannot be found (for the pendulum, of course, such an expression can be found) and numerical calculations must be performed, unavoidable small errors (e.g., roundoff) in the calculation need not grow with time and one can, with confidence, predict the long-range evolution of the system. Since two calculated trajectories that start close together will remain close, one can use a single calculation to predict the behavior of a group of trajectories which all start out in the same neighborhood. If the original trajectory comes close to its initial point after a time t elapses, one can then predict that during the next time interval t, its trajectory will be very similar; in that sense a long-range pattern may be discerned. (The periodic behavior of the pendulum is a special case of this.)

Systems which are sensitive to initial data do not behave so nicely. Small changes in initial conditions lead to large and unpredictable changes in the long-range evolution of the system. Two trajectories which start close to each other do not remain close. Indeed *Ruelle* suggests [4.1, 2] that the difference between them will generally grow exponentially with time so that they rapidly lose any relationship to each other. This means that computations of long-range behavior will be seriously affected by small errors. Furthermore, when a trajectory winds close to its initial point in phase space, sensitivity to initial data implies that its future evolution will, in general, be completely different from its past; in that sense, no long-range pattern can be discerned. An experimentalist who tries to describe the detailed behavior of such a system will look in vain for reproducible trajectories, for unavoidable errors in initial data as well as noise during the experiment will conspire to produce a different pattern during each run. Both experimentally and computationally such a system would be described as chaotic or irregular.

Although sensitive systems are as widespread in nature and as important as insensitive ones, they are difficult to characterize experimentally and relatively costly to model numerically, so their study is in its infancy. From the practical point of view, the most important sensitive system may be the atmosphere. *Lorenz* [4.3] has suggested that its dynamics are severely sensitive to initial data. The implications of this for weather prediction are serious because even if massive improvements in modelling and data collection were made, it would be unlikely that accurate detailed weather forecasts could ever be made a few weeks in advance.

4.2 The Lorenz Equations

One set of models was introduced by *Saltzman* [4.4] and was elaborated upon by *Lorenz* [4.3] in an investigation of the instability of convection of a fluid between two parallel plates. He found that, under some severe approximations, the behavior of the fluid would be described by three ordinary differential equations

$$\frac{dx}{dt} = \sigma y - \sigma x ,$$
$$\frac{dy}{dt} = -xz + rx - y , \qquad (4.1)$$
$$\frac{dz}{dt} = xy - bz .$$

As our derivation will show, $x(t)$ represents the fluid's angular velocity while $y(t)$ and $z(t)$ are mode amplitudes. Depending upon the values of the parameters σ, b, and r, the solutions $x(t)$, $y(t)$, $z(t)$ can show steady-state behavior with no motion, $(x, y, z) = (0, 0, 0)$, or steady state convective flow $(x, y,$ and z time independent but nonzero). Linear stability analysis of the Lorenz equations shows that the zero flow state is stable for $r < 1$. At $r = 1$ a bifurcation occurs and two additional steady-state solutions develop, and these can be interpreted as steady convective flow (clockwise or counterclockwise). Steady convective flow is stable for $r < r_2$ where $r = \sigma(\sigma + b + 3)/(\sigma - b - 1)$. The general solutions $x(t)$, $y(t)$, $z(t)$ cannot be written as combinations of elementary functions of t and the initial data, but an understanding of their behavior can be obtained through computer experimentation, that is, by choosing some values of the parameters σ, b, and r and then choosing several sets of initial data $x(0)$, $y(0)$, $z(0)$ and observing graphically how the solution $x(t)$, $y(t)$, $z(t)$ evolves as t increases. For $r > r_2$ there are no stable steady-state solutions and for a substantial range of $r > r_2$ no stable periodic solutions are observed. Instead, irregularly oscillating solutions are observed in computer experiments. These do not settle down to steady flow or to periodic oscillations. They do not show any long-range pattern which can be predicted (without actually performing the numerical integrations), and small changes in initial conditions result in solutions which differ substantially. For a small range of r just below r_2 (i.e., $r_1 < r < r_2$), (x, y, z) space can be divided into two regions which display very different long-term behavior. In one of these regions, the two convective steady state solutions are stable and so any initial condition chosen near one of these steady states gives a solution which decays to that state. In the other, it is found that solutions starting in that region will remain in it, oscillating irregularly forever. For the commonly used values $\sigma = 10$, $b = 8/3$, it is found numerically that $r_1 \approx 24.06$ and $r_2 \approx 24.74$. See [4.5, 6] for a more detailed interpretation of the transition at 24.06. Even for $r < r_1$ there are still some exceptional orbits which oscillate irregularly forever. However, almost any choice of initial data yields a solution which eventually settles down, possibly after running through a long irregular set of oscillations. Here we say the chaotic behavior is "metastable". We discuss the significance of metastable behavior in Sect. 5.5.

The Lorenz equations are sensitive to initial data and they are sufficiently simple so that their trajectories in phase space may be studied numerically. If we examine two solutions $[x_1(t), y_1(t), z_1(t)]$ and $[x_2(t), y_2(t), z_2(t)]$ which are close to each other at time zero, we see that their difference increases until after a short time the pattern of their oscillation is quite different. The trajectories in phase space, that is, in (x, y, z) space, slowly move apart, then back near each other again, only to later move apart again. The system is predictable in the short run but not predictable over long time intervals because of the growth in the uncertainties of the initial data. Periodic solutions can exist but their sensitivity implies that they will not be observed numerically or experimentally since initial data chosen close to but not on a periodic solution will generally produce a trajectory which diverges from it.

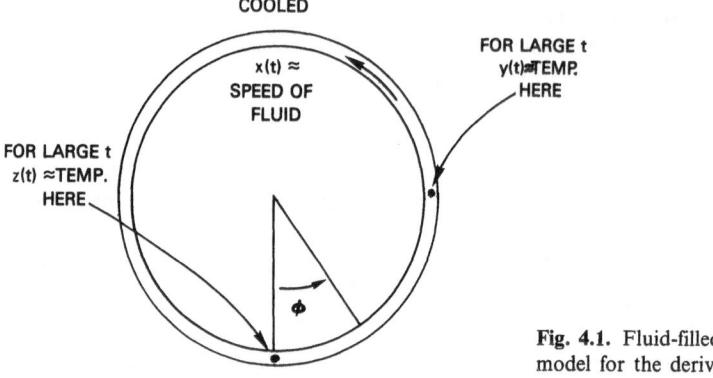

Fig. 4.1. Fluid-filled tube, a physical model for the derivation of Lorenz's equations

Calculations [4.7, 8] have illustrated that the approximations needed to obtain the Lorenz equations for Rayleigh-Bénard convection (two-dimensional convection between a pair of parallel surfaces) are of course rather severe and that they do not adequately describe the instability of convective flow in that geometry. We shall describe another physical system from which the Lorenz equations can emerge under some circumstances and we shall then discuss the behavior of the solutions of those equations in more detail. A similar derivation was described by *L. Howard* and *W. Malkus* (see [4.9]) with few mathematical details. For a more general account see [4.10].

The system we consider is similar to that described by *Welander* [4.11]. A toroidal tube of cross-sectional area A and mean radius l is filled with an incompressible fluid which has density ϱ_0 at temperature T_0. The density of the fluid is assumed to depend upon temperature as $\varrho = \varrho_0[1 - \alpha(T - T_0)]$ where, in the system, the thermal expansion coefficient α is positive. The tube is oriented vertically in a uniform gravitational field g. Angular distance around the tube, ϕ, will be measured counterclockwise from the direction of g. The walls of the tube will be maintained at some time-independent (but angle-dependent) temperature $T_w(\phi)$ which we shall specify. In particular the bottom will be warmer than the top. Figure 4.1 is a sketch of the situation.

Before writing down the equation of motion of the system and the accompanying heat transfer equation, we make several approximations:

a) Except in determining the buoyant force on an element of fluid, the variation in the material properties of the fluid with temperature (and therefore ϕ) will be neglected.

b) The fluid flow through the tube is laminar and in the ϕ direction. This condition, together with incompressibility, means that the flow rate $q(t) = \int v \cdot dA = \int v_\phi dA$ is angle independent, where integration is over the cross section.

c) The fluid flow is opposed by a frictional force which is proportional to the instantaneous flow rate $q(t)$ where R is a constant,

$$F_{fr} = \int f_{fr} dA = -R\varrho_0 q(t)\hat{\phi} \, . \tag{4.2}$$

d) The temperature $T(\phi, t)$ and velocity of the fluid are uniform across a cross section of the tube except for a negligibly thin boundary layer.

e) Heat transfer between the walls of the tube and the fluid is given by $K[T_w(\phi) - T(\phi, t)]$ where K is a constant.

The equation of motion of the fluid is

$$\varrho_0 \frac{dv}{dt} = -\nabla P + [\varrho_0 + \alpha\varrho_0(T_0 - T)]g + f_{fr} , \tag{4.3}$$

where P is the pressure. Evaluating $\varrho_0 \int\limits_0^{2\pi} d\phi \int (dv/dt) \cdot dA$ we find

$$\frac{dq(t)}{dt} = \frac{A\alpha g}{2\pi} \int\limits_0^{2\pi} T(\phi, t) \sin\phi \, d\phi - Rq(t) . \tag{4.4}$$

The heat transfer equation, ignoring thermal conduction in the fluid compared to convection, is

$$\frac{\partial T(\phi, t)}{\partial t} + \frac{q}{Al} \frac{\partial T(\phi, t)}{\partial \phi} = K[T_w(\phi) - T(\phi, t)] . \tag{4.5}$$

Let T_w and T be expanded as a Fourier series in ϕ. The coefficients in the expansion for T are time dependent. In general

$$T(\phi, t) = C_0(t) + \sum_{n=1}^{\infty} [S_n(t) \sin n\phi + C_n(t) \cos n\phi] ,$$

and $\tag{4.6}$

$$T_w(\phi) = W_0 + \sum_{n=1}^{\infty} (V_n \sin n\phi + W_n \cos n\phi) .$$

Inserting the Fourier expansions into (4.4) and (4.5) and solving for the coefficients yields an infinite set of ordinary differential equations,

$$\frac{dq}{dt} = \frac{A\alpha g}{2} S_1 - Rq , \tag{4.7}$$

$$\begin{cases} \dfrac{dS_1}{dt} - \dfrac{q}{Al} C_1 = K(V_1 - S_1) , \\[2ex] \dfrac{dC_1}{dt} + \dfrac{q}{Al} S_1 = K(W_1 - C_1) , \end{cases} \tag{4.8}$$

and

$$\begin{cases} \dfrac{dS_n}{dt} - \dfrac{nq}{Al} C_n = K(V_n - S_n) , \\[2ex] \dfrac{dC_n}{dt} + \dfrac{nq}{Al} S_n = K(W_n - C_n) . \end{cases} \tag{4.9}$$

Equations (4.7) and (4.8) are coupled to each other and these three equations determine the flow rate q. The flow rate appears only as a time-dependent coefficient in (4.8). In this sense, the flow of the fluid is governed by the $n=1$ Fourier components.

This is an example of a technique which is often used in attempting to approximate solutions of partial differential equations. Unfortunately, in most cases the differential equations for the Fourier series coefficients do not decouple (though they clearly do in our case), and the infinite set of equations for the derivatives of any finite collection of the Fourier series coefficients involves other coefficients outside its collection so the "closure problem" – that is, the question of how to obtain a closed set of equations – must be faced.

A change of variables converts (4.7–9) to dimensionless form. We let $\gamma = g\alpha/2KRl$ and introduce dimensionless variables $\sigma = R/K$, $\tau = Kt$, $x = q/KAl$, $y = \gamma S_1$, $r = \gamma W_1$, $r' = \gamma V_1$, and $z = r - \gamma C_1$; for $n \neq 1$, $y_n = \gamma S_n$, $r_n = \gamma W_n$, $r'_n = \gamma V_n$, and $z_n = \gamma C_n$. Then we find

$$
\begin{cases}
\dfrac{dx}{d\tau} = \sigma(y - x) , \\[2mm]
\dfrac{dy}{d\tau} = -y + rx - zx + r' , \\[2mm]
\dfrac{dz}{d\tau} = xy - z ,
\end{cases}
\tag{4.10}
$$

$$
\begin{cases}
\dfrac{dy_n}{d\tau} = nxz_n + r'_n - y_n , \\[2mm]
\dfrac{dz_n}{d\tau} = -nxy_n - z_n + r_n ,
\end{cases}
\quad n > 1
\tag{4.11}
$$

$$
dz_0/d\tau = r_0 - z_0 .
\tag{4.12}
$$

A particularly simple situation results when only W_1 and V_1 are nonzero. Under these conditions (4.11) become homogeneous. Letting $A_n^2 = y_n^2 + z_n^2$, $n > 1$, (4.11) yield

$$
(dA_n^2/d\tau)/2 = -A_n^2 .
\tag{4.13}
$$

Each Fourier component corresponding to $n > 1$ decays exponentially to zero leaving x, y, and z as the only variables with nontransient dynamics. Furthermore, the time dependence of z gives the variation with time of the temperature in the fluid at the top ($\phi = \pi$) of the tube while the time dependence of y is the variation of the temperature at $\phi = \pi/2$.

If $r' = 0$, (4.10) reduce to the Lorenz equations with $b = 1$. For $1 < r < r_2$ there is a constant speed circulation (which can be either clockwise or counterclock-

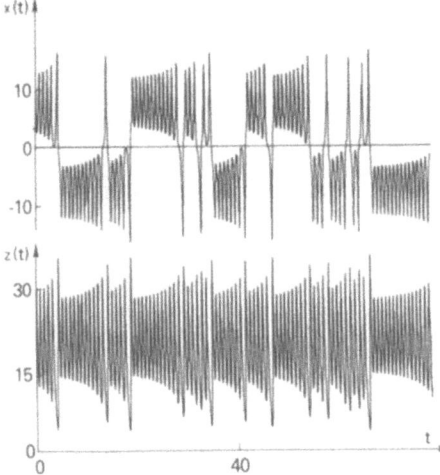

Fig. 4.2. A sample solution of Lorenz's system (4.1)

wise) which is stable. For $r > r_2$ all solutions are sensitive to initial data and the fluid circulates with irregularly varying speed, sometimes clockwise and sometimes counterclockwise. Figure 4.2 shows the sort of flow observed in a computer experiment. In conjunction with this behavior of the fluid motion the temperature difference between the fluid at the bottom and top of the tube [measured by $2C_1 = 2(r - z)/\gamma$] fluctuates irregularly as seen in Fig. 4.2. Particularly large values of z (negative values of C_1) are associated with reversal of the flow direction. In Sect. 4.4 a neat one-dimensional representation of the data involving successive maxima of z will be discussed. Experimentally observed fluid flow in a closed loop under conditions somewhat different from those which lead to the Lorenz equations is qualitatively similar to the flow predicted by these equations [4.12].

A brief investigation of the situation with r' nonzero shows that the fluid circulates preferentially so that it rises on the heated side (determined by the sign of r'). If r' is too large compared to r, steady flow in the preferred direction is stable. For a small range of values of r' and for $r > r_2$, the solutions are sensitive to the initial data but with a bias toward circulation in the preferred direction.

Lorenz investigated a typical "chaotic" value, $r = 28$ (with $\sigma = 10$, $b = 8/3$). There is a complicated region in (x, y, z) space that attracts all nearby solutions. This "attractor" set looks roughly like a two-dimensional surface, though close examination reveals a number of layers. A solution picked at random will wander through this set and, given sufficient time, will eventually pass arbitrarily close to every point of the attractor. Dynamical processes which exhibit sensitivity to initial data and have the property that most of these trajectories have are sometimes called "chaotic".

The geometric shapes of this and related attractors have received considerable attention. *Lorenz* noted that while the attractor at $r = 28$ has the appearance of a surface, if we were to "drill" down through it, we would hit an infinite number of layers. This attractor is indeed strange, and *Ruelle* and *Takens* [4.13] used the term "strange attractor" for any attractor of a dynamical process which is i) bounded, ii) connected, and iii) looks strange; that is, it is neither a point nor a periodic orbit nor a surface. They were unaware of Lorenz's attractor when they suggested the term, but Lorenz's attractor is definitely strange. It is not clear how strange an attractor must be to deserve the name; *Moser* [4.14] has pointed out that differential equations can have a badly distorted torus as an attractor (actually the three-dimensional analog) and it can be nowhere differentiable even if the right-hand sides of the differential equations are analytic. This nowhere differentiability can be expected whenever the rate of attraction to the torus is slow while the flow on the torus has high sensitivity to initial data. Nearly turbulent fluids may satisfy such a combination of high sensitivity and low viscosity, permitting errors in initial data to grow rapidly while having waves damp out quite slowly. Simple numerical studies [4.15] have led to conjecture estimating the dimension of the strange attractor in terms of the sensitivity and stability levels. *Curry* and *Yorke* [4.16] have investigated how an attracting torus can become infinitely fringed with wrinkles so that it becomes a strange attractor and is no longer a torus.

Guckenheimer [4.17] and *Guckenheimer* and *Williams* [4.18] have investigated how Lorenz's strange attractor changes as the parameters change. They have argued convincingly that the topological nature of the Lorenz attractor is different for each value of r. That is, attractors for different values of r are so different that even if they were made of rubber, one could not be squeezed or stretched into the shape of another. The fine structure of how the edges of the infinitely many sheets are interwound changes with r. In our derivation we have introduced a parameter r' which breaks the clockwise vs counterclockwise symmetry; *Guckenheimer* and *Williams* have argued that for such a model even more topological types appear. It appears that for each r and r' (in the chaotic range) the topological type of the attractor is distinct. No two are alike. But any small perturbations to the Lorenz equations – that is, the addition of *any* extra terms that are sufficiently small – can only produce strange attractors which are topologically equivalent to one of those already obtained. That is, the differences will only be a matter of compression of stretching.

4.3 Landau's Idea: A Continuous Transition to Turbulence via an Infinite Cascade of Bifurcations

The Lorenz equations exhibit one type of irregular behavior but there may be other ways in which a flow can become chaotic as a Rayleigh or Reynolds number R is increased. *Landau* and *Lifschitz* [4.19] suggested a model of a transition to chaotic behavior which is quite different from what is observed in the differential equations model of *Lorenz*. (See the discussion by *Joseph* in Sect. 3.3.) The model of *Lorenz* exhibits a sudden transition as R is increased, while *Landau* was interested in situations whose dynamics gradually became more and more irregular as a parameter is increased, and he proposed a qualitative scheme that he felt would be applicable to many flows.

Landau started from the Navier-Stokes equation and the equation of continuity. Because he believed that the nature of the transition to turbulence would be quite general, he did not further specify the system, except to require time-independent boundary conditions. For many situations, steady-state solutions can be found and, experimentally, for sufficiently low Reynolds number the steady flow is stable. The stability of the steady flow can be investigated as the Reynolds number is increased. Such a program had already been carried out theoretically as well as experimentally for several situations including Couette flow. A solution to the Navier-Stokes equation is written as $v_0(x) + v_1(x, t)$, $p_0(x) + p_1(x, t)$, and if the deviations from steady flow are small, the equations can be linearized. That is, one starts from the equations

$$\partial v/\partial \tau + (v \cdot V)v = -(Vp/\varrho) + (V^2 v)/R ,\tag{4.14}$$

$$V \cdot v = 0 .\tag{4.15}$$

The velocity v is prescribed on the boundaries. If deviations from a steady flow, $[v_0(x), p_0(x)]$, are assumed to be small, the deviations approximately satisfy the linear equations

$$\frac{\partial v_1}{\partial \tau} + (v_0 \cdot V)v_1 + (v_1 \cdot V)v_0 = -\frac{Vp_1}{\varrho} + \frac{1}{R} V^2 v_1 ,\tag{4.16}$$

$$V \cdot v_1 = 0 ,\tag{4.17}$$

where $v_1 = 0$ on boundaries.

In principle, separation of variables can be used to find the time dependence of v_1. Solutions of the form

$$v_1 = e^{i(\omega + i\gamma)t} v_{1\omega}(x)\tag{4.18}$$

are examined and the linearized equations are used to determine the complex eigenfrequencies $\omega + i\gamma$. If all the eigenfrequencies have positive imaginary parts, any small disturbance will decay and v_0 is stable. However, if, as R is increased, one eigenfrequency $\omega_1 + i\gamma_1$ crosses the real axis, v_0 will be unstable against perturbations of real frequency ω_1. Beyond the critical Reynolds number R_c at which this occurs, the flow will be different from v_0 and will be time dependent with a frequency ω_1. If ω_1 is zero, one steady flow is replaced by another; this is what happens in Couette flow (see Sect. 6.2).

The speculative part of Landau's scheme involves the behavior of the system as R is increased beyond R_c. His hypothesis is that, if a similar program could be carried out for even higher values of R, a second value, R_2, would be reached at which the solution with one characteristic frequency becomes unstable to perturbations with a characteristic frequency ω_2. Above R_2 the flow has two characteristic frequencies, ω_1 and ω_2. The relative phases with which these frequencies enter the sum are arbitrary; in principle they depend on conditions at the time that R_2 is reached but precise knowledge of these conditions is beyond experimental control. As R is increased beyond R_2, yet a third frequency, ω_3, is introduced at R_3 and so on. Each time a new frequency enters, a new arbitrary phase is introduced. *Landau* further assumed as R is increased above the initial transition from the steady flow of low R "the range of Reynolds numbers between successive appearances of new frequencies diminishes rapidly in size. The new flows themselves are on a smaller and smaller scale". The irregular behavior is thus hypothesized to result from an infinite *cascade* of bifurcations. It is clear (from observing even a two-frequency situation as simple as Lissajous figures) that very involved motions can occur – in Landau's words "for $R > R_c$, therefore, the flow rapidly becomes complicated and confused". The behavior of the system appears to be aperiodic.

No matter how "complicated and confused" the system may appear to the casual observer, there are fundamental differences between Landau's model for chaos and that of a system which is sensitive to initial data. The flows that *Landau* described are *not* sensitive to initial data. Two trajectories which start out close will remain close. And if an experimenter measures the spectrum of some physical quantity related to the flow, the Landau-type flow will yield sharp peaks at the frequencies ω_1, ω_2, ω_3, etc. (and their harmonics), while the spectrum of a system which is sensitive to initial conditions will yield a flat noiselike spectrum. If the frequencies are sufficiently close, it might be possible to temporarily interpret a Landau flow as a truly chaotic one – but that would be a problem of improving instrumental resolution. In principle, Landau flows are really more "ordered" than the types of flows exemplified by the Lorenz equations. In the Landau case an experimental variable $x(t)$ will generally be (positively or negatively) correlated with $x(t+s)$ even if s is large, while if the correlation tends to 0 as s increases, the system is chaotic. For autonomous (i.e., time-independent) differential equations which have sensitivity to initial data, we would expect the correlation of $x_i(t)$ and $x_i(t+s)$ to tend to 0 as $t \to \infty$ for each coordinate x_i. If the system has periodic coefficients (or if we are examining the iterates of a map), the correlation does not have to go to 0.

Landau's hypothesis of the introduction of an ever-increasing (but always finite) number of characteristic frequencies and an extended region of Reynolds numbers over which the transition to turbulence occurs does not seem to correspond to any experimentally observed situation in fluid dynamics. The mathematical validity is discussed in [4.20]. However, a simple mathematical model which has applications in ecology [4.21, 22] has some qualitative similarities to Landau's description of the transition to turbulence. The behavior of this model is discussed in the following section, and the relation of the model to ecology is discussed in Sect. 9.3.

4.4 One-Dimensional Maps: A Continuous Transition to Chaos via an Infinite Cascade of Bifurcations

The difference equation

$$x_{n+1} = rx_n(1 - x_n) , \tag{4.19}$$

where r is a constant, has been studied by mathematicians for at least 50 years. If the subscripts n and $n+1$ are assumed to represent two successive instants of time, (4.19) is a dynamical process. For example, a related equation has been used to model the dependence of the size of the $(n+1)$st generation of insects on that of the preceding nth generation (see Sect. 9.3). Equation (4.19), with its quadratic dependence on x_n, is extremely simple. Yet for different values of the parameter r the sequence generated by successive iterations can show steady-state behavior (for $r < 3$), behavior similar to the Landau transition to turbulence (as r increases to $r_c \approx 3.57$) and even chaotic behavior which is sensitive to the initial conditions (for many values of $r > 3.57$). We shall restrict r to $0 < r \leq 4$ so that the interval $[0, 1]$ will be mapped into itself.

For all values of r, there is a fixed point (a solution to $x_{n+1} \equiv x_n$) at $x = 0$, and for $r > 1$ there is an additional fixed point at $x = 1 - 1/r$. If (4.19) is regarded as a dynamical process, the fixed points represent steady states. Notice that

$$x_{n+1} = rx_n(1 - x_n) \leq rx_n \quad \text{if} \quad x_n \text{ is in } [0, 1] , \tag{4.20}$$

so for $r < 1$ we have $x_n \to 0$ if the initial x_1 is in $[0, 1]$. For $1 < r < 3$ linear analysis shows that if x_n is sufficiently near the fixed point $1 - 1/r$, then x_{n+1} will be closer to the fixed point by a factor of approximately $|2 - r|$. Numerically we observe that indeed for every point x_1 in $(0, 1)$, $x_n \to 1 - 1/r$ for $1 < r \leq 3$. For $r > 3$ a "time-dependent" solution of period 2 occurs. That is, the equation $x_{n+2} = x_n$ develops two real roots p and q in addition to the steady states, and beginning at point p successive iterations of (4.19) will yield the sequence $p, q, p, q, p, \ldots$. The period 2 solution is stable and attracting for r ranging from $r = r_2 = 3.0$ until $r = r_3 \approx 3.45$. At r_3 a period 4 solution develops and the period 2 solution, while still existing, loses its stability. Between r_3 and $r_\infty \approx 3.56998$, stable solutions of period 2^k are successively supplanted by stable solutions of period 2^{k+1}. The lower period solutions remain but are no longer stable. The higher the value of k, the smaller is the interval of r value over which the period 2^k solution is stable. The behavior of iterates of (4.19) is reminiscent of Landau's infinite sequence of transitions in that the bifurcations to higher and higher periodicities, and thus to more complex behavior, occur with ever smaller increases in r. Beyond the period 4 solution the transitions are followed numerically, leading to Fig. 4.3. At $r_\infty \approx 3.57$ there are no stable periodic solutions. The oscillations all settle down to what appears to be period 2 if the sequence of x_n is not examined very closely. If we examine only the first 2 digits,

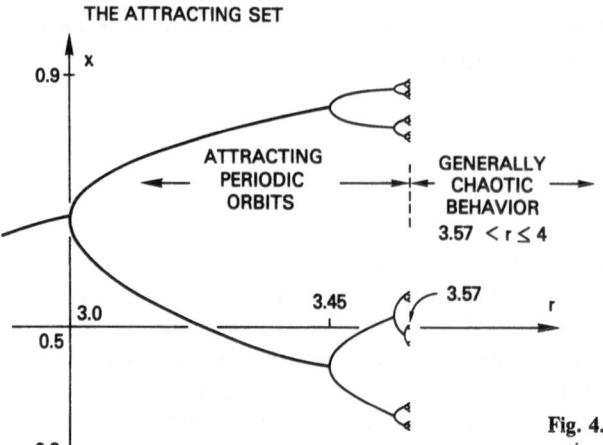

Fig. 4.3. An infinite cascade of bifurcations on the way to chaos

the sequence x_n appears to have period 32, repeating periodically forever, but if 3 digits are observed, it appears to have period 256. The system is not periodic and yet it is highly predictable over large numbers of iterations. If two points x_1 and y_1 are chosen close together, then the thousandth iterates x_{1001} and y_{1001} will be close to each other. Therefore, the system at r_∞ is *not* sensitive to initial data. This long-term predictability is what we would also expect from Landau's model at his corresponding parameter value r_∞. On the other hand, for most values of r in $(r_\infty, 4)$, the sequences of iterates of (4.19) appear highly sensitive to initial data: small changes in initial x eventually make large differences in the x_n's. In that way the sequences are similar to the solutions of the Lorenz equations above r_2. The sensitivity is not merely an artifact of numerical iterations, but can be shown to be a rigorous property of (4.19). A similar cascade of period doubling bifurcations can occur in ordinary differential equations though the phase space must be at least three-dimensional. See for example the "first equation" in [4.23] and for a plasma physics occurrence see [4.24].

In a paper on chaotic dynamics, *Li* and *Yorke* [4.21] investigated the dynamics of

$$x_{n+1} = F(x_n) , \tag{4.21}$$

where F is any continuous function defined on an interval. ([4.25] has many related results.) If there exists some initial point x_0 such that $x_1 > x_0$ and $x_2 > x_1$ but $x_3 \leqq x_0$, then they show that there is an uncountable subset of initial conditions for which the resulting sequence never settles down to a constant or periodic behavior, and furthermore, any two points in this subset lead to sequences which in the long run are very different, regardless of how close these two initial points are. There can in addition be one or more stable periodic orbits but these seem to be somewhat rare, and when they do exist, they are

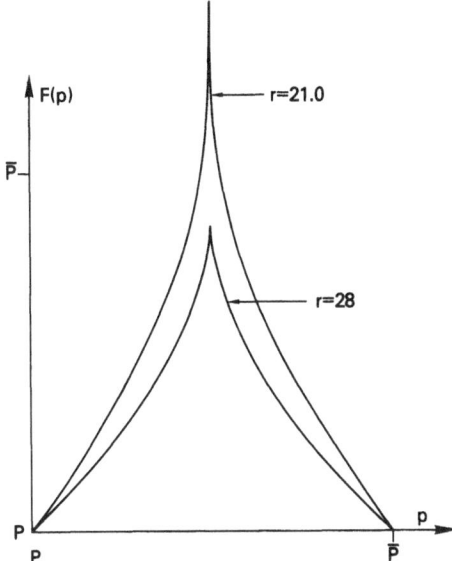

Fig. 4.4. Two examples are shown of curves obtained from Lorenz's system. If p is a local peak value of $z(t)$ (see Fig. 4.2), an excellent approximation of the next peak is $F(p)$. These functions are scaled to fit on a single graph, letting P be a point satisfying $F(P) = P$. For $r = 28$, $P \approx 27.0$ and for $r = 21$, $P = 27.8$. The point $\bar{P}$ denotes the point such that $P = F(\bar{P})$, $\bar{P} \neq P$. For $r = 28$, $\bar{P} \approx 53.5$ while for $r = 21$, $\bar{P} \approx 32.5$. Notice that for $r = 28$, F maps $[P, \bar{P}]$ into itself, while this is not true for $r = 21.0$

totally disrupted by small random perturbations. For example, (4.19) with $r = 3.83$ satisfies the above hypothesis with $x_1 = 0.16$. There is also a stable periodic orbit of period 3 into which most trajectories of x_n will become locked. (There is still an uncountable set of initial conditions whose trajectories do not get locked in.) In preparing this chapter we decided to calculate how fragile this periodic behavior was. We found in fact that the slightly noisy equation

$$x_{n+1} = 3.83 x_n(1 - x_n) \pm 0.001 \tag{4.22}$$

has a different behavior when the signs $\pm$ are chosen randomly for each iteration. No semblance of periodicity is observed, and the trajectories x_n will not stay near the original period 3 orbit, should x_n pass near. The stability is destroyed.

The sensitivity to initial data observed in (4.19) for certain r makes it a model which is even more irregular and chaotic than Landau's model. It is remarkable that this simplest of nonlinear models has behavior worse than that envisioned by *Landau* who even called upon the infinite number of degrees of freedom available.

Coincidentally (4.21) can be used to describe some properties of time-dependent solutions to the Lorenz equations. Numerical integration of the equations for $r = 28$ will lead to solutions for $z(t)$ like those of Fig. 4.2 which show no apparent pattern. Yet *Lorenz* found that if the height of the $(n+1)$st maximum, z_{n+1}, is plotted against the height of the nth maximum, z_n, the points fall on a cusped curve like that of Fig. 4.4. The peaks of $z(t)$ approximately satisfy a difference equation like (4.21) and the approximation is very good.

It is possible to fit a curve F to the set of points so that $|F(z_n) - z_{n+1}| \leq 10^{-3}$ for all n. This error of 10^{-3} is difficult to determine precisely, but there is some point below which the error cannot be reduced. In fact the existence of the functional relationship that appears in Fig. 4.4 is somewhat illusory. A microscopic examination of the points (z_n, z_{n+1}) would reveal a more complicated structure. The points would appear to lie along two close nearly parallel curves separated by about 10^{-3} and which of the curves a pair (z_{n+1}, z_n) was on would be found to depend on z_{n-1}. A submicroscopic examination would show a further splitting, a dependence on z_{n-2} and so on. Notice $|dF/dx| > 1$ for any reasonably fitted curve, and this fact can be shown to have a surprising worthwhile property: if $\{z_n\}$ is an infinite sequence of peaks generated from the original differential equation, then there will exist an x_1 such that for $x_{n+1} = F(x_n)$ we have

$$|x_n - z_n| \leq \varepsilon \quad \text{for all} \quad n = 1, 2, \ldots, \tag{4.23}$$

where $\varepsilon \approx 10^{-3}$. It is unlikely that the correct choice of x_1 would be exactly equal to z_1, and only one choice of x_1 would produce a good fit for all n. However, the existence of an approximate relation like (4.21) for the Lorenz system is useful in that it permits a less expensive study of the statistical properties of the sensitive solutions.

4.5 Long-Term Average Behavior

The combination of sensitivity to initial data and only approximate information about initial data leads to an inability to make accurate long-term predictions. Compare this situation with a long sequence of flips of a fair coin. The coin situation is much worse in that it lacks even short-term predictability. Nonetheless, we can still speak about the average behavior in the long run, expecting about half the flips to yield heads, and the longer the run we examine, the closer to 1/2 we expect the "head" fraction of the flips to be.

In the same spirit we can study a differential equation $dx/dt = f(x)$ on an open subset B of $\mathbb{R}^n$ and examine long-term average behavior along a trajectory $x(t, p)$ which starts at point $p \in \mathbb{R}^n$. Let $g(x)$ be some continuous function of the state, such as the energy or position or even the cube of the first coordinate. For each initial point in $\mathbb{R}^n$, we write

$$\langle g, p \rangle = \lim_{\tau \to \infty} \tau^{-1} \int_0^\tau g[x(t, p)] dt . \tag{4.24}$$

Suppose we examine a Hamiltonian system where the vector x is the vector of the generalized coordinates and momenta. Let g represent the energy of the system. Since energy is conserved, $g[x(t, p)]$ is independent of time so that

$\langle g, p \rangle$ exists. However, given different initial conditions the system will in general have different energies so $\langle g, p \rangle = g(p)$. This is quite different from the coin flipping situation where almost any infinite sequence of flips will yield the same long-term average.

The term *ergodic* is used to describe dynamical situations where the long-term average of any continuous function is essentially independent of p. If two points p and q are chosen at random from the phase space $\mathbb{R}^n$ (using a uniform probability distribution in making the choice), then $\langle g, p \rangle = \langle g, q \rangle$ — at least for almost any choice of p and q in $\mathbb{R}^n$. There can be exceptional (i.e., unlikely) choices of p giving different values or having no long-term average provided that the chance of choosing such an x is 0. See [4.26] for a discussion of this definition. There is a frequently used definition of ergodic behavior which is more restrictive. Here it is assumed that trajectories must take each subset of B to a set of the same volume (though rather generalized concepts of volume are allowed) any time t later. When such a generalized concept of volume is available it is called an "invariant measure". The broader definition extends the concept of ergodicity to situations where there is a relatively small set that attracts almost all trajectories. It even includes the extreme case where there is a single rest point to which all trajectories tend. It should also be noted that whatever definition of ergodic is used, verification that a particular system is ergodic is very difficult. But many dynamical situations appear ergodic and the concept is quite useful for describing the way systems appear to act. Lorenz's system appears ergodic for $r = 28$, the value *Lorenz* studied. *Lucke* [4.27] calculates trajectory averages for the Lorenz model and shows how these averages depend on r. *Di Prima* and *Swinney* describe in Chap. 6 the results of long time averages in experiments. The difficulty in verifying ergodicity for general differential equations leads us to examine simpler systems which are inherently easier to deal with to give us some intuition by analogy for the more complicated systems.

A similar definition of "ergodic" can be made for discrete time system. Let $B \in \mathbb{R}^n$ and $F : B \to B$. For each continuous $g : B \to R$,

$$\langle g, p \rangle = \lim_{N \to \infty} N^{-1} \sum_{n=1}^{N} g[F^n(p)] . \tag{4.25}$$

The map F, or the process (4.21), is called *ergodic* if $\langle g, p \rangle$ is essentially independent of the choice of p for every continuous $g : B \to R$.

That is, there is a real number denoted $\langle g \rangle$ such that when p is chosen at random using a uniform distribution on B, the probability is one that limit $\langle g, p \rangle$ exists and furthermore equals $\langle g \rangle$.

The study of real-valued functions that are defined on an interval allows us to investigate further dynamic and ergodic behavior which is similar to some observed behavior of higher dimensional situations. Suppose $F : [0, 1] \to [0, 1]$. Then given $x_1 \in [0, 1]$ we can iteratively define

$$x_{n+1} = F(x_n) \qquad n = 1, 2, \dots . \tag{4.26}$$

A major theme of this chapter is the importance of sensitivity to initial data. We can guarantee sensitivity to initial data if for each x and y sufficiently close to each other, we have $|F(x) - F(y)| > L|x - y|$ for some $L > 1$.

If F were continuous and differentiable, we would have $|F(1) - F(0)| > L$, so $F([0, 1])$ could not be $[0, 1]$. We therefore assume F is piecewise differentiable. Assume that F is twice differentiable on $[0, p]$ and on $[p, 1]$, where p is some point in $(0, 1)$, and this point can even be a point of discontinuity. In addition assume $|dF/dx| > L$ for some $L > 1$. Assume further that F, dF/dx, and d^2F/dx^2 have left-hand limits and right-hand limits as $x \to p$. *Li* and *Yorke* [4.28] and *Kosygin* and *Sandler* [4.29] have described results which imply that this iterative process is ergodic. In addition there is a set A to which almost every point in $[0, 1]$ is attracted and almost every point in $[0, 1]$ comes arbitrarily close to every point in $\{A\}$. The set $\{A\}$ is the union of a finite number of intervals, and indeed must be a single interval if $L > 2^{1/2}$. Note however that $\{A\}$ is often strictly smaller than $[0, 1]$.

The function F which best represents the sequence of peaks of z for the Lorenz equations has infinite slope at the midpoint of the interval. *Pianigiani* [4.30] and *Wong* [4.31] independently showed how the above ergodic result can be extended to such infinite slope maps.

When there are several discontinuities, it is possible that the interval will have two or more regions, on each of which the map is ergodic. For continuous g, the long-term average $\langle g, x \rangle$ will be the same for almost every x in such a region, but the value of $\langle g \rangle$ will vary from region to region. Higher dimensional analogs exist, though the assumption $|dF/dx| \geq L > 1$ that we used to guarantee sensitivity to initial data does not have a single analog in higher dimensions. *Bowen* and *Ruelle* [4.32] have found higher dimensional analogs in which the analog to $|dF/dx| \geq L$ is valid in some directions but not in others, and they are able to establish that their situations are ergodic.

4.6 Metastable Chaotic States

Some physical processes have one type of state for a substantial period of time and then suddenly switch to another. A burning candle or any system that consumes a fuel is such a process. A quite different type of state – which is sometimes called "metastable" – persists an indeterminant time. The super-cooled state of a fluid is metastable with the onset of freezing abrupt and indeterminant. Also the initial state of a radioactive atom is metastable. By a metastable state we mean a transitory state whose decay times have an exponential distribution. See [4.33] for a mathematical theory.

For $r = 3.83$ the difference equation (4.19) has metastable chaotic behavior. Almost every trajectory eventually settles down to a periodic oscillation after some initial period of irregular oscillation. The mean duration of this chaotic regime is not long, about 50 iterations, but initial conditions can be found by trial and error which remain chaotic much longer.

For certain ranges of parameter values, the system of *Lorenz* has a metastable chaotic state. In this regime there is a substantial region of (x, y, z) space for which a trajectory starting in the region appears to be of the chaotic sort described in Sect. 4.1 for many oscillations. However, eventually, with little warning, the trajectory begins to spiral into one of the steady "convective" states. Just as it is not possible to tell when an individual radioactive atom will decay, so it is not possible to predict when, for a specific trajectory, the change from chaotic behavior will occur. However, if an ensemble of initial points is observed, there is seen to be a well-defined half-life for the chaotic state. That is, at least for any strictly positive and smooth distribution of initial points, the fraction of trajectories which are chaotic for time t decreases as a function of t. For sufficiently large values of t, the decrease is an exponential decay and the decay constant depends only on the Lorenz parameters σ, b, and r; see [4.6]. It has been found that for r slightly less than $r_1 \simeq 24.06$ ($b = 8/3$, $\sigma = 10$), the mean "kick out" time (number of chaotic oscillations before decay sets in) is approximately $300(r_1 - r)^{-K}$, where K is approximately 4. This means that for $r = 23.5$, the transient chaotic behavior would probably exceed the time one observed the system, and chaotic behavior would improperly be inferred.

The system of *Lorenz* was the first continuous chaotic model system to be investigated in detail. Perhaps there are other situations in which an apparently turbulent fluid is actually exhibiting a metastable state, that is, states that are transitory and have *exponentially distributed decay times. L. Reith* and *H. Swinney* have recently found experimental situations which appear to be metastable in this sense. Only the simplest possible metastable chaotic transition is illustrated by the Lorenz system. In other situations a metastable chaotic state might decay to a state or "regime" which is periodic or to one which is also chaotic but which has significantly different behavior. Or the new state might also be metastable, eventually decaying to a third regime, or possibly back to the original metastable state.

The richness of behaviors possible in the simplest nonlinear dynamical systems is awesome. Some interesting work has categorized a number of types of chaotic dynamics. (See in particular the studies of *Rössler* [4.23].) We may ask which of these chaotic behaviors and transition to chaotic behavior occur in fluid dynamics. Landau's philosophy leads to a possible answer, for he says that turbulence is a mathematical phenomenon arising from oscillatory behaviors in spaces with sufficiently many degrees of freedom, which can be understood in principle without detailed analyses of pressure, vorticity, viscosity, and the like and without tedious analyses of general solutions of partial differential equations. Lorenz's severely truncated system seems similarly to have lost all its wetness, all the properties of a fluid except chaos; the same equations could conceivably arise in many situations that have nothing to do with fluid dynamics. The great distance of these equations from fluid dynamics emphasizes the abstract mathematical nature of chaos. A natural extension of Landau's mathematical philosophy, an extension we feel is valid, is that every kind of chaotic behavior and every type of transition to chaotic behavior exhibited by simple systems will be realized in some fluid-dynamical situation. The variety of possible experimental geometries and the choices of physical parameters seem quite rich enough to accommodate all species of chaos.

Acknowledgement. The research for this paper was partially supported by NASA grant NSG 5209.

References

4.1   D. Ruelle: "Dynamical Systems with Turbulent Behavior", in *Mathematical Problems in Theoretical Physics*, Lecture Notes in Physics, Vol. 80 (Springer, Berlin, Heidelberg, New York 1978) p. 341

4.2   D. Ruelle: Sensitive dependence on initial condition and turbulent behavior of dynamical systems. Ann. N.Y. Acad. Sci. **316**, 408 (1979)

4.3   E. N. Lorenz: Deterministic nonperiodic flow. J. Atmos. Sci. **20**, 130 (1963)

4.4   B. Saltzman: Finite amplitude free convection as an initial value problem, I. J. Atmos. Sci. **19**, 329 (1962)

4.5   J. L. Kaplan, J. A. Yorke: Preturbulence: a regime observed in a fluid flow model of Lorenz. Commun. Math. Phys. **67**, 93–108 (1979)

4.6   J. A. Yorke, E. D. Yorke: Metastable chaos: the transition to sustained chaotic behavior in the Lorenz model. J. Statist. Phys. **21**, 263–277 (1979)

4.7   J. B. McLaughlin, P. C. Martin: Transition to turbulence in a statically stressed fluid system. Phys. Rev. A **12**, 186 (1975)

4.8   J. H. Curry: Ph. D. Dissertation, University of California, Berkeley (1976)

4.9   W. V. R. Malkus: Mém. Soc. Royale de Sci. de Liège, 6th Ser. **4**, 125 (1972)

4.10   L. A. Rubenfeld, W. L. Siegman: Nonlinear dynamic theory for a double-diffusive convection model. SIAM J. Appl. Math. **32**, 871 (1977)

4.11   P. Welander: On the oscillatory instability of a differentially heated fluid loop. J. Fluid Mech. **29**, 17 (1967)

4.12   H. F. Creveling, J. F. dePaz, J. Y. Baladi, R. J. Schoenhals: Stability characteristics of a single-phase free convection loop. J. Fluid Mech. **67**, 65 (1975)

4.13   D. Ruelle, F. Takens: On the nature of turbulence. Commun. Math. Phys. **20**, 167 (1971); **23**, 343 (1971)

4.14   J. K. Moser: On a theorem of Anasov. J. Differ. Equat. **5**, 411 (1969)

4.15   J. L. Kaplan, J. A. Yorke: "Chaotic behavior of multidimensional difference equations", in *Functional Differential Equations and Approximations of Fixed Points*, Lecture Notes in Mathematics, Vol. 730, ed. by H. O. Peitgen, H. O. Walther (Springer, Berlin, Heidelberg, New York 1979) pp. 228–237

4.16   J. Curry, J. A. Yorke: "A Transition from Hopf Bifurcation to Chaos: Computer Experiments with Maps in $\mathbb{R}^2$", in *The Structure of Attractors in Dynamical Systems*, Springer Notes in Mathematics, Vol. 668 (Springer, Berlin, Heidelberg, New York 1977) p. 48

4.17   J. Guckenheimer: "Structural stability of the Lorenz attractor", Institut des Hautes Études Scientifiques, Publications Mathématiques No. **50** (1980)

4.18   J. Guckenheimer, R. F. Williams: The structure of Lorenz attractors. Appl. Math. Sci. **19**, 368–381 (1976)

4.19   L. Landau, L. Lifschitz: *Fluid Mechanics* (Pergamon, Oxford 1959)

4.20   J. L. Kaplan, J. A. Yorke: The onset of turbulence in a fluid flow model of Lorenz. Ann. N.Y. Acad. Sci. **316**, 400 (1979)

4.21   T. Y. Li, J. A. Yorke: Period three implies chaos. Amer. Math. Mon. **82**, 985 (1975)

4.22   R. M. May: Simple mathematical models with very complicated dynamics. Nature **261**, 459 (1976). See also R. M. May: Bifurcations and dynamic complexity in ecological systems. Ann. N.Y. Acad. Sci. **316**, 517 (1979)

4.23 O. Rössler: Continuous chaos – four prototype equations. Ann. N.Y. Acad. Sci. **316**, 376 (1979)

4.24. J.-M. Wersinger, J. M. Finn, E. Ott: Bifurcation and "Strange" Behavior in Instability Saturation by Nonlinear Three Wave Mode Coupling. Phys. Fluids **23**, 1142 (1980)

4.25 A. N. Sharkovskii: Coexistence of the cycles of a continuous mapping of the line into itself (Russian). Ukr. Math. J. **16**, No. 1, 61 (1964)

4.26 R. Bowen: "A Model for Couette Flow Data", in *Berkeley Turbulence Seminar*, Lecture Notes in Mathematics, Vol. 615 (Springer, Berlin, Heidelberg, New York 1977) p. 117

4.27 M. Lucke: Statistical dynamics of the Lorenz model. J. Stat. Phys. **15**, 455 (1976)

4.28 T. Y. Li, J. A. Yorke: Ergodic transformations from an interval into itself. Trans. Am. Math. Soc. **235**, 183 (1978)

4.29 A. A. Kosygin, E. A. Sandler: Izv. Vyss. Ucebn. Zaved. Mat. **118**, 32 (1972)

4.30 G. Pianigiani: Existence of Invariant Measures for Piecewise Continuous Transformations. Ann. Polon. Math. (to appear)

4.31 S. Wong: Some Metric Properties of Piecewise Monotonic Mappings of the Unit Interval. Trans. Math. Soc. **252**, 351 (1979)

4.32 R. Bowen, D. Ruelle: The ergodic theory of axiom A attractors. Invent. Math. **29**, 181 (1975)

4.33 A. Lasota, J. A. Yorke: The Law of Exponential Decay for Expanding Maps. Rendiconti Padova (in press)

5. Transition to Turbulence in Rayleigh-Bénard Convection

F. H. Busse

With 13 Figures

The present knowledge of convection in a layer heated from below is reviewed in this chapter with emphasis on those properties that provide some insights into the general problem of turbulence in fluid systems. From the experimental as well as the theoretical point of view convection in a layer heated from below is a particularly simple and accessible case, in which the development of turbulence can be studied. Some properties such as the discrete transitions in the measured heat transport have not yet been observed in other turbulent systems. For the study of time dependence, convection offers unique opportunities because nonlinear advection terms for both a vector quantity (momentum) and a scalar quantity (temperature) are present in the basic equations and can be distinguished by the Prandtl number dependence. In order to provide a comprehensive review of the basic physical properties of convection, attention is restricted to fluid layers which satisfy the Boussinesq approximation.

5.1 Overview

Thermal convection in a fluid layer heated from below represents the simplest example of hydrodynamic instability and transition to turbulence in fluid systems. In the idealized case of an infinitely extended layer with prescribed constant temperatures at the boundaries, the physical conditions are homogeneous and isotropic with respect to the horizontal dimensions. Since under stationary conditions the problem is also homogeneous in time, it exhibits the minimal dependence on a single dimension of a process that dissipates energy and thus requires an energy flux.

From the experimenter's point of view, thermal convection is also a particularly simple subject of research because of the absence of a mean flow. In contrast to most other cases of hydrodynamic instability, the eddies of the secondary motion are not swept past the observer by a mean flow, but stay approximately at the same place such that their properties can be readily observed. In addition, the relative ease with which thermal convection can be realized and visualized in the laboratory has made it a favored subject of experimental research, and it is not surprising that more distinct structures have been discovered in turbulent convection than in any other turbulent fluid systems.

Because the onset of convection occurs as the instability of a primary static state and because the temperature appears as a variable in the problem, thermal convection is often not recognized as a typical hydrodynamic instability. This superficial impression is misleading, however. It can be shown, for example, that the nonlinear solution of the Navier-Stokes equations describing axisymmetric Taylor vortex flow between coaxial cylinders rotating with nearly the same angular velocity is mathematically identical in the small gap limit with the solution describing two-dimensional convection in a fluid of Prandtl number one. The azimuthal component of the Taylor vortex flow corresponds to the temperature field of the convection rolls. With respect to the three-dimensional properties exhibited by secondary instabilities, the two problems differ (see Chap. 6), but this does not detract from the fact that thermal convection cannot be separated from other hydrodynamic instabilities on theoretical grounds.

Both convection and Taylor vortex flow share the property that the principle of exchange of stability is valid, i.e., the onset of instability occurs in the form of steady motions. That both flows lack a time dependence in the initial stage distinguishes them from other hydrodynamic instabilities which lead immediately to a turbulent state with a statistically fluctuating time dependence. It is generally recognized today that the onset of turbulence is delayed in the cases of thermal convection, Taylor vortex flow, and related forms of hydrodynamic instability. The intermediate steps in the development towards fully turbulent flow exhibited by convection and Taylor vortices make them ideal subjects for the study of the mechanisms causing the transition to turbulence. It is with this objective in mind that the properties of convection are considered in this article.

There is no generally accepted definition of turbulent fluid flow, and it may be worthwhile to discuss different possibilities of a definition. For simplicity, the attention will be restricted to systems with steady exterior conditions. It will become clear from the following discourse that no single one of the four attempted definitions is satisfactory since it either does not coincide with the intuitive physical concept of turbulence or it cannot be applied unambiguously in the experimental situations. But the discussion may help to set the stage for some interesting questions that are being asked in the study of thermal convection.

From the mathematical point of view it is natural to separate laminar and turbulent states of a fluid system by the property of uniqueness of the stationary solution of the problem. The conductive static state of a fluid layer heated from below and the Couette flow between differentially rotating cylinders are unique solutions when the forcing parameters represented by the Rayleigh and the Reynolds numbers, respectively, are sufficiently small. Beyond a certain value of these parameters, more than one solution becomes possible, resulting in the instability of the basic solution, at least if a sufficiently large perturbation is introduced. The identification of turbulence with the domain of nonunique solutions is mathematically convenient because necessary criteria for non-uniqueness can readily be derived, and in some cases it is possible to determine analytically the parameter value separating "laminar" and "turbulent" regimes. This definition of turbulence appears to be too general since it does not imply a time dependence. Taylor vortices and convection rolls are the best examples of nonunique steady solutions.

The realization of time-dependent flows for given steady exterior conditions is certainly one of the most startling aspects of turbulence. Thus, one is led to require the property of time dependence to restrict the manifold of nonunique states which correspond to turbulence. Indeed, Taylor vortices, as well as convection rolls in a low Prandtl number fluid, exhibit a distinct transition to a three-dimensional time-dependent state. This transition depends in general on the wavelength of the two-dimensional solution and thus on the history of the particular experiment. For this reason the onset of time dependence can only approximately be predicted on the basis of the exterior parameters of the system. A more serious shortcoming of this definition of turbulence is that the time dependence often occurs in the form of periodic waves, as in the cases mentioned above, which contradicts the aperiodic time dependence one intuitively associates with turbulence. Moreover, the answer to the question whether a fluid flow is time dependent may depend on the observer. Wavy Taylor vortices appear as a steady phenomenon to the observer positioned in a system rotating with a suitable angular velocity about the axis of the cylinders.

Since most scientists associate turbulence with randomness, one may attempt to define turbulence as the realization of solutions of the Navier-Stokes equations with random properties. A simple example is provided by convection in a horizontal layer of large aspect ratio. Theory predicts that convection assumes the form of two-dimensional rolls. But because of the horizontal isotropy of the layer, all directions of the convection rolls are equally likely. Starting from infinitesimal random disturbances, the onset of convection occurs in the form of patches of short convection rolls with different orientations. As time proceeds, some patches grow at the expense of others owing to the tendency of convection rolls to align with each other. As this process continues, the rate of change in time of the convection pattern decays because it is determined by the time of horizontal diffusion over an increasing length scale.

Since the random spatial organization of convection is a relatively weak property which disappears when the aspect ratio decreases to a value of the order 30 or less, and since the corresponding fluid flow is nearly steady, low amplitude convection lacks typical properties of turbulence. A more satisfactory definition for the onset of turbulence in fluid system is the first appearance of random elements in the time dependence of the system. Although this definition does not exclude without more detailed specifications the case of spatially random convection in a large aspect ratio layer, it seems to work well when the random behavior is preceded by a periodic or quasi-periodic time dependence. In their experimental investigation of Taylor vortices, *Gollub* and *Swinney* [5.1] have found a fairly distinct onset of broadband noise in the frequency spectrum preceded by periodic and quasi-periodic states (see Chap. 6). Similar phenomena have been found in some low Prandtl number convection experiments [5.2–5]. For a detailed discussion of the related mathematical models, we refer to Chap. 3 by *Joseph*.

Before completing our discussion, we wish to comment on the nature of the apparently random time dependence. *Gollub* and *Swinney* [5.2] and *Martin* [5.6] have expressed the view that aperiodic time dependence indicates the presence of a strange attractor in the phase space of the system. Indeed, the sequence of transitions leading to the aperiodic state strongly supports the strange attractor concept of *Ruelle* and *Takens* [5.7]. While the time dependence would not be truly random in this case, in the sense of a stochastic process, since the system still obeys deterministic equations, the alternative possibility that minute amounts of experimental noise affect the macroscopic properties of the system in a truly random sense cannot be easily excluded. An example of this latter form of random time dependence is described in Sect. 5.6 in the context of convection in a rotating layer [5.8].

In this chapter, the present knowledge of convection in a layer heated from below is reviewed with special emphasis on those aspects which are relevant for the understanding of the development of turbulence in fluid systems. The attention is focussed on the simplest problem of convection. Many interesting effects, such as deviations from the Oberbeck-Boussinesq approximation, the influence of sidewalls, penetrative convection, and others, are either mentioned only briefly or are not discussed at all. More general information can be obtained from the books of *Chandrasekhar* [5.9] and *Gershuni* and *Zhukovitskii* [5.10], which are mostly restricted to the linear theory of the onset of convection, and from the recent review articles of *Spiegel* [5.11, 12], *Koschmieder* [5.13], *Palm* [5.14], *Normand* et al. [5.15], and *Busse* [5.16]. Useful chapters on thermal convection can be found in the books of *Turner* [5.17] and *Joseph* [5.18]. The latter is especially recommended to readers interested in the mathematical aspects of the theory of convection.

5.2 Linear Theory

5.2.1 Basic Equations

Most of hydrodynamic stability theory is restricted to fluids of constant density for which a solenoidal velocity field can be assumed. Thermal convection is caused by temperature-induced variations of density, but the theoretical advantages of constant density fluids can be retained if the Oberbeck-Boussinesq approximation of the Navier-Stokes equations of motion is used. In this approximation all material properties are assumed to be constant with the exception of the temperature dependence of the density which is taken into account in the gravity term only. For derivations of the Oberbeck-Boussinesq approximation we refer to *Spiegel* and *Veronis* [5.19], *Mihaljan* [5.20], *Malkus* [5.21], and *Gray* and *Giorgini* [5.22]. It is easy to satisfy the assumptions of the Oberbeck-Boussinesq approximation to a high degree of accuracy in laboratory experiments. To a lesser degree of accuracy, the theoretical description based on this approximation can be applied to convection in the atmosphere and other large-scale systems if the temperature difference across the convection layer is replaced by the excess over the adiabatic temperature difference.

To obtain dimensionless equations it is convenient to introduce the thickness d of the layer as the length scale and the thermal diffusion time d^2/κ as the time scale, where κ is the thermal diffusivity. In addition, we shall use $(T_2 - T_1)/R$ as scale of the temperature where T_2 and T_1 are the mean temperatures of the lower and the upper boundaries, respectively, and R is the Rayleigh number defined below. In theoretical treatments of the problem it is often assumed that the upper and lower boundaries are infinitely conducting such that the temperature fluctuations induced by convection vanish there. But in experiments this condition is usually only approximately satisfied, which sometimes causes discrepancies between experimental measurements and theoretical predictions.

Using the Oberbeck-Boussinesq approximation, the nondimensional equations for the velocity vector u and the heat equation for the deviation θ from the static temperature distribution can be written in the form

$$P^{-1}\left(\frac{\partial}{\partial t}u + u \cdot \nabla u\right) = -\nabla \pi + \theta \lambda + \nabla^2 u , \tag{5.1a}$$

$$0 = \nabla \cdot u , \tag{5.1b}$$

$$\frac{\partial}{\partial t}\theta + u \cdot \nabla \theta = R\lambda \cdot u + \nabla^2 \theta , \tag{5.1c}$$

where λ is the unit vector in the direction opposite to gravity. The dependence of the problem on the material properties and the exterior condition has been reduced to two dimensionless parameters, the Rayleigh number and the Prandtl number,

$$R \equiv \frac{\gamma g(T_2 - T_1)d^3}{\nu \kappa} , \qquad P \equiv \frac{\nu}{\kappa} . \tag{5.1d}$$

Here g, ν, and γ denote the acceleration of gravity, kinematic viscosity, and the coefficient of thermal expansion, respectively. In (5.1a) all terms that can be written in the form of a gradient have been combined in the term $\nabla \pi$. Since only the curl of (5.1a) enters into the analysis of the problem, it is not necessary to specify this term in more detail. The particular choice (5.1d) of the dimensionless parameters of the problem offers the advantage that the onset of convection depends only on the Rayleigh number. The Prandtl number measures the relative importance of the advection of momentum and heat and thus affects the nonlinear properties of convection strongly. Values of P for some fluids used in laboratory experiments are listed in Table 5.1.

Most theoretical work has focussed on two sets of boundary conditions to be used in conjunction with (5.1). The assumption of stress-free boundaries has the advantage that simple solutions can be obtained, at least in the limit of small convection amplitudes. In most respects, the results differ only quantitatively when the more realistic no-slip condition of a rigid boundary is applied. The equation of continuity allows the conditions involving the tangential components of the velocity to be expressed in terms of the normal component. Thus the conditions

$$u_z = \partial_{zz}^2 u_z = \theta = 0 \quad \text{at} \quad z = \pm \tfrac{1}{2} \tag{5.2a}$$

are obtained in the case of stress-free boundaries, while the conditions

$$u_z = \partial_z u = \theta = 0 \quad \text{at} \quad z = \pm \tfrac{1}{2} \tag{5.2b}$$

hold in the case of rigid boundaries. When the finite horizontal extent of the layer is important, additional boundary conditions must be considered at the sidewalls. Unless indicated otherwise, the limit of an infinitely extended layer will be assumed in the following. In writing conditions (5.2) a Cartesian system of coordinates has been assumed with the origin in the center of the layer and the z coordinate in the direction of the unit vector λ. This choice will be retained throughout this chapter.

5.2.2 The Onset of Convection

A physical understanding of the onset of convection can be gained by considering the energy balances of (5.1a) and (5.1c). By multiplying (5.1a) and (5.1c) with u and θ, respectively, and averaging the result over the entire fluid layer, the relationships

$$P^{-1}(d\langle u \cdot u \rangle/dt)/2 = \langle u \cdot \lambda \theta \rangle - \langle |\nabla u|^2 \rangle , \tag{5.3a}$$

$$(d\langle \theta \theta \rangle/dt)/2 = R\langle u \cdot \lambda \theta \rangle - \langle |\nabla \theta|^2 \rangle , \tag{5.3b}$$

are obtained, where the property has been used that all expressions which can be written as a divergence vanish because of the boundary conditions and because the contributions from the sidewalls become negligible in the limit of an infinitely extended layer. For steady or statistically stationary convection the left-hand sides of (5.3) are zero. In that case, (5.3a) describes the balance between the work done by the buoyancy force and viscous dissipation, while (5.3b) expresses an analogous relationship between the convective heat transport and the entropy production by convection. That $\langle u \cdot \lambda \theta \rangle$ represents the convective part of the heat transport is easily seen from the horizontal mean of (5.1c),

$$\frac{\partial}{\partial t}\,\bar{\theta}+\partial_z\overline{u_z\theta}=\partial_{zz}^2\bar{\theta}\;. \tag{5.4}$$

In contrast to the average over the entire layer used in (5.3), the average over the x, y dependence at given value of z is indicated by a bar. By integrating (5.4) for a steady temperature field $\bar{\theta}$ and using the boundary conditions for the determination of the constant of integration,

$$\partial_z\bar{\theta}=\overline{u_z\theta}-\langle u_z\theta\rangle \tag{5.5}$$

is obtained. Since the normal component of the velocity vanishes at the boundary, the entire heat flux is carried by conduction at the boundary. The expression

$$-\partial_z\bar{\theta}|_{z=1/2}=\langle u_z\theta\rangle \tag{5.6}$$

thus represents the convective contribution to the heat transport in addition to that resulting from conduction alone. The ratio of the heat transports with and without convection is called the Nusselt number.

$$\mathrm{Nu}=(R+\langle u_z\theta\rangle)/R\;. \tag{5.7}$$

From (5.3a) it is evident that $\mathrm{Nu}\geq 1$ under stationary conditions.

By shifting the terms with $\langle u\cdot\lambda\theta\rangle$ to the left-hand sides of (5.3) and multiplying the resulting equations with each other, we obtain the relationship

$$R=\frac{\langle|\nabla u|^2\rangle\,\langle|\nabla\theta|^2\rangle}{\langle u\cdot\lambda\theta\rangle^2}\;. \tag{5.8}$$

This expression for R clearly indicates that steady or statistically stationary convection can exist only above a certain positive value of R. The right-hand side of (5.8) can be interpreted as a functional of the trial fields u and θ. When this functional is minimized subject to the constraints of the continuity equation (5.1b) and the boundary conditions (5.2), the energy stability limit R_E is obtained, i.e., no steady or statistically stationary form of convection is possible for $R<R_E$.

It can be shown that the Euler equations corresponding to the variational problem which determines R_E are mathematically identical to the linearized steady version of (5.1) in which the left-hand sides of (5.1) vanish. This identity has two important consequences [5.23]. First, Boussinesq convection represents one of the few cases of hydrodynamic instability where a criterion of the form $R\geq R_E$ provides not only a necessary condition for instability but a sufficient one as well. Second, there is no need to look for convection modes with an oscillatory time dependence, since they can exist only at higher values of the Rayleigh number if they exist at all. After neglecting the left-hand sides of (5.1), the following equation for u_z can be obtained by taking the vertical component of the curl curl of (5.1a) and by eliminating θ,

$$(\nabla^6 - R\Delta_2)u_z^{(0)} = 0 ,\qquad(5.9)$$

where Δ_2 denotes the Laplacian with respect to the horizontal coordinates. Equation (5.9) is solved by solutions of the form

$$u_z^{(0)} = f(z, \alpha) \sum_n c_n \exp(i k_n \cdot r) ,\qquad(5.10)$$

where r is the position vector and the vectors k_n satisfy the conditions

$$|k_n| = \alpha , \quad k_n \cdot \lambda = 0 \quad \text{for all } n \qquad(5.11)$$

but are arbitrary otherwise. This arbitrariness reflects the horizontal isotropy of the layer and causes the degeneracy of the eigenvalue R, which depends only on the length of the vectors k_n but not on their directions. The function $f(z, \alpha)$ is determined by an ordinary differential equation which yields

$$f(z, \alpha) = \sqrt{2} \cos \pi z , \quad R(\alpha) = (\pi^2 + \alpha^2)^3 / \alpha^2 \qquad(5.12)$$

in the case of stress-free boundaries. This result was obtained by *Rayleigh* [5.24] who did the first theoretical analysis of thermal convection. Of physical significance is the minimum value of $R(\alpha)$ at which convection first sets in when the temperature difference across the layer is slowly increased,

$$R_c = 27\pi^4/4 \approx 657 , \quad \alpha_c = \pi/\sqrt{2} .\qquad(5.13)$$

In the case of rigid boundary conditions (5.2b), the function $f(z, \alpha)$ is not independent of α and the problem is slightly more difficult to solve. The calculations of *Pellew* and *Southwell* [5.25], and *Reid* and *Harris* [5.26] have yielded

$$R_c = 1707.76 , \quad \alpha_c = 3.117 .\qquad(5.14)$$

The interpretation of R_c in terms of laboratory parameters in Table 5.1 indicates that the onset of convection can be observed easily for a wide variety of fluids. The fact that α_c is close to π indicates that convection cells with a nearly square cross section are preferred. In cells with $\alpha < \alpha_c$ the potential energy released by the vertical motion is too small in comparison with viscous dissipation associated with the horizontal motion. Cells with $\alpha > \alpha_c$ are less

Table 5.1. Properties of fluids in convection experiments at 20 °C

	Mercury	Air	Water	Glycerin
Prandtl number P	0.027	0.66	6.94	11,460
Depth d [cm] needed for $R_c = 1708$ for $T_2 - T_1 = 1\,°C$	0.773	2.54	0.497	3.47

favored because heat conduction between up- and down-going fluid parcels diminishes the available buoyancy.

There is one case of convection in which α_c actually tends to zero. Since this case is also distinguished by a simple analytical solution of (5.9), it is worth mentioning here. When the constant temperature boundary condition is replaced by a constant heat flux condition and the limit is assumed where the conductivity of the boundaries becomes small in comparison with that of the fluid, the condition $\theta = 0$ must be replaced by $\partial_z \theta = 0$ at the boundaries. $R(\alpha)$ reaches its lowest value in the limit $\alpha \to 0$ in this case and θ becomes independent of z. Straightforward calculations [5.27, 28] yield

$$f(z, \alpha)|_{\alpha = 0} \propto (z^2 - \tfrac{1}{4})^2 , \qquad R_c = 720 \tag{5.15a}$$

in the case of rigid boundaries and

$$f(z, \alpha)|_{\alpha = 0} \propto (z^2 - \tfrac{1}{4})(z^2 - \tfrac{5}{4}) , \qquad R_c = 120 \tag{5.15b}$$

in the case of free boundaries. The relatively low values of R_c and the vanishing values of α_c are caused by the fact that the vertical length scale of the temperature perturbation θ is no longer constrained by the thickness of the convection layer and increases without limits.

The main problem posed by the results of the linear theory is the infinite degeneracy of the eigenvalue R_c. This degeneracy is twofold. First there is the orientational and translational degeneracy. A solution of the form (5.10) with a given set of coefficients c_n remains a solution if the position vector r_i is replaced by

$$r'_i = a_i + b_{ik} r_k , \tag{5.16}$$

where a_i is an arbitrary constant vector and b_{ik} is an arbitrary orthogonal tensor. This degeneracy reflects the homogeneity and isotropy of the infinitely extended layer and thus cannot be removed without abandoning those properties. When the aspect ratio of the layer is decreased to finite values, the continuous function $R(\alpha)$ splits into a large number of discrete values, and a cluster of solutions competes for the lowest possible value of R_c of the Rayleigh number. But an orientational degeneracy may still persist, depending on the geometrical configuration of the side walls.

The second degeneracy is the pattern degeneracy associated with the arbitrary choice of the coefficients c_n. In order that (5.10) represent a real expression it is appropriate to impose the conditions

$$c_{-n} = c_n^+ , \qquad \mathbf{k}_{-n} = -\mathbf{k}_n , \tag{5.17}$$

where the superscript $^+$ indicates the complex conjugate. It is convenient to make the additional assumption that N different $\mathbf{k}$ vectors correspond to non-vanishing coefficients c_n in (5.10), where N may tend to infinity. The fact that the amplitude of solution (5.10) is not determined by linear theory suggests the introduction of the normalization condition,

$$\sum_{n=-N}^{N} |c_n|^2 = 1 , \tag{5.18}$$

But the conditions (5.17, 18) hardly constrain the arbitrary choice of coefficients c_n. While part of this arbitrariness reflects the orientational and translational degeneracy mentioned above, the more interesting freedom of choice represents the large variety of different forms of three-dimensional convection flows. It is this part of the degeneracy problem which can be resolved by considering the nonlinear problem and the stability of its solutions.

5.3 Nonlinear Theory

5.3.1 The Perturbation Approach

The only systematic method for analyzing the manifold of three-dimensional nonlinear steady solutions of (5.1) is the perturbation approach based on the amplitude ε of convection as small parameter. This approach is particularly appropriate in the case of convection because the instability occurs in the form of infinitesimal disturbances. Obviously, the perturbation expansion is of limited usefulness when the Rayleigh number is increased much beyond its critical value. Thus direct numerical methods have been used to solve the problem of fully nonlinear convection. Because these computations are usually restricted to the two-dimensional case, the numerical approach is to some extent complementary to the perturbation analysis and quite different questions can be answered by the two methods.

There are two questions suggested by the results of the linear theory to which the perturbation approach can be addressed. First, do all the different patterns of convection represented by the general form (5.10) of the linear solution correspond to steady solutions of the nonlinear problem in the limit $\varepsilon \to 0$? If this is not the case, is there a unique form of convection determined by the stationary problem in the limit $\varepsilon \to 0$? The answer to both questions is no. But the second question can be answered in the affirmative sense if the analysis of the steady problem is supplemented by a stability analysis.

The perturbation approach was first used by *Gorkov* [5.29] and *Malkus* and *Veronis* [5.30]. Here we follow the general analysis of the problem by *Schlüter* et al. [5.31]. The method proceeds by introducing the series expansions

$$\boldsymbol{u} = \varepsilon[\boldsymbol{u}^{(0)} + \varepsilon^2 \boldsymbol{u}^{(1)} + \varepsilon^3 \boldsymbol{u}^{(2)} + \ldots], \tag{5.19a}$$

$$R = R_c + \varepsilon R^{(1)} + \varepsilon^2 R^{(2)} + \ldots, \tag{5.19b}$$

and analogous expressions for θ and π into (5.1) and by solving successively the linear equations corresponding to each power of ε. Since only steady solutions are considered, the $\partial/\partial t$ terms in (5.1) vanish and in the order ε^1 the problem becomes identical to the linear problem. In the order ε^2, inhomogeneous linear equations are obtained with the left-hand sides given in terms of expression (5.10). A solution of the inhomogeneous system of equations exists if and only if the left-hand side is orthogonal to all solutions of the adjoint homogeneous problem. That the problem is self-adjoint is easily seen by multiplying (5.1a) and (5.1c) by $R_c \boldsymbol{u}^{(0)*}$ and $\theta^{(0)*}$, respectively, where $u_n^{(0)*}$ is given by expression (5.10) with arbitrary coefficients c_n^* replacing c_n. After adding the equations, averaging them over the fluid layer, and using the equation of continuity, the right-hand side vanishes because $\boldsymbol{u}^{(0)*}$ and $\theta^{(0)*}$ satisfy the homogeneous equations.

The solvability condition thus imposed on the inhomogeneous part of the equations determines the coefficients $R^{(n)}$ and provides constraints on the choice of coefficients c_n. But because of the vertical symmetry of the problem, the solvability condition of the order ε^2 is trivially satisfied with $R^{(1)} = 0$. In the order ε^3 the following system of equations is obtained:

$$R^{(2)}|c_l|^2 - \sum_{n=-N}^{N} A(\boldsymbol{k}_l \cdot \boldsymbol{k}_n)|c_n|^2 \cdot |c_l|^2 = 0 \quad \text{for} \quad l = 1, \ldots, N \tag{5.20}$$

when all independent choices in the form $c_n^* = c_l \delta_{ln}$ with $n = 1, \ldots, N$ are applied. The general manifold of solutions of (5.20) has not been explored, but, in the case when each vector $\boldsymbol{k}_n$ assumes the same set of angles with all other $\boldsymbol{k}$ vectors, it is easily seen that

$$|c_1|^2 = \ldots = |c_N|^2 = 1/2N \tag{5.21}$$

represents a solution of (5.20) together with the normalization condition (5.18).

The case of a regular k vector distribution with a constant angle of π/N between neighboring vectors obviously fulfills the requirements of (5.21). The requirements are also met in the semiregular case when each vector k_n encloses the angle $2\pi/N$ with its two second nearest neighbors without satisfying the more restrictive conditions of the regular distribution. It can be shown that the solutions (5.21) are not further restricted by higher order solvability conditions as long as the symmetry of the problem with respect to the plane $z=0$ is retained.

Only two of the solutions (5.21) are of physical importance. The simplest solution is given by $N=1$ and corresponds to convection in the form of two-dimensional rolls. It has the property that a translation by half a wavelength takes place when c_n is replaced by $-c_n$ for all n. The simplest solution that does not have this property is the hexagon solution corresponding to $N=3$. Thus there are really two hexagon solutions, one with upward motion and one with downward motion, in the center of the cells. This property is the basic reason for the preference of the hexagonal pattern in cases when the physical conditions of the layer are not symmetric with respect to its midplane.

By evaluating expression (5.20) the increase of the heat transport due to convection can be determined. In the case of two-dimensional rolls, different dependences on the Prandtl number are obtained for the two cases of boundary conditions (5.2). In the case of stress-free boundary conditions [5.30]

$$\text{Nu}-1=2(R-R_c)/R \tag{5.22a}$$

while in the case of rigid boundary conditions [5.31]

$$\text{Nu}-1=\frac{(R-R_c)}{R(0.69942-0.00472P^{-1}+0.008321P^{-2})} \tag{5.22b}$$

is found. Since in the limit of vanishing Prandtl number thermal conduction dominates, the convective heat transport is expected to decrease in that limit as shown by expression (5.22b). Indeed, the Prandtl number independent expression (5.22a) represents a singular case. Even slight changes in the boundary conditions lead to a dramatic decrease of the heat transport at low Prandtl numbers. The sensitivity of expressions (5.22) to small changes in the physical conditions of the problem indicates that the radius of convergence of the expansion (5.19) becomes rather small in the limit $P\to0$. This property is reflected in the formal convergence proofs [5.32, 33] for the expansion.

The solvability condition (5.20) restricts the number of possible solutions of the form (5.10) significantly. But there still exist infinitely many solutions of the nonlinear steady problem, and a stability analysis is required to select the solutions which are physically realizable. The stability analysis proceeds by superimposing arbitrary infinitesimal disturbances $\tilde{u}$, $\tilde{\theta}$ on the steady solution. By subtracting the steady equations from the equations for $u+\tilde{u}$, $\theta+\tilde{\theta}$ the following stability problem is obtained:

$$P^{-1}(\sigma\tilde{u} + u\cdot\nabla\tilde{u} + \tilde{u}\cdot\nabla u) = -\nabla\tilde{\pi} + \lambda\tilde{\theta} + \nabla^2\tilde{u} \ , \tag{5.23a}$$

$$0 = \nabla\cdot\tilde{u} \ , \tag{5.23b}$$

$$\sigma\tilde{\theta} + \tilde{u}\cdot\nabla\theta + u\cdot\nabla\tilde{\theta} = R\tilde{u}\cdot\lambda + \nabla^2\tilde{\theta} \ . \tag{5.23c}$$

Since the amplitude of the disturbances is infinitesimal, the stability problem is linear and a time dependence of the form $\exp(\sigma t)$ can be assumed. Here σ represents the eigenvalue of the boundary value problem given by (5.23) and the corresponding condition (5.2). When the real parts of all eigenvalues σ are negative, the steady solution u, θ is stable. It is unstable when an eigenvalue with a positive real part exists.

The eigenvalue problem (5.23) can be solved by expanding $\tilde{u}$, θ, and σ into series in powers of ε analogous to (5.19). By considering coefficients up to $\sigma^{(2)}$ in the series of the eigenvalue σ, *Schlüter* et al. [5.31] were able to demonstrate that all steady solutions of (5.1) are unstable with the exception of the two-dimensional solution in the form of rolls. Moreover, it could be shown that at small but finite values of $R - R_c$, rolls corresponding to a finite range of wave numbers α are stable. But this range represents only a small fraction of the available range of wave numbers [5.31, 34]. For a detailed discussion of the instabilities restricting the region of stability, see Sect. 5.5. The main conclusion to be drawn from these results is that even if the orientational and translational degeneracy of the problem is ignored, a spectrum of different steady convection modes is physically realizable and the asymptotic state of a convection layer will in general depend on the initial conditions. The statistical analysis of the three-dimensional initial value problem has not been done, but some typical features such as the tendency towards the realization of a single wave number spectrum are already apparent in the two-dimensional case [5.35].

5.3.2 Numerical Computations

The fact that convection rolls represent physically realizable solutions of the convection problem, at least for Rayleigh numbers sufficiently close to the critical value, together with the relative ease with which problems of two-dimensional flow can be solved on modern computers, has led to a large number of numerical investigations of convection rolls. Because little can be learned about the transition to turbulence from two-dimensional computations, they will be discussed only briefly here.

Of special interest is the dependence of the heat transport on the Rayleigh number. In the less restrained case of stress-free boundaries, the computations of *Moore* and *Weiss* [5.36] suggest an asymptotic dependence

$$\mathrm{Nu} = 1.8(R/R_c)^{0.365} \ . \tag{5.24a}$$

In the case of rigid boundaries, computations have not been carried out to quite as high Rayleigh numbers as in the stress-free case. A good representation for Rayleigh numbers between 10^4 and 10^6 is given by [5.37, 38]

$$Nu = 1.56(R/R_c)^{0.296} \qquad (5.24b)$$

Both expressions (5.24) have been obtained for $\alpha = \alpha_c$ and $P = 1$. Equation (5.22b) suggests that the maximum Nusselt number is reached in the limit of infinite Prandtl number. But the numerical computations at high Rayleigh numbers show a distinct maximum near $P = 1$ in the case of rigid boundaries. In the case of stress-free boundaries there is a slight monotonic increase of Nu as P decreases.

Numerical difficulties have prohibited computations of the convective heat transport at high Rayleigh and low Prandtl numbers except in the atypical case of stress-free boundaries. Experimental measurements of the Nusselt number in mercury [5.39] indicate that the efficiency of convection increases remarkably at supercritical Rayleigh numbers and that the range of $R - R_c$ for which expression (5.22b) is approximately valid decreases rapidly with decreasing P. The increased efficiency may be connected with the onset of time-dependent convection, which also occurs at decreasing values of $R - R_c$ as P decreases (see Sect. 5.5). Or it could be a manifestation of the phenomenon of inertial convection which was discovered by *Jones* et al. [5.40] in the case of convection in a cylinder with stress-free boundaries. In this case the axisymmetric convection characterized by a heat transport dependence of the form (5.22b) was replaced by a different axisymmetric mode with a Nusselt number dependence close to that given by (5.22a) as the Rayleigh number exceeded a second critical value about 33% above R_c. The new solution exhibits a "flywheel" character, i.e., its kinetic energy far exceeds the potential energy gained within each circulation period. An illuminating analytical study of this phenomenon in the case of convection in a horizontal tube has been given by *Proctor* [5.41]. But it is not yet well understood in which form "inertial" convection may occur in the case of an infinite layer.

Next to the Prandtl number dependence, the wave number dependence of the convective heat transport has received the most attention. The wave number at which the Nusselt number reaches a maximum increases with the Rayleigh number. On the other hand, the wave number of convection rolls observed in experiments tends to decrease. This contradicts the intuitive hypothesis first applied by *Malkus* [5.30, 42] that the convection mode with maximum heat transport should be physically preferred. The Nusselt number as a function of the wave number is shown in Fig. 5.1 in a special case. Besides the heat transport, other properties such as the potential energy or the kinetic energy of the convection flow may be related to the property of stability. The dimensionless expression E_p for the decrease of the potential energy of the layer from that of the static state is given by

$$E_p = \int_{-1/2}^{1/2} \bar{\theta} dz \qquad (5.25)$$

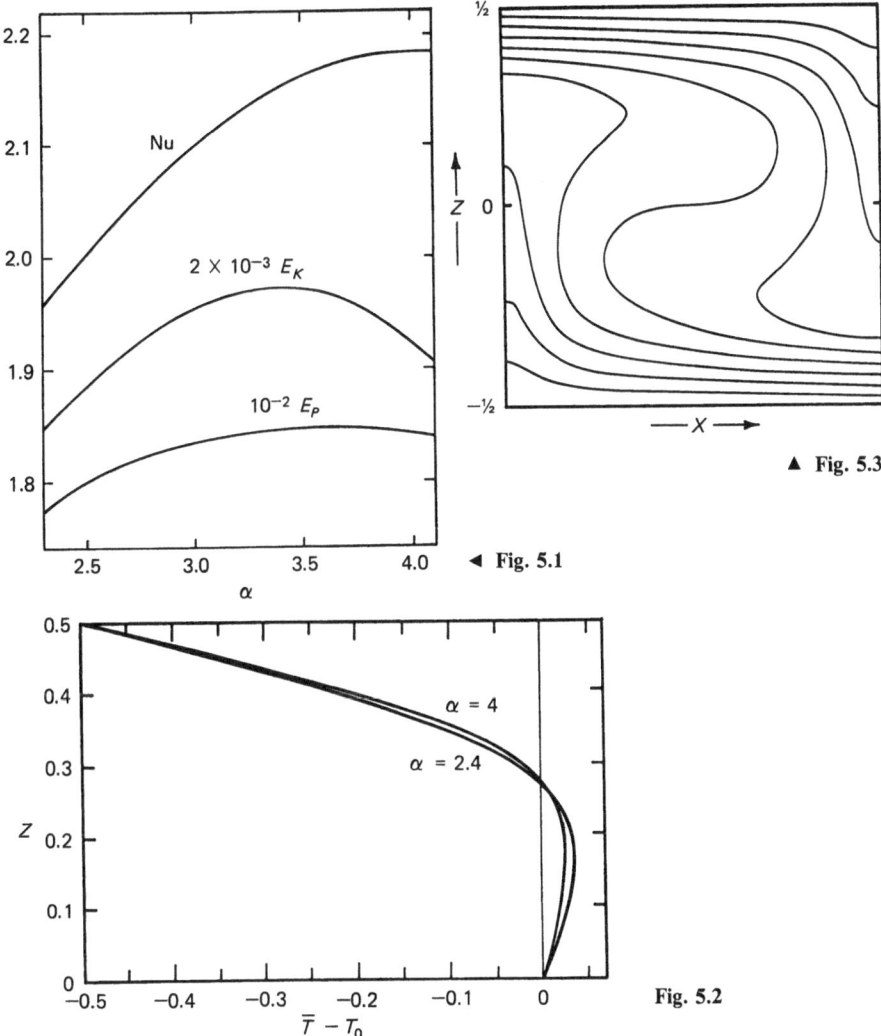

◀ **Fig. 5.1**

▲ **Fig. 5.3**

Fig. 5.2

Fig. 5.1. Nusselt number Nu, kinetic energy E_k (given by $E_k \cdot \pi^3$), and decrease in potential energy E_p (given by $E_p \cdot 100 \pi^2$) as a function of α for $R = 2 \cdot 10^4$ and $P = \infty$. The data of [5.45] have been used. (Note that $b_{\lambda v} \cdot \pi^{-3}$ is given in the figures of that paper instead of $b_{\lambda v} \cdot \pi^{-4}$ and the heat transport in Table 1 is too large by 10^3 for $\alpha = 2.4$, 2.7 in the case $R = 2 \cdot 10^4$)

Fig. 5.2. The variation of the mean temperature, $-z + \bar{\theta}/R$, throughout the upper half of a convection layer in the case $R = 2 \cdot 10^4$, $P = \infty$ for two different values of α

Fig. 5.3. Isotherms of a convection roll in the case $R = 2 \cdot 10^4$, $P = \infty$ for $\alpha = 3.117$

and is also shown in Fig. 5.1. Both E_p and the kinetic energy E_k have a maximum at a wave number larger than α_c. Thus it must be concluded that simple physical properties show little relationship with the property of stability. This view is confirmed by the rigorous stability analysis described in Sect. 5.5.

It is worth mentioning, however, that the stable solution in the form of rolls is distinguished among the unstable three-dimensional patterns of convection by a maximum heat transport [5.31, 53]. In the order ε^2 in which this result holds, a maximum heat transport is equivalent to a maximum of E_p or E_k.

The variation of the mean temperature profile as a function of depth is another interesting aspect of convection. In large amplitude convection rolls, the mean temperature gradient tends to reverse in the interior of the layer, while sharp gradients develop in the boundary layers as shown in Fig. 5.2 in the case of $P = \infty$. The reversal of the mean temperature gradient is caused by the mushroomlike spreading of the hot ascending and the cool descending fluid sheets at the opposite boundaries, as shown in Fig. 5.3. According to the computations of *Plows* [5.44], the reversal becomes slightly weaker as the Prandtl number decreases.

The steepening of the mean temperature profile at the boundary reflects the efficiency of the convective heat transport. In the interior of the layer, convection carries the entire heat flux and even a reversed conductive heat flux does not reduce the efficiency. At the boundary, the heat flux must be carried by conduction, since the normal component of the velocity vanishes. This restriction is most severe in the case of rigid boundaries, where the normal derivative of the vertical component vanishes as well. The thermal boundary layer is nearly static, and it is not unreasonable to apply the stability criterion for the onset of convection. Defining δ as the ratio between the boundary layer thickness and d, the criterion for instability becomes

$$1/2 R \delta^3 \geqq R_c , \tag{5.26}$$

where R_c is a number of the order 10^3 between the values given by (5.13) and (5.14). $R/2\delta$ is the total heat transport and the criterion (5.26) can be expressed in terms of the Nusselt number,

$$\mathrm{Nu} \leqq (R/2R_c)^{1/3}/2 . \tag{5.27}$$

Thus the thermal boundary tends to become unstable when the Nusselt number increases with less than the 1/3 power of the Rayleigh number. In accordance with relationships (5.24) the rigorous stability analysis shows indeed that convection rolls become unstable in the case of rigid boundaries [5.45], while they remain stable in the stress-free case [5.46].

Two-dimensional numerical calculations do not exhibit this instability because it assumes the form of rolls at a right angle to the given rolls. This effect is caused by the presence of the horizontal velocity component near the boundary. It is well known from the analysis of the stability of a static layer [5.47, 48] that the presence of a plane parallel horizontal shear flow has a stabilizing influence for all convection modes except for those which do not depend on the coordinate in the direction of the shear. Only when the Reynolds number of the shear exceeds a critical value may the stabilizing influence be reversed. Since the Reynolds number is small in the case considered here, the criterion (5.26) remains approximately valid even if the presence of the horizontal velocity is taken into account.

There are a few three-dimensional numerical computations of convection available in the literature. The most extensive work is that of *Lipps* [5.49]. But the imposition of a low aspect ratio periodicity interval prevents realistic simulation of the time-dependent forms of convection in a layer of large aspect ratio. Despite the rapid progress of computational facilities, the numerical simulation of fully developed turbulent convection is likely to remain an elusive goal. It may not even be a desirable goal because the large amounts of data produced by the computations would show an even more complex picture than that provided by optical observations of convection realized in the laboratory.

5.3.3 The Optimum Theory of Turbulent Convection

Experimental measurements of the heat transport by turbulent convection give well-defined and reproducible values at high Rayleigh numbers. But· the intricacies of the nonlinear problem of convection prevent the derivation of a theoretical prediction for the Nusselt number–Rayleigh number relationship. The theoretical difficulties are characteristic of the general problem of turbulent fluid flow and can be circumvented only by the introduction of heuristic assumptions of doubtful reliability such as the mixing length hypothesis. For a detailed application of the latter hypothesis to the problem of turbulent convection, we refer to *Kraichnan*'s paper [5.50].

In this section an alternative approach will be considered. It is possible to avoid the introduction of arbitrary assumptions if the more restricted goal of bounds on the mean transport of a physical quantity by turbulent fluid flow is pursued. The derivation of useful upper and lower bounds is accomplished by the optimum theory of turbulence. But even more interesting than the bounds are the corresponding optimizing vector fields which offer some remarkable insights into the mechanisms of turbulent transport.

The basic concept of the optimum theory of turbulence was developed by *Howard* [5.51] in the case of thermal convection long before the theory was applied to more general cases of turbulence [5.52–54]. Today convection in a layer heated from below still exhibits the closest correspondence between theory and experimental observations, and further experimental research on the discrete structures of turbulent convection is likely to emphasize this relationship. A lower bound on the heat transport is given by the static solution of pure conduction according to (5.7), and the main objective is to find an upper bound. By determining the maximum of the heat transport among a manifold of fields u, θ that includes all statistically stationary solutions of the basic equations (5.1), an upper bound for the physically realized heat transport can be obtained. The particular manifold of fields for which the variational problem will be formulated is given by all fields $\mathbf{u}$, θ that satisfy the relationships

$$\langle \mathbf{u} \cdot \lambda \tilde{\theta} \rangle - \langle |\nabla \mathbf{u}|^2 \rangle = 0 , \tag{5.28a}$$

$$R\langle \mathbf{u} \cdot \lambda \tilde{\theta} \rangle - \langle |\nabla \tilde{\theta}|^2 \rangle - \langle (\mathbf{u} \cdot \lambda \tilde{\theta} - \langle \mathbf{u} \cdot \lambda \tilde{\theta} \rangle)^2 \rangle = 0 , \tag{5.28b}$$

the equation of continuity (5.1b), and the boundary conditions (5.2b). Equations (5.28) are identical to (5.3) in the statistically stationary case when the left-hand side of (5.3) vanishes. By defining

$$\tilde{\theta} \equiv \theta - \bar{\theta}$$

and using (5.5), $\bar{\theta}$ has been eliminated from the energy balances (5.28). The Euler equations obtained as necessary conditions for an extremum of $\langle \mathbf{u} \cdot \lambda \tilde{\theta} \rangle$ among the above-defined manifold of vector fields differ from the basic equations (5.1) essentially in two respects. Since the variational problem does not involve derivatives with respect to time, only the three spatial coordinates appear as independent variables in the Euler equations. Secondly, although the Euler equations are strongly nonlinear, the nonlinearity is of the simplest possible form in that it depends on the z coordinate only. Thus the Euler equations can be satisfied by superposition of wavelike dependences with respect to the horizontal dimensions.

$$\tilde{\theta} = \sum_{n=1}^{N} \theta_n(z) w_n(x, y) , \tag{5.29}$$

where $w_n(x, y)$ represents an arbitrary solution of the equation

$$\Delta_2 w_n(x, y) = -\alpha_n^2 w_n(x, y) . \tag{5.30}$$

An expression analogous to (5.29) can be assumed for u_z, from which the other components of $\mathbf{u}$ can be derived since it can be shown that the vertical vorticity, $\lambda \cdot \nabla \times \mathbf{u}$, vanishes. Solutions of the form (5.29) have been called multi-α solutions [5.55] in extension of the case $N = 1$ considered by Howard [5.51]. The functions $\theta_n(z)$ can be determined by boundary layer analysis. In the limit $R \to \infty$ a multiple boundary layer structure develops. Assuming the ordering $\alpha_n > \alpha_m$ for $n > m$, the function $\theta_N(z)$ differs from zero only in the immediate neighborhood of the boundaries. In the region where $\theta_N(z)$ decays towards the interior of the layer, $\theta_{N-1}(z)$ grows. As $\theta_{N-1}(z)$ decays towards the interior, $\theta_{N-2}(z)$ grows and so on. Thus a hierarchy of boundary layers is established with $\theta_1(z)$ describing the interior dependence of the field $\tilde{\theta}$.

Each solution of the variational problem corresponding to a given integer value of N describes a local maximum of the quantity $\langle u_z \theta \rangle$. To obtain the absolute maximum $\mu(R)$, the relative maxima of $\langle u_z \theta \rangle$ given by the different multi-α solutions must be compared. It turns out that for low Rayleigh numbers $\mu(R)$ is given by the single-α solution with $N = 1$. At $R \approx 3.2 \cdot 10^4$ the

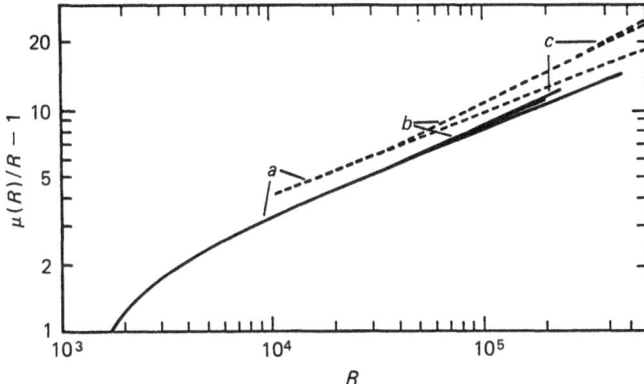

Fig. 5.4. The upper bound $\mu(R)$ of the convective heat transport. The dashed lines show the results of boundary layer analysis [5.55], while the solid lines indicate the numerically computed bounds [5.56]. Curves a, b, and c refer to the single-α, two-α, and three-α solutions, respectively. The solid curve c represents a guess, since only the point of bifurcation from the curve b has been determined

two-α solution corresponding to $N = 2$ in (5.29) begins to provide the absolute maximum $\mu(R)$. At Rayleigh numbers exceeding $6.5 \cdot 10^4$, $\mu(R)$ corresponds to $N = 3$, etc. [5.56]. Thus the upper bound $\mu(R)$ is produced piecewise by different extremalizing vector fields as shown in Fig. 5.4. The kinks in the upper bound curve show a striking resemblance with those exhibited by the actual heat transport measurements and to some extent the causes for the kinks are similar as we shall discuss in Sect. 5.5.

While the upper bound $\mu(R)$ tends to zero at $R = R_c$ and closely approximates the measured heat transport at low values of R, it rises substantially above the measured values as the Rayleigh number becomes large. For $R \to \infty$, $\mu(R)$ is given by

$$\mu(R) \approx 0.0309 R^{3/2} \tag{5.31}$$

which represents a much stronger dependence than the power law derived from experimental measurements [see (5.32)]. But it must be kept in mind that the experimentally realized values of R are still small compared to values of the order 10^{22} where a $R^{3/2}$ power law for the heat transport is expected to be realized according to mixing length arguments [5.50].

The upper bound (5.31) is Prandtl number independent, and it may be expected that improvements of the upper bound can be expected if the analysis is restricted to certain ranges of the Prandtl number. Only in the case $P = \infty$ has it been possible to derive a more restrictive upper bound. In the limit $R \to \infty$, $\mu(R)$ is given by

$$\mu(R) \approx 0.152 R^{4/3}$$

according to *Chan* [5.57].

5.4 Experimental Observations

5.4.1 Steady Convection

Convection in a layer heated from below is often called Bénard convection in honor of the man who did the first quantitative investigation of the phenomenon. *Bénard* [5.58] used thin oil layers with a free upper surface. Although *Bénard* was aware of the important role played by the temperature-dependent surface tension in his experiments, it did not become evident until much later [5.59, 60] that surface tension inhomogeneity rather than thermal buoyancy was the dominant driving force for the cellular motions. Surface tension effects become negligible is sufficiently thick layers or symmetric boundary conditions are used and we shall not consider them further.

The onset of convection has been observed by a number of early investigators [5.61, 62] and good agreement with the theoretical prediction (5.14) has been found. As the photographs by *Silveston* [5.62] show, at the onset convection tends to assume a cellular form corresponding to nearly hexagonal cells. This is caused by small asymmetries of the layer such as those caused by the temperature-dependent material properties, notably the viscosity of the fluid. An extension of the analysis of Sect. 5.3.1, including these effects [4.43, 63], shows that the coefficient $R^{(1)}$ in (5.19b) differs from zero for the hexagon solution corresponding to the case $N = 3$ of (5.21). Since the $R^{(1)}$ contribution dominates at Rayleigh numbers close to the critical value, hexagonal convection cells are observed at the onset of convection even when the Oberbeck-Boussinesq approximation is well approximated.

As has been mentioned before, the hexagon solution is the simplest among the solutions of the form (5.10) that correspond to two physically different forms of convection. Cells with upward motion in the center usually appear in liquids, while convection in gases exhibits downward motion in the center. Experimental [5.64] and theoretical [5.65, 66] evidence has indicated that this difference is caused by the property that the viscosity of liquids decreases with increasing temperature while the viscosity of gases increases.

It is noteworthy that it has been possible to realize experimentally even the onset of convection in the case of stress-free boundaries. By using a mercury layer below and a helium layer above the convecting oil, *Goldstein* and *Graham* [5.67] have been able to approximate the boundary condition (5.2a) and found reasonable agreement with Lord Rayleigh's prediction (5.13).

While the predictions of the linear theory are well confirmed by the experimental observations, the comparison between the measured heat transport and the theoretically derived Nusselt number has been a more complex task. The discrepancy between the experimental data and the calculated values for $\alpha = \alpha_c$ was resolved when *Wills* et al. [5.68] used the observed values of α for the theoretical computations. Some of the reasons for the significant increase of the observed wavelength of convection rolls with increasing Rayleigh number will be discussed in Sect. 5.5.2.

When controlled initial conditions are used, as in the experiments of *Busse* and *Whitehead* [5.69, 70] and nearly exactly two-dimensional rolls of a given wavelength are established, this wavelength does not change as the Rayleigh number increases unless an instability occurs. Usually experiments are started without controlled initial conditions and the isotropy of the layer causes the onset of roll-like cells of different orientation. When the aspect ratio of the layer is sufficiently large, the rearrangement of these cells is responsible for a weak time dependence of the convection pattern for many thermal diffusion times after a stationary value of the Rayleigh number has been reached. This behavior may account for the observation of *Ahlers* and *Behringer* [5.4] that in a layer of aspect ratio 57 and a Prandtl number of the order unity, the time dependence never does disappear, in contrast to the experiments carried out with lower aspect ratios. In layers with low aspect ratios of the order 10 or 20, the influence of the sidewalls is apparent already at the onset of convection since the small inhomogeneities associated with the sidewalls cause finite amplitude disturbances. The steady pattern usually reflects the geometrical configuration of the sidewalls. A circular boundary gives rise to a roll pattern in the form of concentric rings, as in *Koschmieder*'s experiments [5.71], while the rolls tend to align with the shorter sides in the rectangular layers used by *Stork* and *Müller* [5.72] in agreement with *Davis*' theoretical prediction [5.73].

A property of steady as well as turbulent convection that can be accurately measured by interferometric methods is the gradient of the mean temperature distribution $\bar{\theta}$. For Rayleigh numbers in excess of 3.8 times the critical value, *Farhadieh* and *Tankin* [5.74] found a reversed temperature gradient such as that theoretically predicted by *Veronis* [5.75]. The shape and the magnitude of the reversal agree with theoretical computations if the wavelength dependence (see Fig. 5.2) is taken into account. A smaller reversal than for rolls is found for three-dimensional forms of convection, and for convection at large Rayleigh numbers the reversal disappears according to the measurements of *Chu* and *Goldstein* [5.76]. Using a special differential interferometer *Oertel* and *Bühler* [5.77] have obtained recently quantitative measurements for horizontal as well as vertical variations of the temperature in convection layers.

The recent development of laser Doppler velocimetry has made it possible to obtain even more detailed quantitative data in the form of the spatial variations of the velocity in convection rolls. In the earlier work [5.78] *Bergé* and *Dubois* compared their measurements with the predictions of the small amplitude theory outlined in Sect. 5.3.1. In a more recent paper [5.79] the authors have extended the measurements to much higher Rayleigh numbers, where the analytical theory is no longer applicable and the results of the numerical analysis [5.45] must be used for a comparison. Good agreement between experimental data and theoretical predictions has been found in both the low and the high Rayleigh number cases for the structure of steady convection.

5.4.2 Transitions

The range of Rayleigh numbers for which the experimentally realized convection can be approximately described by steady two-dimensional rolls extends to about $13 \cdot R_c$ in high Prandtl number fluids, but narrows considerably as the Prandtl number decreases to values less than unity. The transition to three-dimensional forms of convection depends strongly on P, and the picture shown by experiments and theory for moderate values of P is a rather complex one. But in the limits of small and very large Prandtl numbers, the problem becomes much simpler, mainly because the nonlinear term of (5.1a) dominates in the former, while the nonlinear term of (5.1c) dominates in the latter limit.

In high Prandtl number fluids convection rolls are replaced by another steady form of convection, called bimodal convetion, when R exceeds a value of the order $2 \cdot 10^4$. This transition is obviously caused by the instability of thermal boundary layers described by criterion (5.26), and, as expected, manifests itself in the form of additional roll-like motions at right angles to the basic rolls. Experimental observations [5.69, 80] of the transition have been found in good agreement with the predictions of the theory [5.45] which will be discussed in Sect. 5.5.2. Since two wave numbers characterize bimodal convection, a large number of possible solutions exists, some of which are stable and can be realized experimentally by using controlled initial conditions [5.81].

The main interest of experimenters and theoreticians alike has always been the transition to time-dependent convection. The onset of oscillations in a convection layer has been observed as a function of Prandtl number by a large number of investigators. The earlier work by *Willis* and *Deardorff* [5.3, 82], *Krishnamurti* [5.83, 84], *Busse* and *Whitehead* [5.70], and *Whitehead* and *Parsons* [5.85] has emphasized visual observations. At low Prandtl numbers of the order one or less, the oscillations assume the form of waves propagating along the convection rolls, but the picture becomes more complicated at higher Prandtl numbers, when the oscillations are preceded by a transition to three-dimensional convection. There is no general agreement about the Prandtl number dependence of the Rayleigh number R_{III} for the onset of oscillations. *Krishnamurti* [5.83, 84] advocates a value of $5.5 \cdot 10^4$ for R_{III} independent of P for $P \gtrsim 50$, while the more recent observations of *Whitehead* and *Parsons* [5.85] indicate a continuing increase of R_{III} as P tends to infinity. This disagreement is related to the fact that oscillations tend to occur first in isolated spots where the convection pattern is particularly inhomogeneous. It requires a considerable increase of the Rayleigh number beyond the value R_{III} to obtain a nearly uniform distribution of oscillations. In the case of a spatially periodic bimodal convection pattern which can be realized by using controlled initial conditions [5.70] the onset of oscillations is much delayed as shown by the upper boundary of the shaded area in Fig. 5.5.

In recent years *Ahlers* [5.3] has introduced high precision heat flux measuring techniques for convection in liquid helium and *Bergé* and *Dubois* [5.86], *Gollub* et al. [5.87]. and *Gollub* and *Benson* [5.5] have used laser

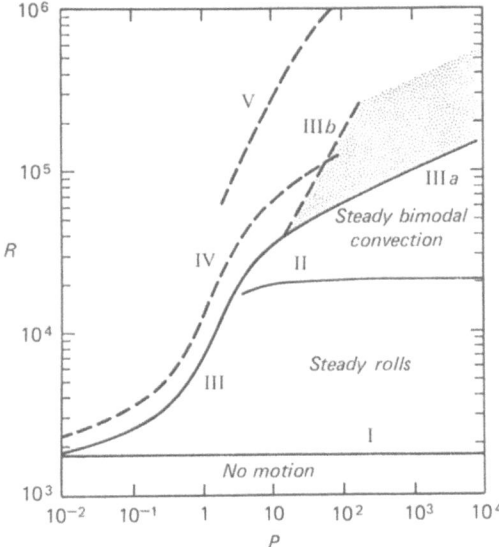

Fig. 5.5. Transitions in thermal convection as a function of Rayleigh and Prandtl numbers according to *Krishnamurti* [5.83] and others

Doppler velocimetry to obtain more detailed information on the time dependence of convection. The picture that emerges from the new experimental data is a complex one because the time dependence does not depend only on the Rayleigh and the Prandtl numbers, but also on the aspect ratio of the convection layer and perhaps even on the geometrical configuration of the sidewalls. In addition, the initial conditions can have a significant effect even in cases when the spatial pattern of convection does not depend much on the history of the experiment. [5.5].

There appears to be general agreement that the evolution of the time dependence of convection is distinctly different in small and in large aspect ratio layers [5.4, 88, 89]. In the latter case, the time dependence of convection probably sets in first in the form of slow but erratic changes of the pattern of convection. Periodic oscillations become clearly recognizable only at significantly higher values of the Rayleigh number. Because of the erratic shifts in the basic convection pattern, the spectral peaks of the oscillations are broadened. In the case of a small aspect ratio layer, on the other hand, sharp peaks are found in the frequency spectrum since a steady convection flow precedes the onset of oscillations. At a somewhat higher Rayleigh number a broadening of the spectral peaks is observed and a low frequency noise becomes noticeable. From these observations it appears that convection layers with small and large aspect ratios differ in respects similar to those of the experiments with controlled and uncontrolled initial conditions mentioned above. The Rayleigh number for the onset of low frequency aperiodic time dependence appears to decrease with increasing aspect ratio. In a layer of liquid helium ($P \approx 1$) of very large aspect ratio (ratio of radius to height of layer $\Gamma = 57$)

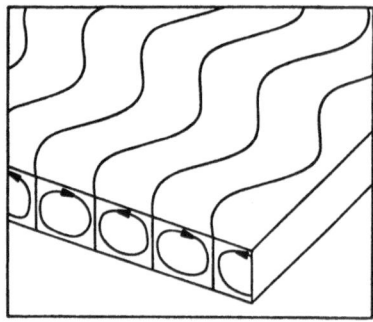

Fig. 5.6. Sketch of oscillatory convection rolls [5.38]

Ahlers and *Behringer* [5.4] were unable to find any regime of strictly steady convection.

In some cases of oscillatory convection in small aspect ratio layers, ranges of periodic and quasi-periodic time dependence can be clearly distinguished [5.5]. The latter is characterized by the appearance of a sharp peak corresponding to an incommensurate frequency in the spectrum of the primary frequency and its higher harmonics. Thus, there is evidence that the evolution of turbulent convection in small aspect ratio layers parallels closely the corresponding phenomenon in the Taylor vortex case [5.1]. But it is not yet clear whether the appearance of a second incommensurate frequency is a general property or depends on the particular geometry of the convection layer. In this respect, it would be of interest to see whether a spectral decomposition of the time dependence measured by *Olson* and *Rosenberger* [5.90] in a cylinder heated from below would reveal a second incommensurate frequency.

The sinusoidal oscillations of convection rolls in large aspect ratio layers with $P \lesssim 2$ can be readily interpreted by the rigorous theory discussed in Sect. 5.5.2. The oscillations are caused by waves in the form of transverse displacements of the rolls as shown in Fig. 5.6. These waves may either propagate along the rolls or form standing patterns. The phase of the waves shows little dependence on the vertical coordinate, a property which has been observed in the case of oscillatory convection in air [5.3] as well as in liquid metals. But as *Hurle* et al. [5.91] have emphasized, the presence of mean horizontal temperature gradients present in most of the reported experiments with liquid metals must be taken into account for a quantitative interpretation of the data.

To some extent the theory of the oscillatory instability of rolls (Sect. 5.5.2) can be applied when the rolls are already replaced by a three-dimensional pattern of convection. Standing waves observed in a bimodal pattern fit the predicted dispersion relation fairly well [5.70] and the transverse motion is clearly evident for $P \lesssim 50$. For higher values of P thermal effects instead of hydrodynamic effects appear to dominate the character of the oscillations. In some respects the oscillation resembles the advection of a hot blob in a circular tube heated from below, a problem whose interesting nonlinear properties have been investigated by *Keller* [5.92] and *Welander* [5.93].

Another model of time-dependent convection has been proposed by *Chang* [5.94] and *Howard* [5.95]. It is based on a periodic instability of the thermal boundary layer and yields an $R^{2/3}$ dependence of the frequency of oscillations which appears to match the observations well. A further development of these models for the purpose of a quantitative comparison with the observations would be of considerable interest.

The convection loop model of *Keller* and *Welander*, as well as the earlier *Lorenz* model [5.96] which corresponds to a highly truncated Fourier representation of convection, exhibits aperiodic nonlinear oscillations which are suggestive of the onset of turbulence in addition to the periodic solutions. But because of the finite number of degrees of freedom in these models, no direct relationship to the experimental observations can yet be drawn. Even the model with 39 Fourier components of *McLaughlin* and *Martin* [5.97] is likely to suffer from this limitation. *McLaughlin* and *Martin*'s model, which describes nonlinear oscillations in low Prandtl number convection, is of special interest since it supports the *Ruelle* and *Takens* [5.7] picture of the onset of turbulence. While the presence of three interacting oscillatory modes still yields periodic solutions, the addition of a fourth harmonic gives rise to an aperiodic time dependence if the Rayleigh number is sufficiently high. The extension of computations of this kind for a sufficiently large number of modes such that numerical convergence can be claimed will be the next step in the theoretical exploration of this fascinating aspect of the onset of turbulence.

5.4.3 Turbulent Convection

Since turbulent convection plays an important role for the heat transport in the atmosphere and in stars, much of the scientific interest in turbulent convection has focussed on its efficiency in transporting heat. In the absence of reliable theoretical estimates, a considerable effort has been devoted to the experimental measurements of the Nusselt number. Typical power laws fitting the data for an extended range of Rayleigh numbers of the order 10^6 up to 10^9 are

$$\text{Nu} = 0.184 \cdot R^{0.281} \quad \text{for} \quad P = 200 \quad [5.39] \,, \tag{5.32a}$$

$$\text{Nu} = 0.130 \cdot R^{0.293} \quad \text{for} \quad P \approx 6 \quad [5.98] \,, \tag{5.32b}$$

$$\text{Nu} = 0.123 \cdot R^{0.294} \quad \text{for} \quad P = 0.7 \quad [5.99] \,, \tag{5.32c}$$

$$\text{Nu} = 0.147 \cdot R^{0.247} \quad \text{for} \quad P = 0.025 \quad [5.39] \,. \tag{5.32d}$$

The results have not been obtained for the same range of Rayleigh numbers and caution must be exercised in interpreting the Prandtl number dependence. But there is evidence for a weak maximum of heat transport for values of P of the order unity similar to that apparent in the Nusselt number dependence of convection rolls. Even quantitatively the expression (5.24b) for the numerically derived Nusselt number fits the measured values quite well. This is surprising, since turbulent convection has little in common with the steady convection rolls of the numerical analysis. The compensating effects of the three dimensionality and of the time dependence of turbulent convection may be responsible for the close coincidence of the heat transport values.

The transition to bimodal convection clearly manifests itself as a kink in the dependence of the heat transport on the Rayleigh number [5.80]. Even the hysteresis property of the position of the kink can be explained by the fact that convection rolls tend to have a shorter wavelength in experiments where R has been increased than in the opposite case [5.80]. More surprising is the fact that an entire sequence of kinks has been observed as the heat transport was carefully measured as a function of the Rayleigh number. *Malkus* [5.100] found five additional kinks for values of R up to 10^7. Just as in the case of the similar kinks in the upper bound curve shown in Fig. 5.4, the kinks probably do not correspond to true discontinuities of the derivative $\partial Nu/\partial R$, at least not in the limit of a large aspect ratio layer. But the second derivative $\partial^2 Nu/\partial R^2$ is discontinuous according to theoretical considerations [5.101]. In the case of small aspect ratios, even breaks in the curve have been observed by *Threlfall* [5.102] which indicates the importance of the geometrical configuration of the convection layer for the phenomenon of kinks. The Malkus transitions have been confirmed by numerous investigators [5.76, 103, 104], although it should be mentioned that some researchers have found only smooth curves for Nu(R) in spite of highly accurate heat flux measurements [5.4, 105]. The agreement of the positions of the kinks found in different experiments is only partially satisfactory. Little attention has yet been paid to the expected Prandtl number dependence of the kinks.

The phenomenon of discrete transitions in a highly turbulent fluid system challenges the traditional view that fully developed turbulence is governed by random processes. Regular structures seem to underlie the chaotic appearance of turbulent convection. The similarity between the measured heat transport dependence and the upper bound suggests that discrete wave numbers α_n characterize the different stages of turbulent convection. Since the relevant wave numbers describe the two-dimensional horizontal dependence of convection, they cannot be expected to be visible in the usual measured spectra obtained from one-dimensional traverses of the layer. In spite of this limitation *Deardorff* and *Willis* [5.106] found evidence for discrete wave numbers in turbulent convection. Perhaps the low aspect ratio of one of the horizontal dimensions of their apparatus led to an alignment of the turbulent eddies which produced the peaks in the measured spectrum. In the square layer used by *Fitzjarrald* [5.107], these results could not be reproduced. The development of optical correlation methods, such as that used by *Somerscales* and *Parsapour* [5.108], may lead to suitable tools for the investigation of discrete peaks in the wave number spectrum.

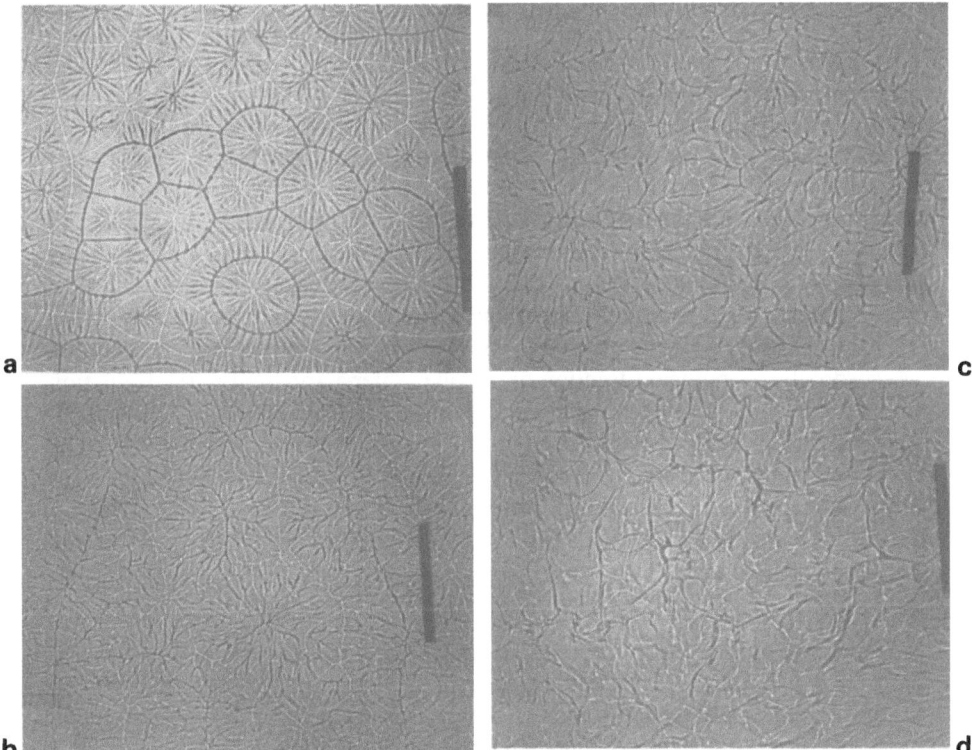

Fig. 5.7a–d. Shadowgraph observations of convection in a layer of methyl alcohol $(P=7)$ corresponding to the Rayleigh numbers. (**a**) $4.2 \cdot 10^4$ $(d=5.6\,\text{mm})$; (**b**) $1.3 \cdot 10^5$ $(d=1\,\text{cm})$; (**c**) $1.9 \cdot 10^6$ $(d=2\,\text{cm})$; (**d**) $1.7 \cdot 10^7$ $(d=4\,\text{cm})$. The rod seen in the photographs is 10 cm long

A visual impression of the development of turbulent convection can be gained from the shadowgraph photographs of Fig. 5.7, which indicate the vertically averaged temperature distribution of the layer (for the experimental details see [5.70]). The form of convection apparent in the first three pictures has been called summarily spoke pattern convection. It is characterized by a large-scale, nearly steady pattern which provides the framework for a small-scale structure that fluctuates strongly in time. The spoke pattern cells seen at Rayleigh numbers of a few times 10^4 resemble to some extent ·the hexagonal cells of low Rayleigh number convection. But there is now no preference for either hot or cold fluid in the centers and both kinds of cells can usually be found in different parts of the layer. At higher Rayleigh numbers the network of bright (cold) and dark (hot) lines becomes more symmetric and the wavelength increases. But the characteristic difference between the nearly steady large-scale and the fluctuating small-scale structure persists. Only for Rayleigh numbers higher than 10^7 does this distinction appear to vanish.

5.5 Instabilities of Convection Rolls

5.5.1 Theoretical Analysis

The analysis of the instabilities of convection rolls has become a useful tool for the theoretical study of the transition to turbulence in convection for two reasons. First, the experimental observations suggest that the transitions correspond to supercritical bifurcations and thus can be described by linear theory. Second, some of the instability mechanisms isolated by the numerical stability analysis of rolls operate with minor modifications in the case of higher transitions from one three-dimensional form of convection to another. Accordingly, a significant fraction of the phenomena apparent in the transition to turbulence in convection can be understood qualitatively if not quantitatively on the basis of the linear stability theory of convection rolls.

The stability analysis is based on (5.23). It is assumed that the steady two-dimensional solution describing convection rolls is given by

$$u_z = \sum_{\lambda, \nu} a_{\nu\lambda} \phi_\nu(z) \cos \lambda \alpha y \tag{5.33}$$

and corresponding expressions for u_y and θ. The functions $\phi_\nu(z)$ represent a complete orthonormal set of functions satisfying the boundary conditions for u_z in (5.2b). The stability equations (5.23) for the solution (5.33) are solved by introducing general three-dimensional disturbances of the form

$$\tilde{u}_z = \left[\sum_{\lambda, \nu} (b_{\nu\lambda} \cos \lambda \alpha y + c_{\nu\lambda} \sin \lambda \alpha y) \phi_\nu(z) \right]$$
$$\cdot \exp(idy + ibx + \sigma t) \tag{5.34}$$

with corresponding expressions for $\tilde{u}_x$, $\tilde{u}_y$, and $\tilde{\theta}$. The linear equations (5.23) constitute an eigenvalue problem with the growth rate σ as eigenvalue. For prescribed values of R, P, and α, σ must be determined as a function of the parameters b and d for all possible solutions of the form (5.34). The steady solution given by (5.33) is unstable if an eigenvalue with positive real part exists; otherwise it is stable. The curve in the R-α plane on which the largest real part of σ vanishes describes the boundary of the domain of stable rolls as shown, for example, in Fig. 5.9.

In the following, the instabilities of rolls that lead to rolls of different wavelength are discussed separately from those that lead to three-dimensional forms of convection. This distinction is not well justified on theoretical grounds, since the theory describes only the initial growth of disturbances, but not the asymptotic stationary state produced by the instability. The distinction is useful, however, for the comparison with the experimental observations and even without this comparison the distinction can usually be made on the basis of the mathematical properties of the growing disturbances.

5.5.2 Wavelength Changing Instabilities

At Rayleigh numbers close to the critical value R_c rolls represent the only stable form of stationary convection. Thus any instability of rolls must ultimately lead to convection in the form of rolls again, though the direction and the wavelength will be changed in general. The sole purpose of instabilities of rolls at low Rayleigh numbers is the restriction of realizable wave numbers to values close to α_c. The zigzag instability accomplishes this very effectively for rolls with wave numbers less than α_c by bending the rolls in a sinusoidal fashion. The lengthening of the boundaries between the rolls between the rolls corresponds to an increase of the effective wave number as shown in Fig. 5.8a. The name zigzag instability derives from the experimentally observed property that new rolls are created in two directions enclosing angles of approximately 40° and 140° with the direction of the original rolls [5.69].

For rolls with wave numbers much larger than α_c, the restricting mechanism of instability is not as effective as in the case $\alpha < \alpha_c$. The cross-roll instability occurs in the form or rolls at a right angle to the given rolls such as to minimize the interaction between the original pattern and its disturbance (see Fig. 5.8b). As expected, the wave number of the cross-roll instability is close to α_c if R is not too large, as shown in Fig. 5.9.

(a)

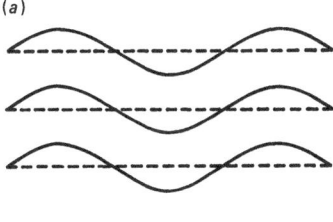

(b)

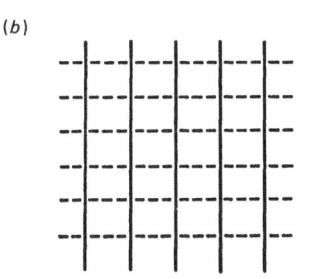

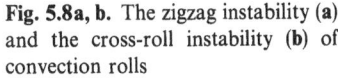

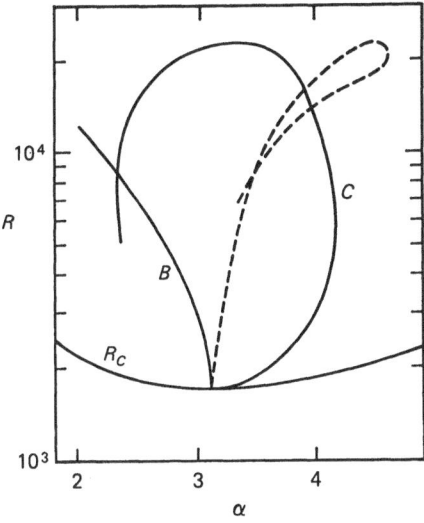

Fig. 5.8a, b. The zigzag instability (a) and the cross-roll instability (b) of convection rolls

Fig. 5.9. The stability domain of convection rolls as a function of the wave number α for $P = \infty$ according to the analysis of [5.45]. The dashed line indicates the wave number b of the cross-roll instability (C). The zigzag instability is responsible for the boundary B and R_c indicates the minimum Rayleigh number for the existence of solutions with wave number α

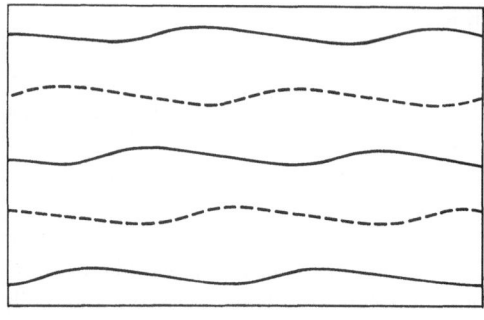

Fig. 5.10. The distortion of rolls owing to the skewed varicose instability (after [5.112]). The solid and dashed lines indicate the hot and the cold sides of a single roll

At low Prandtl numbers, a purely two-dimensional instability mechanism becomes important [5.34]. The growing disturbances of the Eckhaus instability consists of two interacting rolls with one wave number slightly less and the other slightly larger than the wave number of the given roll. Because the disturbing rolls and the original rolls are parallel, the third spatial dimension is not involved, and the theory of the Eckhaus instability applies in cases of Taylor vortices as well [5.34, 109]. For a critique of the original work of *Eckhaus* [5.110] and for general discussion of the theory, we refer to the recent paper of *Stuart* and *DiPrima* [5.111].

Most important among the instabilities that change the wavelength of convection rolls is the skewed varicose instability [5.112, 113] because it affects rolls with the wave number α_c. A sketch of the exaggerated deformation of a pattern of straight rolls owing to this instability is shown in Fig. 5.10. The experimental shadowgraph observation of this instability shown in Fig. 5.11 indicates that the nonlinear development of this instability does not correspond to a monotonous growth of the initially visible disturbances. Instead, a subharmonic wave develops corresponding to the pairing of two wavelengths of the original disturbances, but this may be caused by the presence of the knot instability (see Sect. 5.5.3) which cannot be easily separated from the skewed varicose instability in fluids of $P \approx 6$. The main result of the instability is the introduction of large-scale convection as shown in the last picture of Fig. 5.11.

The skewed varicose instability appears to be the main cause of the generally observed increase of the average wavelength of convection with increasing Rayleigh number in experiments without controlled initial conditions. This effect is especially pronounced in low Prandtl number fluids such as air [5.68]. The general shift towards lower wave numbers of the domain of stable rolls shown in Fig. 5.12 at $P = 0.7$ is in good agreement with the observations [5.113].

5.5.3 Pattern Changing Instabilities

From the dependence of the wave number b of the cross-roll instability shown in Fig. 5.9 it is evident that this instability cannot lead to stable two-

Fig. 5.11. Skewed varicose instability (modified by the knot instability) in a layer of methyl alcohol ($P \approx 7$). The time interval between the photographs is about 5 min, $d = 0.84$ cm, and $R \approx 4 \cdot 10^4$. The rod seen in the fourth photograph is 10 cm long

dimensional rolls for Rayleigh numbers above $1.5 \cdot 10^4$ since the dashed line lies outside the domain of stability. Instead the cross-roll instability leads to three-dimensional convection in the form of bimodal cells. The monotonic growth of the instability suggests a steady form of convection in agreement with the observations mentioned in Sect. 5.4.2.

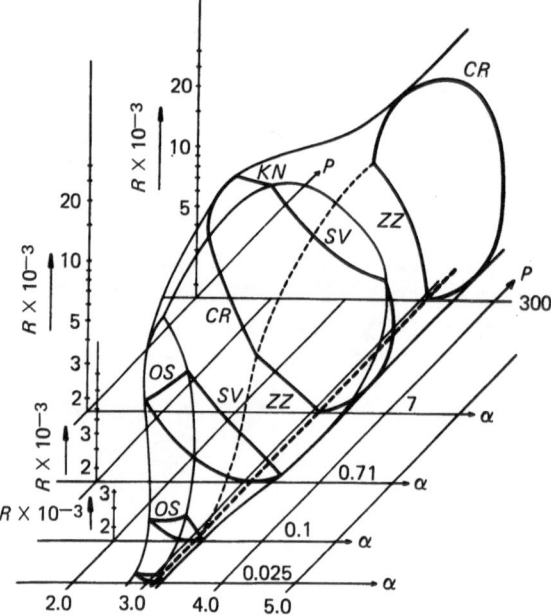

Fig. 5.12. Stability domain of convection rolls in the $R-P-\alpha$ space. The thick curves represent computed stability boundaries for the oscillatory (OS), the skewed varicose (SV), the cross-roll (CR), the knot (KN), and the zigzag (ZZ) instabilities after [5.38, 112, 113]. For $P=300$, the results for $P=\infty$ [5.45] have been used

At lower Prandtl numbers the cross-roll instability develops a new branch exhibiting a wave number b significantly lower than α_c. This instability manifests itself in the form of knots appearing on the boundaries of the convection rolls in the shadowgraph observations [5.112]. It competes with the skewed variose instability, and the instability developing in the first few pictures of Fig. 5.11 appears to be actually a combination of the two instabilities. At Prandtl numbers of the order 10 and larger, the knot instability is preceded by the transition to bimodal convection. The collective instability of the bimodal pattern of convection observed experimentally [5.70] resembles the knot instability in all basic properties. As indicated by its large wavelength, the knot instability tends to develop into spoke pattern cells. These cells differ from those shown in the first picture of Fig. 5.7 in that only tiny spokes are visible initially, which show little time dependence. As the Rayleigh number increases, the spoke pattern cells seem to evolve gradually into those shown in Fig. 5.7.

Of particular interest is the oscillatory instability which corresponds to the only growing solution of the form (5.34) with a finite imaginary part of σ. It was first analyzed for a convection layer with stress-free boundaries [5.114]. The Rayleigh number R_{III} for the onset of this instability is given in this case by

$$R_{\mathrm{III}}=R_{\mathrm{c}}(1+0.31P^2) \tag{5.35a}$$

if $\alpha = \alpha_c$ is assumed and the Prandtl number is sufficiently low. The associated period of oscillation is given by

$$t = [2R_c/3(R-R_c)b^2]^{1/2} d^2/\kappa . \tag{5.35b}$$

In the case of rigid boundaries, $R_{III} - R_c$ remains finite in the limit $P \to 0$ [5.28]. Like the zigzag instability, the oscillatory instability can be described as a modification of the translation disturbance given by

$$\tilde{u} = j \cdot \nabla u , \qquad \tilde{\theta} = j \cdot \nabla \theta , \tag{5.36}$$

where j is the horizontal unit vector normal to the axis of the steady rolls. It is easily seen that (5.36) solves (5.23) for $\sigma = 0$. The translation disturbance (5.36) does not describe any instability since it becomes stabilized by the presence of sidewalls. But small modifications of the solution (5.36) do indeed lead to positive values of σ. Because of their relationship to the translational disturbance, the curve of marginal instability corresponds to the limit $b \to 0$. But since beyond the stability boundary the wave number b of the fastest growing disturbance increases proportionally to the third power of $R - R_{III}$, only disturbances with finite values of b are observed experimentally.

Figure 5.6 exhibits the close relationship of the oscillatory instability to the translational disturbance in that the rolls are moved forward and backward in a periodic fashion. The period is essentially given by the circulation time of a convection roll. When measured in multiples of d^2/κ, the period depends primarily on R and is essentially independent of P.

5.6 Convection in a Rotating Layer

Convection in a horizontal layer rotating about a vertical axis differs in many respects from convection in a nonrotating layer. In low Prandtl number fluids the onset of convection can occur in the form of oscillatory motions and steady finite amplitude convection can exist at Rayleigh numbers below the values required for the onset of convection in the form of infinitesimal motion [5.115, 116]. The proof for the nonexistence of these phenomena in nonrotating layer mentioned in Sect. 5.2.2 does not hold once the Coriolis force $-2\Omega\lambda \times u$ is added in (5.1a). Since the Coriolis force does not contribute to the energy relations, (5.3) remains unchanged, and the energy stability limit R_E is actually independent of the rotation rate of the system. But the critical Rayleigh number R_c determined by the linearized version of (5.1) is a monotonically increasing function of the rotation parameter Ω. In contrast to the time scale defined in Sect. 5.2.1, the dimensionless angular velocity Ω of the rotating layer traditionally is measured in multiples of v/d^2. Often the Taylor number $T = 4\Omega^2$ is used as a measure of the effects of rotation.

Instead of reviewing the general properties of convection in a rotating system, the attention is focussed in this section on a particular instability phenomenon that can be described within the framework of the small amplitude perturbation approach outlined in Sect. 5.3.1. Because the rotating layer retains the properties of isotropy and homogeneity in the limit in which the centrifugal force can be neglected in comparison with gravity, the analysis proceeds in the same way as in the nonrotating case of Sects. 5.2.2 and 5.3.1. Starting with solutions of the form (5.10), the linear problem is solved by (5.12) if the stress-free boundary condition (5.2a) is assumed and if the term $4\Omega^2\pi^2/\alpha^2$ is added in the expression for $R(\alpha)$. Accordingly, the minimum Rayleigh number R_c for the onset of convection in the form of steady infinitesimal motions increases proportionally to $\Omega^{4/3}$ and the corresponding wave number α_c increases proportionally to $\Omega^{1/3}$ as Ω tends to infinity.

Using the series expansion (5.19), the perturbation approach to the nonlinear problem yields $R^{(1)}=0$ in the second order and the solvability condition

$$R^{(2)}|c_l|^2 - \sum_{n=-N}^{N} [\hat{A}(k_l \cdot k_n) + B(k_l \cdot k_n)\lambda \cdot k_l \times k_n]|c_n|^2|c_l|^2 = 0$$

$$\text{for} \quad l=1,\dots,N \tag{5.37}$$

in the third order. These equations differ from (5.20) principally by appearance of a term proportional to $\lambda \cdot k_l \times k_n$. Owing to this term, only solutions of the form (5.21) with a regular distribution of k vectors satisfy (5.37). *Küppers* and *Lortz* [5.117], who obtained this result, continued the perturbation approach outlined in Sect. 5.3.2 by analyzing the stability of all small amplitude steady solutions. As in the nonrotating case, they found that only two-dimensional rolls represent a stable solution, but even rolls become unstable when Ω exceeds a critical value Ω_c. In the case of stress-free boundaries this value is $\Omega_c = 23.9$ for $P = \infty$.

The Küppers-Lortz instability of rolls poses a paradoxical problem. Since all steady solutions are unstable for $\Omega \geq \Omega_c$ and since oscillatory forms of convection cannot exist at Rayleigh numbers close to R_c for $P \geq 1$, it is difficult to imagine what kind of solution describes the physically realized convection. The mathematical description of the Küppers-Lortz instability does not reveal any singular properties. It occurs in the form of monotonically growing rolls with an angle near 60° with respect to the steady rolls. The angle is measured in the direction of rotation. To obtain further information and to resolve the paradoxical situation, the evolution in time of the instability must be investigated.

This problem was addressed recently [5.8] by assuming that the coefficients c_v in the general expression (5.10) for the solution of the linear problem are time dependent. Close to the critical value R_c of the Rayleigh number, the time derivative $c_v^{-1}\partial c_v/\partial t$ must be of the order ε^2, and the solvability condition (5.37)

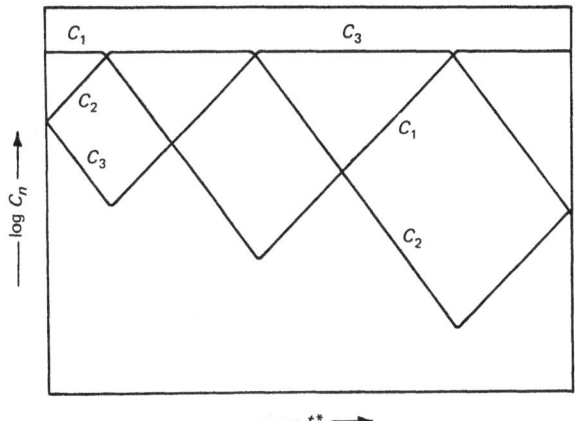

Fig. 5.13. Sketch of the time dependence of a solution of (5.38) in the case $N = 3$

can be generalized to the weakly time-dependent case as

$$M(\partial/\partial t^*)|c_l|^2 = \left\{ R^{(2)} - \sum_{n=-N}^{N} [\hat{A}(\mathbf{k}_l \cdot \mathbf{k}_n) + B(\mathbf{k}_l \cdot \mathbf{k}_n)\boldsymbol{\lambda} \cdot \mathbf{k}_l \times \mathbf{k}_n]|c_n|^2 \right\} |c_l|^2$$

$$\text{for} \quad l = 1, \ldots, N , \tag{5.38}$$

where $t^* \equiv \varepsilon^{-2}t$ and M is a constant. The general behavior of the solutions of (5.38) is most clearly apparent in the case $N = 3$. Figure 5.13 shows a sketch of solution in the case when c_1 assumes the steady value of the roll solution and c_2 and c_3 have significantly lower values initially. Because of the instability of the roll solution, the disturbance c_2 grows until it reaches a value comparable to that of c_1, at which point the latter starts to decay. As c_2 replaces c_1 in representing the steady equilibrium value of the roll solution, c_3 assumes the same position with respect to c_2 that c_2 had initially with respect to c_1. But since c_3 decayed while c_2 was growing, it takes a longer time for c_3 to reach the equilibrium value. Thus c_1, c_2, and c_3 continue to replace each other cyclically, but the periods of the cycles grow without limit.

Although the time-dependent solutions of (5.38) resolve the paradoxical result of *Küppers* and *Lortz*, they are not physically acceptable because they are inhomogeneous in time. The origin of the unphysical behavior of the solutions is the assumption of the presence of disturbances initially, but not throughout the evolution of the system. In an experimental situation the continuous presence of noise does not allow disturbances to decay to arbitrarily low amplitudes as shown by the solutions of (5.38). If a lower bound for the function $c_n(t)$ is introduced artificially, the solution of (5.38) becomes asymptotically periodic. Because of the random nature of experimental noise, the problem is a statistical one, and the period of the cycles becomes a stochastic function. Thus the weakly nonlinear problem of convection in a rotating system exhibits a typical feature of turbulent fluids in that its evolution in time is governed by random effects.

5.7 Concluding Remarks

There are two major aspects of thermal convection that are of importance for the understanding of the general phenomenon of fluid turbulence. One of these aspects is the existence of discrete structure in turbulent convection as indicated by the kinks in the heat transport dependence and by the evidence for discrete wave numbers in the horizontal spectrum of convection. The most startling evidence for the discrete structures of highly turbulent systems comes actually from convection phenomena observed on the sun and in the earth's atmosphere. The phenomena of granulation, supergranulation, and giant cells in the solar convection zone, each separated by a factor of about 10 from the next in its characteristic horizontal scale, are difficult to understand on the basis of any statistical theory of turbulence. Rather well-defined wave numbers are also observed for mesoscale convection and other patterns of convection of smaller scale in the earth's atmosphere. The understanding of the mechanisms by which turbulent systems seem to develop a hierarchy of scales is probably the most challenging problem posed by the observations of thermal convection.

The other aspect in which thermal convection provides a model for our understanding of turbulent systems is the onset of apparent randomness of fluid motions. There are at least three different frameworks in which this problem has been discussed from the mathematical point of view, even if the possibilities of inverted bifurcations leading to the onset of turbulence by "snap-through" instability [5.18] are not considered. The traditional concept is the Landau-Hopf picture of subsequent bifurcations leading to increasingly complex solutions. *Hopf* [5.118] considered problems described by ordinary differential equations, but it is obvious that in problems of fluid mechanics the number of bifurcating solutions increases extremely fast, and highly complex states corresponding to turbulent motion can be reached rapidly. There is no particular point in the Landau-Hopf picture that marks the onset of turbulence because the system becomes turbulent gradually. Such a point has been introduced in the concept of *Ruelle* and *Takens* [5.7]. These authors found in their study of systems of ordinary differential equations that typically at the fourth bifurcation the solutions tend towards a "strange attractor" in the phase space of the system, with the consequence that an aperiodic time dependence occurs. A third concept is the noise amplification mechanism for the onset of turbulence as described in the example of convection in a rotating layer in Sect. 5.6. The idea that a fluid system is capable of amplifying microscopic disturbances to a macroscopic level is not surprising. Any picture of the onset of convection in an extended horizontal layer without controlled initial conditions demonstrates the irreproducible random nature of the pattern. In the case where the boundary temperatures are varied sinusoidally such that the Rayleigh number oscillates about its critical value, the time dependence of convection exhibits, in general, random features [5.119]. But convection in a rotating layer appears to be the simplest case that exhibits a time dependence governed by random effects under steady external conditions.

Other mathematical concepts for the description of the onset of turbulence can be constructed and the variety of models becomes even larger if the possibilities of inverted bifurcations are included. But a common property of all models appears to be that they are highly sensitive to small perturbations, at least in certain stages of their time evolution. Thus it may be concluded that the macroscopic chaos of turbulent fluids reflects to some extent the microscopic randomness of the physical conditions of the respective system.

Acknowledgement. The author's research on thermal convection has been supported by the Atmospheric Science Section of the U.S. National Science Foundation.

References

5.1   J.P.Gollub, H.L.Swinney: Onset of turbulence in a rotating fluid. Phys. Rev. Lett. **35**, 927–930 (1975)

5.2   G.E.Willis, J.W.Deardorff: The oscillatory motions of Rayleigh convection. J. Fluid Mech. **44**, 661–672 (1970)

5.3   G.Ahlers: Low temperature studies of the Rayleigh-Bénard instability and turbulence. Phys. Rev. Lett. **33**, 1185–1188 (1974)

5.4   G.Ahlers, R.P.Behringer: Evolution of turbulence from the Rayleigh-Bénard instability. Phys. Rev. Lett. **40**, 712–716 (1978)

5.5   J.P.Gollub, S.V.Benson: Many routes to turbulent convection. J. Fluid Mech. **100**, 449–470 (1980)

5.6   P.C.Martin: "The Onset of Turbulence: A Review of Recent Developments in Theory and Experiment", in *Proc. Int. Conf. Stat. Mech.*, ed. by L.Pàl and P. Szĕpfalusy (North-Holland, Amsterdam 1975)

5.7   D.Ruelle, F.Takens: On the nature of turbulence. Commun. Math. Phys. **20**, 167 (1971

5.8   F.H.Busse, R.M.Clever: "Nonstationary Convection in a Rotating System", in *Recent Developments in Theoretical and Experimental Fluid Mechanics*", ed. by U.Müller, K.G. Roesner, B.Schmidt (Springer, Berlin, Heidelberg, New York 1979) pp. 376–385

5.9   S.Chandrasekhar: *Hydrodynamic and Hydromagnetic Stability* (Clarendon Press, Oxford 1961)

5.10   G.Z.Gershuni, E.M.Zhukovitskii: *Convection Stability of Incompressible Fluids*, transl. from the Russian by D.Louvish (Keter Publications, Jerusalem 1976)

5.11   E.A.Spiegel: Convection in stars: I. Basic Boussinesq convection. An.. Rev. Astron. Astrophys. **9**, 323–352 (1971)

5.12   E.A.Spiegel: Convection in stars: II. Special effects. Ann. Rev. Astron. Astrophys. **10**, 261–304 (1972)

5.13   E.L.Koschmieder: Bénard convection. Adv. Chem. Phys. **26**, 177–212 (1974)

5.14   E.Palm: Nonlinear thermal convection. Ann. Rev. Fluid Mech. **7**, 39–61 (1975)

5.15   C.Normand, Y.Pomeau, M.G.Velarde: Convective instability: A physicist's approach. Rev. Mod. Phys. **49**, 581–624 (1977)

5.16   F.H.Busse: Nonlinear properties of convection. Rep. Prog. Phys. **41**, 1929–1967 (1978)

5.17   J.S.Turner: *Buoyancy Effects in Fluids* (Cambridge University Press, London 1973)

5.18   D.D.Joseph: *Stability of Fluid Motions*, Springer Tracts in Natural Philosophy, Vols. 27, 28 (Springer, Berlin, Heidelberg, New York 1976)

5.19   E.A.Spiegel, G.Veronis: On the Boussinesq approximation for a compressible fluid. Astrophys. J. **131**, 442–447 (1960)

5.20   J.M.Mihaljan: A rigorous exposition of the Boussinesq approximation applicable to a thin fluid layer. Astrophys. J. **136**, 1126–1133 (1962)

5.21   W. V. R. Malkus: Boussinesq equations. Woods Hole Oceanographic Inst. Rep. 64–46 (1964) pp. 1–12

5.22   O. D. Gray, A. Giorgini: The validity of the Boussinesq approximation for liquids and gases. Int. J. Heat Mass Transfer **19**, 545–551 (1976)

5.23   V. S. Sorokin: Variational method in the theory of convection. Prikl. Mat. Mekh. **17**, 39 (1953)

5.24   Lord Rayleigh: On convection currents in a horizontal layer of fluid, when the higher temperature is on the under side. Philos. Mag. **32**, 529–546 (1916)

5.25   A. Pellew, R. V. Southwell: On maintained convective motion in a fluid heated from below. Proc. R. Soc. (London) A **176**, 312–343 (1940)

5.26   W. H. Reid, D. L. Harris: Some further results on the Bénard problem. Phys. Fluids **1**, 102–110 (1958)

5.27   E. M. Sparrow, R. J. Goldstein, V. H. Jonsson: Thermal instability in a horizontal fluid layer: effect of boundary conditions and non-linear temperature profile. J. Fluid Mech. **18**, 513–528 (1964)

5.28   E. Jakeman: Convective instability in fluids of high thermal diffusivity. Phys. Fluids **11**, 10–14 (1968)

5.29   L. P. Gorkov: Stationary convection in a plane liquid layer near the critical heat transfer point. Zh. Eksp. Teor. Fiz. [English transl.: Sov. Phys. – JETP] **6**, 311–315 (1957)

5.30   V. W. R. Malkus, G. Veronis: Finite amplitude convection. J. Fluid Mech. **4**, 225–260 (1958)

5.31   A. Schlüter, D. Lortz, F. Busse: On the stability of steady finite amplitude convection. J. Fluid Mech. **23**, 129–144 (1965)

5.32   V. I. Yudovich: On the origin of convection. J. Appl. Math. Mech. **30**, 1193 (1966)

5.33   D. H. Rabinowitz: Existence and nonuniqueness of rectangular solutions of the Bénard problem. Arch. Ration. Mech. Anal. **29**, 32 (1968)

5.34   F. H. Busse: "Stability Regions of Cellular Fluid Flow", in *Instability of Continuous Systems*, ed. by H. Leipholz (Springer, Berlin, Heidelberg, New York 1971) pp. 41–47

5.35   A. C. Newell, C. G. Lange, P. J. Aucoin: Random convection. J. Fluid Mech. **40**, 513–542 (1970)

5.36   D. R. Moore, N. O. Weiss: Nonlinear penetrative convection. J. Fluid Mech. **58**, 289–312 (1973)

5.37   J. E. Fromm: Numerical solutions of the nonlinear equations for a heated fluid layer. Phys. Fluids **8**, 1757–1769 (1965)

5.38   R. M. Clever, F. H. Busse: Transition to time dependent convection. J. Fluid Mech. **65**, 625–645 (1974)

5.39   H. T. Rossby: A study of Bénard convection with and without rotation. J. Fluid Mech. **36**, 309–335 (1969)

5.40   C. A. Jones, D. R. Moore, N. O. Weiss: Axisymmetric convection in a cylinder. J. Fluid Mech. **73**, 353 (1976)

5.41   M. R. E. Proctor: Inertial convection at low Prandtl number. J. Fluid Mech. **82**, 97–114 (1977)

5.42   W. V. R. Malkus: The heat transport and spectrum of thermal turbulence. Proc. R. Soc. (London) A **225**, 196–212 (1954)

5.43   F. H. Busse: The stability of finite amplitude cellular convection and its relation to an extremum principle. J. Fluid Mech. **30**, 625–649 (1967)

5.44   W. H. Plows: "Numerical Studies of Laminar, Free Convection in a Horizontal Fluid Layer Heated from Below", Ph. D. Thesis, University of California, Berkeley (1971)

5.45   F. H. Busse: On the stability of two-dimensional convection in a layer heated from below. J. Math. Phys. **46**, 149–150 (1967)

5.46   J. M. Straus: Finite amplitude doubly diffusive convection. J. Fluid Mech. **56**, 353–374 (1972)

5.47   J. W. Deardorff: Gravitational instability between horizontal plates with shear. Phys. Fluids **8**, 1027–1030 (1965)

5.48   A. P. Gallagher, A. McD. Mercer: On the behavior of small disturbances in plane Couette flow with a temperature gradient. Proc. R. Soc. (London) A **286**, 117–128 (1965)

5.49   F. B. Lipps: Numerical simulation of three-dimensional Bénard convection in air. J. Fluid Mech. **75**, 113–148 (1976)

5.50   R.H.Kraichnan: Turbulent thermal convection of arbitrary Prandtl number. Phys. Fluids **5**, 1374–1389 (1962)
5.51   L.N.Howard: Heat transport by turbulent convection. J. Fluid Mech. **17**, 405–432 (1963)
5.52   F.H.Busse: Bounds on the transport of mass and momentum by turbulent flow between parallel plates. J. Appl. Math. Phys. (ZAMP) **20**, 1–14 (1969)
5.53   F.H.Busse: Bounds for turbulent shear flow: J. Fluid Mech. **41**, 219–240 (1970)
5.54   F.H.Busse: The optimum theory of turbulence. Adv. Appl. Mech. **18**, 77–121 (1978)
5.55   F.H.Busse: On Howard's upper bound for heat transport by turbulent convection. J. Fluid Mech. **37**, 457–477 (1969)
5.56   J.M.Strauss: On the upper bounding approach to thermal convection of moderate Rayleigh numbers, II. Rigid boundaries. Dyn. Atmos. Oceans **1**, 77–90 (1976)
5.57   S.-K.Chan: Infinite Prandtl number turbulent convection. Stud. Appl. Math. **50**, 13–49 (1971)
5.58   H.Bénard: Les tourbillons cellulaires dans une nappe liquide transportant de la chaleur par convection en régime permanent. Ann. Chim. Phys. **7**, Ser. 23, 62 (1901)
5.59   J.R.A.Pearson: On convection cells induced by surface tension. J. Fluid Mech. **4**, 489–500 (1958)
5.60   M.J.Block: Surface tension as the cause of Bénard cells and surface deformation in a liquid film. Nature **178**, 650–651 (1956)
5.61   R.J.Schmidt, S.W.Milverton: On the instability of a fluid when heated from below. Proc. R. Soc. (London) A **152**, 586–594 (1935)
5.62   P.L.Silveston: Wärmedurchgang in waagerechten Flüssigkeitsschichten. Forsch. Ingenieurwes. **24**, 29–32, 59–69 (1958)
5.63   S.H.Davis, L.A.Segel: Effects of surface curvature and property variation on cellular convection. Phys. Fluids **11**, 470–476 (1968)
5.64   H.Tippelskirch: Über Konvektionszellen, insbesondere in flüssigem Schwefel. Beitr. Phys. Atmos. **29**, 37–54 (1956)
5.65   E.Palm: On the tendency towards hexagonal cells in steady convection. J. Fluid Mech. **8**, 183–192 (1960)
5.66   L.A.Segel, J.T.Stuart: On the question of the preferred mode in cellular thermal convection. J. Fluid Mech. **13**, 289–306 (1962)
5.67   R.J.Goldstein, D.J.Graham: Stability of a horizontal fluid layer with zero shear boundaries. Phys. Fluids **12**, 1133–1137 (1969)
5.68   G.E.Willis, J.W.Deardorff, R.C.Somerville: Roll-diameter dependence in Rayleigh convection and its effect upon the heat flux. J. Fluid Mech. **54**, 351–367 (1972)
5.69   F.H.Busse, J.A.Whitehead: Instabilities of convection rolls in a high Prandtl number fluid. J. Fluid Mech. **47**, 305–320 (1971)
5.70   F.H.Busse, J.A.Whitehead: Oscillatory and collective instabilities in large Prandtl number convection. J. Fluid Mech. **66**, 67–79 (1974)
5.71   E.L.Koschmieder: On convection on a uniformly heated plane. Beitr. Phys. Atmos. **39**, 1–11 (1966)
5.72   K.Stork, U.Müller: Convection in boxes: experiments. J. Fluid Mech. **54**, 599–611 (1972)
5.73   S.H.Davis: Convection in a box: linear theory. J. Fluid Mech. **30**, 465–478 (1967)
5.74   R.Farhadieh, R.S.Tankin: Interferometric study of two-dimensional Bénard convection cells. J. Fluid Mech. **66**, 739–752 (1974)
5.75   G.Veronis: Large-amplitude Bénard convection. J. Fluid Mech. **26**, 49 (1966)
5.76   T.Y.Chu, R.J.Goldstein: Turbulent convection in a horizontal layer of water. J. Fluid Mech. **60**, 141–159 (1973)
5.77   H.Oertel, Jr., K.Bühler: A special differential interferometer used for heat convection investigations. Int. J. Heat Mass Transfer **21**, 1111–1115 (1978)
5.78   P.Bergé, M.Dubois: Convective velocity field in the Rayleigh-Bérnard instability: experimental results. Phys. Rev. Lett. **32**, 1041–1044 (1974)
5.79   M.Dubois, P.Bergé: Experimental study of the velocity field in Rayleigh-Bénard convection. J. Fluid Mech. **85**, 641–653 (1978)

5.80 R. Krishnamurti: On the transition to turbulent convection. Part 1. The transition from two-to three-dimensional flow. J. Fluid Mech. **42**, 295–307 (1970a)

5.81 J. A. Whitehead, G. L. Chan: Stability of Rayleigh-Bénard convection rolls and bimodal flow at moderate Prandtl number. Dyn. Atmos. Oceans **1**, 33–49 (1976)

5.82 G. E. Willis, J. W. Deardorff: Development of short-period temperature fluctuations in thermal convection. Phys. Fluids **10**, 931–937 (1967)

5.83 R. Krishnamurti: On the transition to turbulent convection. Part 2. The transition to time-dependent flow. J. Fluid Mech. **42**, 309–320 (1970b)

5.84 R. Krishnamurti: Some further studies on the transition to turbulent convection. J. Fluid Mech. **60**, 285–303 (1973)

5.85 J. A. Whitehead, B. Parsons: Observations of convection at Rayleigh numbers up to 760,000 in a fluid with large Prandtl number. Geophys. Astrophys. Fluid Dyn. **9**, 201–217 (1978)

5.86 P. Bergé, M. Dubois: Time dependent velocity in Rayleigh-Bénard convection: a transition to turbulence. Opt. Commun. **19**, 129–133 (1976)

5.87 J. P. Gollub, S. L. Hulbert, G. M. Dolny, H. L. Swinney: "Laser Doppler Study of the Onset of Turbulent Convection at low Prandtl Number", in *Photon Correlation Spectroscopy and Velocimetry*, ed. by H. Z. Cummins, E. R. Pike (Plenum Press, New York, London 1977) pp. 425–439

5.88 P. Bergé: "Experimental Evidence of the Behaviour of Convective Velocity in Rayleigh-Bénard Instability: Different Transitions Toward Turbulence", Paper presented at the European Mechanics Colloquium, No. 106, Instabilities and Convection in Fluid Layers, Grenoble (1978)

5.89 A. Libchaber, J. Maurer: Local probe in a Rayleigh-Bénard experiment in liquid helium. J. Physique Lett. **39**, 369–372 (1978)

5.90 J. M. Olson, F. Rosenberger: Convective instabilities in a closed vertical cylinder heated from below. Part 1: Monocomponent gases. J. Fluid Mech. **92**, 609–629 (1979)

5.91 D. T. J. Hurle, E. Jakeman, C. P. Johnson: Convective temperature oscillations in molten gallium. J. Fluid Mech. **64**, 565–576 (1974)

5.92 J. B. Keller: Periodic oscillations in a model of thermal convection. J. Fluid Mech. **26**, 599–606 (1966)

5.93 P. Welander: On the oscillatory instability of a differentially heated fluid loop. J. Fluid Mech. **29**, 17–30 (1967)

5.94 Y. P. Chang: A theoretical analysis of heat transfer in natural convection and in boiling. Trans. Am. Soc. Mech. Engrs. **79**, 1501–1513 (1957)

5.95 L. N. Howard: "Convection at High Rayleigh Number", in *Proc. 11th Internat. Congress of Appl. Mech.*, Munich, 1964, ed. by H. Görtler (Springer, Berlin, Heidelberg, New York 1966) pp. 1109–1115

5.96 E. N. Lorenz: Deterministic nonperiodic flow. J. Atmos. Sci **20**, 130 (1963)

5.97 J. B. McLaughlin, P. C. Martin: Transition to turbulence in a statically stressed fluid system. Phys. Rev. A **12**, 186–203 (1975)

5.98 R. J. Goldstein, T. Y. Chu: Thermal convection in a horizontal layer of air. Progr. Heat Mass Transfer **2**, 55–75 (1971)

5.99 A. M. Garon, R. J. Goldstein: Velocity and heat transfer measurements in thermal convection. Phys. Fluids **16**, 1818–1825 (1973)

5.100 W. V. R. Malkus: Discrete transitions in turbulent convection. Proc. R. Soc. (London) A **225**, 185–195 (1954)

5.101 F. H. Busse, D. D. Joseph: Bounds for heat transport in a porous layer. J. Fluid Mech. **54**, 521–543 (1972)

5.102 D. C. Threlfall: Free convection in low-temperature gaseous helium. J. Fluid Mech. **67**, 17–28 (1975)

5.103 G. E. Willis, J. W. Deardorff: Confirmation and renumbering of the discrete heat flux transitions of Malthus. Phys. Fluids **10**, 1861–1866 (1967)

5.104 J. J. Carroll: "The Structure of Turbulent Convection", Ph.D. Dissertation, Dept. of Meteorology, University of California, Los Angeles (1971)

5.105 E. L. Koschmieder, S. G. Pallas: Heat transfer through a shallow, horizontal convecting fluid layer. Int. J. Heat Mass Transfer **17**, 991–1002 (1974)

5.106 J. W. Deardorff, G. E. Willis: Investigation of turbulent thermal convection between horizontal plates. J. Fluid Mech. **28**, 675–704 (1967)

5.107 D. E. Fitzjarrald: An experimental study of turbulent convection in air. J. Fluid Mech. **73**, 693–719 (1976)

5.108 E. F. C. Somerscales, H. Parsapour: A new approach to the detection of Malkus transition in free convection. Bull. Am. Phys. Soc. **21**, 1236 (1976)

5.109 S. Kogelman, R. C. DiPrima: Stability of spatially periodic supercritical flows in hydrodynamics. Phys. Fluids **13**, 1–11 (1970)

5.110 W. Eckhaus: *Studies in Non-Linear Stability Theory*, Springer Tracts in Natural Philosophy, Vol. 6 (Springer, Berlin, Heidelberg, New York 1965)

5.111 J. T. Stuart, R. C. DiPrima: The Eckhaus and Benjamin-Feir resonance mechanism. Proc. R. Soc. (London) A **362**, 27–41 (1978)

5.112 F. H. Busse, R. M. Clever: Instabilities of convection rolls in a fluid of moderate Prandtl number. J. Fluid Mech. **91**, 319–335 (1979)

5.113 R. M. Clever, F. H. Busse: Large wavelength convection rolls in low Prandtl number fluids. J. Appl. Math. Phys. (ZAMP) **29**, 711–714 (1978)

5.114 F. H. Busse: The oscillatory instability of convection rolls in a low Prandtl number fluid. J. Fluid Mech. **52**, 97–112 (1972)

5.115 G. Veronis: Large-amplitude Bénard convection in a rotating fluid. J. Fluid Mech. **31**, 113–139 (1968)

5.116 R. M. Clever, F. H. Busse: Nonlinear properties of convection rolls in a horizontal layer rotating about a vertical axis. J. Fluid Mech. **94**, 609–627 (1979)

5.117 G. Küppers, D. Lortz: Transition from laminar convection to thermal turbulence in a rotating fluid layer. J. Fluid Mech. **35**, 609–620 (1969)

5.118 E. Hopf: A mathematical example displaying features of turbulence. Commun. Pure Appl. Math. **1**, 303 (1948)

5.119 R. G. Finucane, R. E. Kelly: Onset of instability in a fluid layer heated sinusoidally from below. Int. J. Heat Mass Transfer **19**, 71–85 (1976)

6. Instabilities and Transition in Flow Between Concentric Rotating Cylinders

R. C. Di Prima and Harry L. Swinney

With 9 Figures

In this chapter we consider the stability and transitions of viscous incompressible flow between concentric rotating cylinders as the speed of one or both cylinders is increased. We shall be concerned primarily with the case with the outer cylinder at rest. The basic flow, now known as Couette flow, was studied in nineteenth century experiments [6.1–3] that were designed to test the Newtonian stress approximation in the Navier-Stokes equation. Then in 1923 *Taylor* [6.4] achieved the first dramatic success in the theory of hydrodynamic stability for any system: he observed and calculated by a linear stability analysis the critical rotation speed at which Couette flow becomes unstable. Taylor used dye to visualize the secondary flow which consists of horizontal toroidal vortices, as shown in Fig. 6.1a [6.5], and his measurements were in striking agreement with his calculations.

The concentric rotating cylinder geometry continues to be a paradigm for experimental and theoretical studies of hydrodynamic stability and transition. This geometry is particularly appealing to experimentalists because experiments can be conducted on small closed systems, and the Reynolds number, proportional to the cylinder rotation rate, can be controlled with high accuracy. Moreover, by using a glass tube for the outer cylinder, the complete flow pattern and transitions in the flow can be observed by flow visualization techniques, as Figs. 6.1b–d [6.5, 6] illustrate.

Following an introduction to the problem in Sect. 6.1 we shall discuss in Sects. 6.2 and 6.3, respectively, the instability of Couette flow (the Taylor instability) and the growth of secondary flow (Taylor vortex flow). As the Reynolds number is increased further, the Taylor vortices become unstable at a second critical Reynolds number, and a new flow is established with travelling azimuthal waves superimposed on the Taylor vortices; the wavy instability, discussed in Sect. 6.4, was first studied systematically in 1965 by *Coles* [6.6] (see Fig. 6.1b, c). Experiments in the past six years have shown that further instabilities occur before the fluid becomes turbulent; the higher instabilities and the transition to turbulence are discussed in Sect. 6.5. The theory in Sects. 6.1–5 assumes that the cylinders are infinitely long. In Sect. 6.6 recent theoretical and experimental work on the effect of finite cylinder length will be discussed.

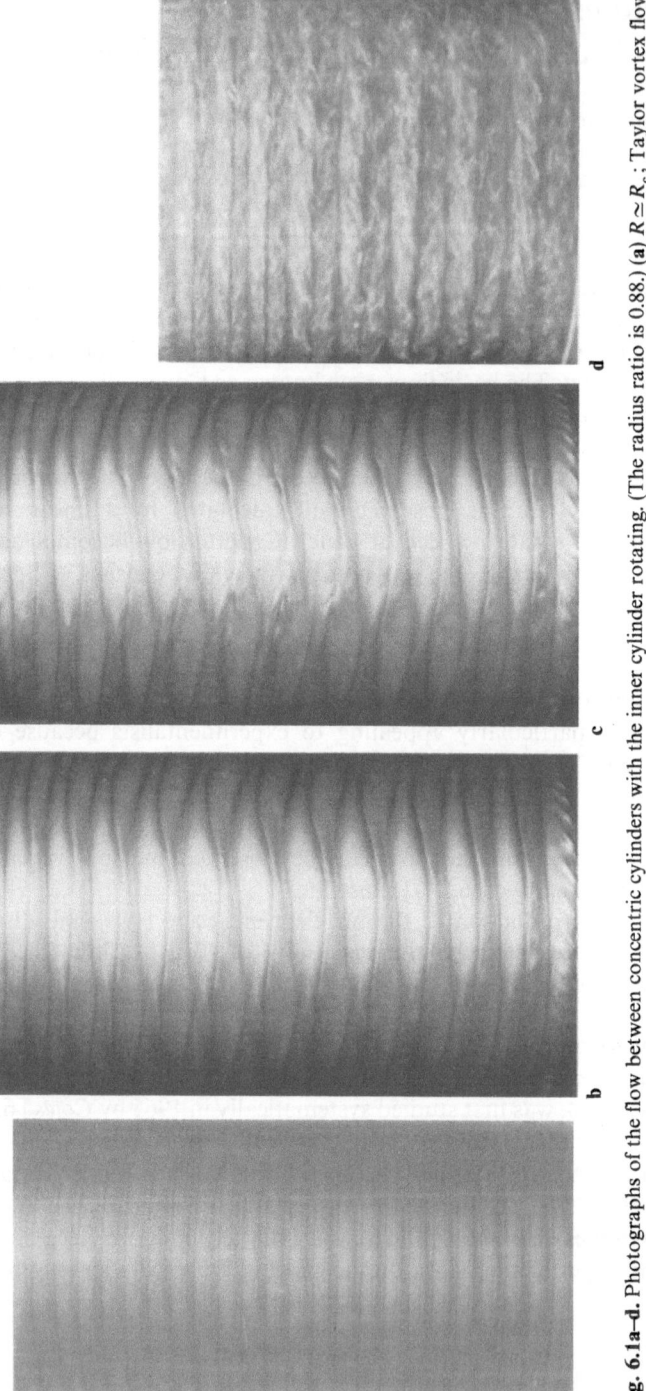

Fig. 6.1a–d. Photographs of the flow between concentric cylinders with the inner cylinder rotating. (The radius ratio is 0.88.) **(a)** $R \simeq R_c$; Taylor vortex flow [6.5]. **(b)** $R/R_c = 10.4$; wavy vortex flow [Ref. 6.6, Fig. 19d]. **(c)** $R/R_c = 12.3$; the "first appearance of randomness" in wavy vortex flow [Ref. 6.6, Fig. 19e]. **(d)** $R/R_c = 23.5$; the azimuthal waves have disappeared and the flow is turbulent, although the axial periodicity remains [Ref. 6.7, Fig. 1d]. The visualization of the flow in these experiments was achieved by suspending small flat flakes in the fluid; the flakes align with the flow, and variations in their orientation are observed as variations in the transmitted or reflected intensity

6.1 Background

Let (r, θ, z) denote the usual cylindrical coordinates, and let a, b and Ω_1, Ω_2 denote the radii and angular velocities of the inner and outer cylinders, respectively. Further, we denote the velocity components in the increasing r, θ, and z directions by u_r, u_θ, and u_z, respectively, and the pressure by p. An exact time-independent solution of the Navier-Stokes equations satisfying the boundary conditions $u_\theta = a\Omega_1$ at $r = a$ and $u_\theta = b\Omega_2$ at $r = b$ is the Couette velocity distribution,

$$u_r = u_z = 0, \quad u_\theta = V(r) = Ar + B/r, \quad \partial p/\partial r = \varrho V^2/r, \tag{6.1}$$

where ϱ is the density and

$$A = \frac{b^2\Omega_2 - a^2\Omega_1}{b^2 - a^2} = -\Omega_1 \frac{\eta^2 - \mu}{1 - \eta^2}, \tag{6.2a}$$

$$B = -\frac{a^2 b^2 (\Omega_2 - \Omega_1)}{b^2 - a^2} = \Omega_1 a^2 \frac{1 - \mu}{1 - \eta^2}, \tag{6.2b}$$

$$\mu = \Omega_2/\Omega_1, \quad \eta = a/b. \tag{6.2c}$$

After dimensionless variables are introduced, the Couette velocity distribution can be characterized by the parameters μ and η and an appropriate Reynolds number.

In this chapter we shall primarily be interested in determing the evolution of infinitesimal disturbances and transitions as the speed of the inner cylinder (Ω_1) is increased while the outer cylinder is at rest ($\mu = 0$). However, before turning to that more specific problem we would like to make a few remarks about the stability problem for arbitrary disturbances and both cylinders rotating.

Serrin [6.8], using energy methods, showed that Couette flow is monotonically and globally stable[1] to arbitrary disturbances that are periodic in the axial direction provided that

$$\frac{|B|}{\nu} = \frac{a^2 b^2 |\Omega_2 - \Omega_1|}{\nu(b^2 - a^2)} < \frac{\pi^2}{(\ln b/a)^2}. \tag{6.3}$$

A slightly sharper, but more complicated, rigorous result has been obtained by *Joseph* and *Munson* [Ref. 6.9, Eq. (4.26)]; or see *Joseph* [Ref. 6.10, Eq. (44.25)]. Critical values of $|B|/\nu$ have been calculated by *Hung* [6.11]. The results are given in [Ref. 6.10, Table 37.1] for several values of η. However, there appears to be an error in the recorded results; possibly the tabulated values are for $|B|/\nu$

---

1 By monotonically and globally stable we mean that the energy associated with an arbitrary (consistent with the Navier-Stokes equations) initial disturbance decays monotonically to zero with increasing time (see [Ref. 6.10, p. 9]).

Table 6.1. Critical values of $|B|/v$ [see (6.2b)] as a function of η for arbitrary disturbances according to energy theory [6.12]

η	$|B|/v$
0.10	11.60
0.25	25.23
0.50	89.71
0.75	503.0
0.95	1.571×10^4

multiplied by some function of η. In Table 6.1 we record the values of $|B|/v$ obtained by *Neitzel* [6.12]. We note that in all of these calculations the actual minimum value of $|B|/v$ occurs for a disturbance that is axisymmetric, that is, independent of the azimuthal coordinate θ.

If it is assumed from the outset that the disturbance is axisymmetric as well as periodic in the axial direction, then stronger results can be obtained. With the additional approximation $\eta \to 1$ (the small gap problem) *Serrin* [6.8] found that the condition for global and monotonic stability is

$$\frac{|\Omega_2 - \Omega_1|a(b-a)}{v} < 2\sqrt{1708} . \tag{6.4}$$

This result has been improved and extended by *Joseph* and *Hung* [6.13]. In that work the appropriate eigenvalue problem was solved numerically for different values of μ and η. In Fig. 6.2 we show the stability boundary for Couette flow for $a/b = 0.75$ as given by *Joseph* and *Hung*. The curve for linear theory is the stability boundary at which infinitesimal axisymmetric disturbances will start to grow according to linear theory. Note that i) the energy theory and the linear theory stability boundaries for axisymmetric disturbances are essentially identical for μ greater than some small negative number, and ii) the stability boundary for arbitrary disturbances is very close to the linear stability theory boundary for $0 < \mu < \eta^2$. The straight line $\Omega_1 a^2 = \Omega_2 b^2$ is $\mu = \eta^2$ ($= 0.5625$). The importance of this line will be made clear in a moment. Also, note that for μ sufficiently negative it can anticipated from the work of *Krueger* et al. [6.14] that the linear theory stability boundary will be given by nonaxisymmetric disturbances. A more complete discussion of the energy stability results can be found in *Joseph* [Ref. 6.10, Chap. 5].

Rayleigh [6.15] and *von Kármán* [6.16] each gave convincing physical arguments that in an *inviscid* flow with circular streamlines a necessary and sufficient condition for stability to axisymmetric disturbances is that the square of the circulation ($r^2 V^2$) increases outwards. Both arguments are described by *Lin* [Ref. 6.17, Sect. 4.2]. We shall briefly summarize the argument due to *von Karman*. Let $V(r)$ be the azimuthal velocity, which in an inviscid fluid need

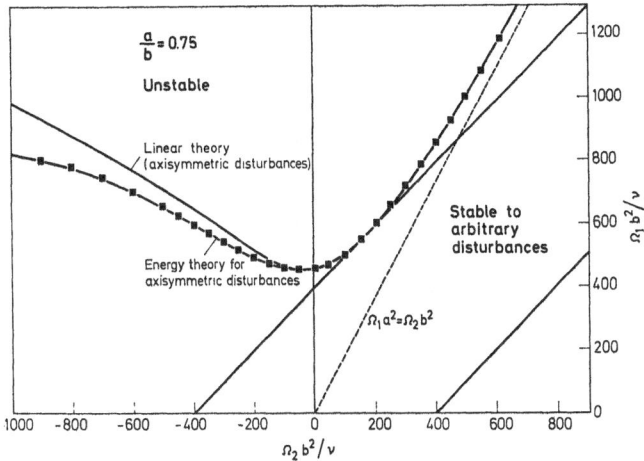

Fig. 6.2. Stability regions for Couette flow between rotating cylinders

not be Couette flow. In the steady state the centrifugal force field $\varrho V^2/r$ must be balanced by the radial pressure gradient $\partial p/\partial r$. According to Kelvin's circulation theorem the angular momentum of a fluid ring must be conserved for an axisymmetric disturbance. A ring of fluid at r_1 with velocity V_1 that moves to r_2 ($>r_1$) will acquire a new velocity $\bar{V}=r_1 V_1/r_2$. The centrifugal force associated with the new fluid ring at r_2 is $\varrho\bar{V}^2/r_2=\varrho(r_1 V_1)^2/r_2^3$. However, the prevailing pressure gradient at r_2 is $\varrho V_2^2/r_2$. Thus, the centrifugal force of the new fluid ring will exceed the prevailing pressure gradient if $(r_1 V_1)^2 > (r_2 V_2)^2$, in which case the fluid ring will continue to move outwards, and the motion will be unstable.

The necessary and sufficient condition $d(rV)^2/dr>0$ for stability is usually referred to as the Rayleigh criterion. It can also be expressed in the form

$$\frac{d}{dr}(rV)^2 = 2r^3\Omega\zeta > 0 , \qquad (6.5)$$

where $\Omega=V/r$ is the angular velocity and $\zeta=r^{-1}d(rV)/dr$ is the vorticity. For inviscid Couette flow, we find as a necessary and sufficient condition for stability that $\Omega_1>0$ and $\mu>\eta^2$. This explains the importance of the line corresponding to $\mu=\eta^2$ in Fig. 6.2. We emphasize again that the Rayleigh criterion (6.5) is for an inviscid fluid and axisymmetric disturbances.

Synge [6.18] derived the Rayleigh criterion mathematically by considering the linear stability problem for axisymmetric disturbances in an inviscid revolving fluid. Later he showed [6.19] that in a viscous fluid the condition $\Omega_1>0$ and $\mu>\eta^2$ was sufficient, but not necessary, for stability according to linear theory. Observe in Fig. 6.2 the difference in the Rayleigh line, $\Omega_1 a^2=\Omega_2 b^2$, and the actual linear theory curve. The Synge result has been extended to finite amplitude disturbances by *Joseph* and *Hung* [6.13]; also see [Ref. 6.10, Sect. 40].

When $\mu \geqq \eta^2$, Couette flow will, of course, be unstable at sufficiently high values of a Reynolds number based on the velocity of the outer cylinder. The linear stability problem for $\Omega_1 = 0$ has been investigated by *Schultz-Grunow* [6.20–22] for disturbances representing a wave propagating in the azimuthal direction. Just as in the case of plane Couette flow, he found that the flow is stable, according to linear theory, for all values of Ω_2.

We now turn our attention to the case $\mu < \eta^2$, and more specifically usually $\mu = 0$ and η near one. It is convenient to introduce a Reynolds number $R = \Omega_1 a(b-a)/v$; then the basic Couette flow is fully described by the values of μ, η, and R (assuming $\Omega_1 \neq 0$).

Taylor [6.4] showed that there exists a critical value of the Reynolds number $R_c(\mu, \eta)$ with the following properties. For $R < R_c(\mu, \eta)$ all initially infinitesimal axisymmetric disturbances that are periodic in the axial direction are damped and decay to zero with increasing time; for $R > R_c(\mu, \eta)$ there are some disturbances that will grow with time. Moreover, his experiments showed that this instability of Couette flow leads to a new steady secondary axisymmetric flow in the form of regularly spaced vortices in the axial direction, commonly referred to as Taylor vortices[2]. A picture of a Taylor vortex flow is shown in Fig. 6.1a. Mathematically, the instability represents a supercritical steady bifurcation from Couette flow to Taylor vortex flow.

For simplicity of explanation and because not all parameter values have been considered, let us take $\mu = 0$ and assume that η is near one, say $\eta \simeq 0.75, 0.8$, etc., or even $\eta \to 1$. *Davey* [6.24], following the method of *Stuart* [6.25] and *Watson* [6.26], was the first to calculate the Taylor vortex flow for values of R slightly greater than R_c. He found that the amplitude of the Taylor vortex flow increases as $(R - R_c)^{1/2}$, and his calculation of the torque on the inner cylinder was in good agreement with experimental measurements. *Davey* et al. [6.27] showed that with increasing R a second critical Reynolds number R_c', which depends on η, is reached. At R_c' the Taylor vortex flow becomes unstable, the instability leading to a wavy vortex flow as shown in Fig. 6.1b. The wavy vortices have a definite frequency and move with a definite wave velocity in the azimuthal direction. Mathematically, this second instability is a time-periodic supercritical bifurcation (a Hopf bifurcation) from Taylor vortex flow to wavy vortex flow. These transitions, and more, have been studied experimentally by *Coles* [6.6]. One of the important observations of his experiments is the nonuniqueness of the spatial structure: different axial and azimuthal wave numbers are observed at the same final Reynolds number. Presumably, these different final states at the same value of $R > R_c$ are functions of the initial conditions and the method by which the final value of R is reached.

---

2 As noted earlier, for μ sufficiently negative and η near one, that the instability will be of a nonaxisymmetric form and, at least for some parameter ranges, of a subcritical nature; see [6.14] and [6.23]. Thus, these statements are not absolutely correct for all values of μ and η with $\mu < \eta^2$.

More recently, *Gollub* and *Swinney* [6.28] and *Fenstermacher* et al. [6.7] have studied the temporal properties of the flow field for increasing R beyond R_c by using an optical heterodyne technique to measure the time dependence of the radial component of velocity at a fixed point. The power spectra obtained by Fourier-transforming the velocity records show i) a transition from Couette flow to steady Taylor vortex flow, ii) a transition from Taylor vortex flow to periodic wavy vortex flow (one time frequency), iii) a transition to a quasi-periodic flow with two frequencies, with the new frequency disappearing at a larger value of R, and iv) a transition to chaotic flow with a continuous spectrum. *Walden* and *Donnelly* [6.29] have observed the same transitions in power spectra obtained by a different technique (see Sect. 6.5). In addition, for large aspect ratios and at large Reynolds numbers, *Walden* and *Donnelly* observed another frequency component. Theoretical work has not kept pace with these experimental studies.

In the rest of this chapter we shall develop in more detail this brief introduction to the transition from Couette flow to Taylor vortex flow, to wavy vortex, to quasi-periodic flow, and eventually to chaos (turbulence) with increasing R.

6.2 Instability of Couette Flow

Let u', v', w', and p' denote perturbations of u_r, u_θ, u_z, and p for the Couette flow described by (6.1). Substituting in the Navier-Stokes equations, we obtain (see [Ref. 6.30, Sects. 69 and 70])

$$\frac{\partial u'}{\partial t} + \frac{V}{r}\frac{\partial u'}{\partial \theta} - \frac{2V}{r}v' + \frac{1}{\varrho}\frac{\partial p'}{\partial r} - \nu\left(\nabla^2 u' - \frac{2}{r^2}\frac{\partial v'}{\partial \theta} - \frac{u'}{r^2}\right)$$

$$= -\left(u'\frac{\partial u'}{\partial r} + \frac{v'}{r}\frac{\partial u'}{\partial \theta} + w'\frac{\partial u'}{\partial z}\right) + \frac{v'^2}{r}, \tag{6.6a}$$

$$\frac{\partial v'}{\partial t} + \frac{dV}{dr}u' + \frac{V}{r}\frac{\partial v'}{\partial \theta} + \frac{V}{r}u' + \frac{1}{\varrho r}\frac{\partial p'}{\partial \theta} - \nu\left(\nabla^2 v' + \frac{2}{r^2}\frac{\partial u'}{\partial \theta} - \frac{v'}{r^2}\right)$$

$$= -\left(u'\frac{\partial v'}{\partial r} + \frac{v'}{r}\frac{\partial v'}{\partial \theta} + w'\frac{\partial v'}{\partial z}\right) - \frac{u'v'}{r}, \tag{6.6b}$$

$$\frac{\partial w'}{\partial t} + \frac{V}{r}\frac{\partial w'}{\partial \theta} + \frac{1}{\varrho}\frac{\partial p'}{\partial z} - \nu\nabla^2 w' = -\left(u'\frac{\partial w'}{\partial r} + \frac{v'}{r}\frac{\partial w'}{\partial \theta} + w'\frac{\partial w'}{\partial z}\right), \tag{6.6c}$$

$$\frac{\partial u'}{\partial r} + \frac{u'}{r} + \frac{1}{r}\frac{\partial v'}{\partial \theta} + \frac{\partial w'}{\partial z} = 0, \tag{6.6d}$$

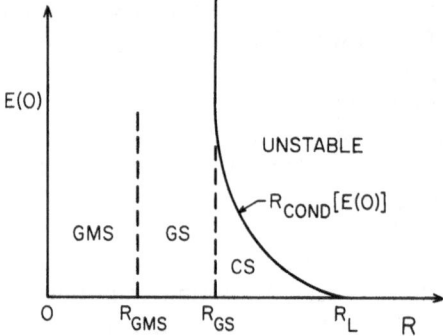

Fig. 6.3. Schematic representation of the stability limits for Couette flow. GMS = globally and monotonically stable; GS = globally stable; CS = conditionally stable

where v is the kinematic viscosity and

$$\nabla^2 = \frac{\partial^2}{\partial r^2} + \frac{1}{r} \frac{\partial}{\partial r} + \frac{1}{r^2} \frac{\partial^2}{\partial \theta^2} + \frac{\partial^2}{\partial z^2} . \tag{6.7}$$

The domain is $a < r < b$, $0 \le \theta < 2\pi$, and $-\infty < z < \infty$. In addition, the condition of no slip at the cylinder walls imposes the boundary conditions

$$u' = v' = w' = 0 \quad \text{at} \quad r = a \quad \text{and} \quad r = b . \tag{6.8}$$

Since the domain is infinite in the axial direction it is necessary to impose some additional condition; we shall assume that the disturbance is periodic in the axial direction. Finally, the initial values of u', v', w', and p' at $t = 0$ must be specified.

Couette flow (or any other flow) is said to be globally and monotonically stable to disturbances if $E(t)/E(0) \to 0$ as $t \to \infty$ and $dE/dt \le 0$ for all t, where $E(t)$ is the disturbance energy (suitably defined for an unbounded region). It is said to be globally stable if $E(t)/E(0) \to 0$ as $t \to \infty$, and conditionally stable if there exists $\Delta > 0$ such that $E(t)/E(0) \to 0$ as $t \to \infty$ provided that $E(0) < \Delta$. Finally, it is said to be linearly stable if, on neglect of the quadratic terms in (6.6), all disturbances die out as $t \to \infty$. These different conditions define a sequence of Reynolds numbers $R_{GMS} \le R_{GS} \le R_L$ as shown schematically in Fig. 6.3. (These definitions and Fig. 6.3 are patterned after *Joseph* [Ref. 6.10, Sects. 1 and 2].)

As we have indicated in Sect. 6.1, our interest here is in i) determining R_L, ii) following the evolution of an infinitesimal growing disturbance for $R > R_L$, and iii) studying the subsequent instabilities and transitions as R is increased further. In this section we consider the problem of determining R_L. Thus, we restrict our attention, as did *Taylor* in his classic paper [6.4], to infinitesimal disturbances that are axisymmetric and periodic in the axial direction. *Krueger* et al. [6.14] have shown that the critical disturbance is axisymmetric for $\mu = 0$.

Thus the disturbance equations are linear, independent of θ, and have coefficients that depend only on r. For disturbances that are periodic in the axial direction we can seek normal mode solutions of the form

$$u'(r,z,t)=u(r)e^{\beta t+i\gamma z} , \qquad v'(r,z,t)=v(r)e^{\beta t+i\gamma z} , \qquad (6.9)$$

with similar expressions for p' and w'. Substituting in the linearized disturbance equations, and using the axial momentum equation and the continuity equation to eliminate p and w, we obtain

$$v\left(\frac{d^2}{dr^2}+\frac{1}{r}\frac{d}{dr}-\frac{1}{r^2}-\gamma^2-\frac{\beta}{v}\right)\left(\frac{d^2}{dr^2}+\frac{1}{r}\frac{d}{dr}-\frac{1}{r^2}-\gamma^2\right)u=2\gamma^2\Omega(r)v , \qquad (6.10a)$$

$$v\left(\frac{d^2}{dr^2}+\frac{1}{r}\frac{d}{dr}-\frac{1}{r^2}-\gamma^2-\frac{\beta}{v}\right)v=2Au , \qquad (6.10b)$$

where $\Omega(r)=V(r)/r$ and we have made use of the fact that $dV/dr+r^{-1}V=2A$, a constant.

The natural length scale is the gap between the cylinders $d=b-a$. If v, the azimuthal velocity, is scaled with respect to $a\Omega_1$, it is clear from (6.10b) that the scale for u is $va\Omega_1/2Ad^2$. Thus, we let

$$r=R_0+dx, \qquad R_0=(a+b)/2$$
$$\delta=d/R_0, \qquad \xi(x)=(1+\delta x)^{-1}$$
$$\Omega=\Omega_1 G(x), \qquad G(x)=[1/(1-\eta^2)]\,[\mu-\eta^2+4\xi^2(x)\eta^2(1-\mu)/(1+\eta)^2] \qquad (6.11)$$
$$\sigma=\beta d^2/v, \qquad \alpha=\gamma d , \qquad T=-4A\Omega_1 d^4/v^2$$
$$D=d/dx, \qquad D^*=D+\delta\xi(x) .$$

On setting $v=a\Omega_1 v_{11}$, $u=(va\Omega_1/2Ad^2)u_{11}$, substituting in (6.10), and using (6.11), we obtain

$$(DD^*-\alpha^2-\sigma)(DD^*-\alpha^2)u_{11}=-\alpha^2TG(x,\mu,\eta)v_{11} , \qquad (6.12a)$$

$$(DD^*-\alpha^2-\sigma)v_{11}=u_{11} . \qquad (6.12b)$$

The boundary conditions, which follow from (6.8) and the continuity equation (6.6d), are

$$u_{11}=v_{11}=Du_{11}=0 \quad \text{at} \quad x=\pm 1/2 . \qquad (6.13)$$

The parameter T that appears in (6.12a) is usually referred to as the Taylor number. It has the form

$$T=\frac{-4A\Omega_1 d^4}{v^2}=\frac{4\Omega_1^2 d^4}{v^2}\left(\frac{\eta^2-\mu}{1-\eta^2}\right)=4\left(\frac{\Omega_1 ad}{v}\right)^2\left(\frac{d}{a}\right)\frac{\eta(1-\mu/\eta^2)}{1+\eta} . \qquad (6.14)$$

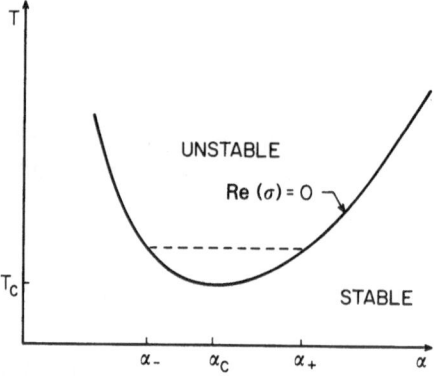

Fig. 6.4. Schematic sketch of the neutral curve for the stability of Couette flow to axisymmetric disturbances for fixed values of μ and η

Notice first that T is only positive if $\mu < \eta^2$. It is precisely the condition $\mu < \eta^2$ that is necessary and sufficient for instability in an inviscid fluid and is necessary for instability in a viscous fluid. Second, the Taylor number has the form of the square of the Reynolds number $R = \Omega_1 a d / v$ times a geometric factor d/a (giving the effect of the curvature) and times a function of μ and η. In the case often of interest in applications and experiments we have the outer cylinder at rest ($\mu = 0$) and the gap between the cylinders very small compared to the radii ($\eta \to 1$). In this case $T = 2R^2(d/a)$.

Equations (6.12) and (6.13) define an eigenvalue problem of the form

$$H(\eta, \mu, T, \alpha, \sigma) = 0 . \tag{6.15}$$

The parameter η describes the geometry, the parameters μ and T (and η) describe the basic flow, the parameter α is the dimensionless wave number of the disturbance in the axial direction, and the value of σ determines the growth rate of this disturbance for the condition given by η, μ, T. If for given values of η, μ and T there exists an $\alpha > 0$ for which σ has a positive real part, then the disturbance grows exponentially in time and the flow is unstable. For given values of $\mu (< \eta^2)$ and η we anticipate that for each value of $\alpha > 0$ there will be a value of T with the property that for T less than this value, all solutions σ of (6.15) have a negative real part, but for T greater than this value, there exists at least one solution σ of (6.15) with a positive real part[3]. This set of points defines a *neutral curve* in the T-α plane as shown schematically in Fig. 6.4. The minimum point on the neutral curve defines a critical value of the Taylor number T_c, below which all axisymmetric disturbances are damped, and above which there is a band of wave numbers ($\alpha_- < \alpha < \alpha_+$) for which the corresponding axisymmetric disturbances will grow.

---

3   This is not the case for all hydrodynamic stability problems. For example, in the linear stability of plane Poiseuille flow there are values of the wave number for which the flow is always stable.

Table 6.2. Critical values of $R(d/a)^{1/2}$ and the dimensionless wave number $\alpha = 2\pi d/\lambda$ for different values of $\eta = a/b$ for $\mu = 0$. Linear stability theory: [6.34–36]. Energy theory: [6.12]. Note that the Reynolds number $R = \Omega_1 ad/v$ and the Taylor number are related by $T = [4\eta/(1+\eta)]$ $[R(d/a)^{1/2}]^2$; see (6.14)

η	Ref.	α	$R(d/a)^{1/2}$	η	Ref.	α	$R(d/a)^{1/2}$
1	[6.35]	3.12	41.18	0.6	[6.35]	3.15	58.56
0.975	[6.36]	3.13	41.79	0.5	[6.36]	3.16	68.19
0.9625	[6.36]	3.13	42.09		[6.34]	3.16	68.19
0.95	[6.36]	3.13	42.44		[6.35]	3.16	68.18
	[6.34]	3.14	42.44		[6.12]	3.17	67.28
	[6.35]	3.12	42.45	0.4	[6.35]	3.17	83.64
	[6.12]	3.12	18.50	0.36	[6.36]	3.19	92.72
0.925	[6.36]	3.13	43.13	0.35	[6.34]	3.21	95.38
0.9	[6.36]	3.13	43.87	0.3	[6.35]	3.20	118.89
	[6.35]	3.13	43.88	0.28	[6.36]	3.22	120.43
0.875	[6.36]	3.13	44.66	0.25	[6.34]	3.24	136.40
0.85	[6.36]	3.13	45.50		[6.12]	3.33	122.91
0.8	[6.35]	3.13	47.37	0.2	[6.36]	3.25	176.26
0.75	[6.36]	3.14	49.52		[6.35]	3.23	176.33
	[6.34]	3.14	49.53	0.15	[6.34]	3.31	250.10
	[6.12]	3.13	42.35	0.10	[6.34]	3.36	423.48
0.7	[6.35]	3.14	52.04		[6.35]	3.30	422.79
0.65	[6.36]	3.14	55.01		[6.12]	3.65	310.07

The solution of the eigenvalue problem is considerably simplified by the so-called (and badly named) principle of exchange of stabilities, which states that when $\operatorname{Re}\{\sigma\} = 0$ then also $\operatorname{Im}\{\sigma\} = 0$. Thus the neutral curve is given by $\sigma = 0$ and the solution of the eigenvalue problem requires only real valued arithmetic. *Yih* [6.31] has given a very intricate proof of the principle of exchange of stabilities for $0 < \mu < \eta^2$. Many numerical calculations confirm this result and show that it probably holds for some negative values of μ, but perhaps not for all values of $\mu \ll 0$. In this regard see the time-dependent finite difference calculations for the linear stability problem by *Astill* and *Chung* [6.32] and *Chung* [6.33].

The eigenvalue problem (6.15) with $\sigma = 0$ can be solved rather easily for not too negative values of μ by standard procedures such as shooting [6.34] or the Galerkin method [6.35]. However, it is worth noting that the original formulas given by *Taylor* [6.4] provide good approximations to the critical value of T for a wide range of values of μ and η. In Table 6.2 we record the critical values of $R(d/a)^{1/2}$ and wave number α for $\mu = 0$ and for several values of η as given by *Roberts* [6.36], *Sparrow* et al. [6.34], and *Walowit* et al. [6.35]. We also record, for the purposes of comparison, the corresponding values of $R(d/a)^{1/2}$ for stability to *arbitrary* axisymmetric disturbances, as given by *Neitzel* [6.12].

It is clear from Table 6.2 that for $\mu = 0$ the wavelength λ of the critical disturbance is approximately $2d$. Thus, the vortices can be expected to have a nearly square cross section which is consistent with the experimental observations dating back to *Taylor* [Ref. 6.4, Fig. 5].

Before continuing we would like to discuss a mathematical detail. We have assumed that the disturbance is periodic in the axial direction with some wave number $\gamma \equiv \alpha/d$. (Within the framework of linear theory we could also assume that the disturbance is the sum of a finite number of periodic disturbances each with a different wave number since there will be no coupling; however, for simplicity we restrict attention to a single disturbance of wave number γ.) In terms of the physical variables, the initial conditions are of the form

$$u'(r, z, 0) = \sum_{n=-\infty}^{\infty} u'_n(r) e^{in\gamma z} \qquad (6.16)$$

with a solution of the linear disturbance equations of the form

$$u'(r, z, t) = \sum_{n=-\infty}^{\infty} f'_n(r, t) e^{in\gamma z} , \qquad (6.17)$$

with similar expressions for v', w', and p'. Since u' is real it follows that f'_{-n} is the complex conjugate of f'_n.

In the following we shall speak schematically referring only to the u' and f'_n with the understanding that we are referring to the actual velocity field. The equations for the $f'_n(r, t)$ are uncoupled and can be solved separately; this is the essence of the normal mode approach. The solutions are of the form

$$f'_n(r, t) = e^{\beta_n t} f_n(r) , \qquad (6.18)$$

with β_n being determined as the solution of an eigenvalue problem and with $f_n(r)$ the corresponding eigenfunction. Ideally we can expect that the eigenvalue problem will have an infinite set of discrete eigenvalues β_{nk} that can be ordered as Re $\{\beta_{n0}\} >$ Re $\{\beta_{n1}\} > \dots$ with no cluster point in the finite plane, and with corresponding eigenfunctions $f_{nk}(r)$. Thus the solution of the initial value problem for the disturbance is

$$u'(r, z, t) = \sum_{n=-\infty}^{\infty} \left[\sum_{k=0}^{\infty} A_{nk} f_{nk}(r) e^{\beta_{nk} t} \right] e^{in\gamma z} , \qquad (6.19)$$

with the arbitrary constants A_{nk} chosen so as to satisfy the initial conditions

$$u'_n(r) = \sum_{k=0}^{\infty} A_{nk} f_{nk}(r) . \qquad (6.20)$$

Hence in order to solve the linearized disturbance equations for the time evolution of a disturbance that is periodic in the axial direction with (arbitrary) wave number γ and arbitrary radial dependence, it is clear that we must have a completeness theorem for the eigenfunctions $\{f_{nk}(r)\}$. The eigenvalue problem (6.12) and (6.13) is non-self-adjoint; however, an appropriate completeness theorem can be established by showing that the differential operator is the sum of a self-adjoint operator plus a bounded operator; see *Di Prima* and *Habetler* [6.37].

Consider the solution (6.19) and suppose that $\gamma = \gamma_c$ and T is slightly greater than T_c. By T slightly greater than T_c we mean T such that $\mathrm{Re}\,\{\beta_{10}\} > 0$ and, for $(n, k) \neq (1, 0)$, $\mathrm{Re}\,\{\beta_{nk}\} = O(1)$ and $\mathrm{Re}\,\{\beta_{nk}\} < 0$. Then for large t we obtain from (6.19)

$$u'(r, z, t) \sim A_{10} f_{10}(r) e^{\beta_{10} t + i\gamma_c z} + \text{c.c.} \quad \text{as} \quad T \to \infty . \tag{6.21}$$

The higher spatial modes $f_{1k}(r)$, $k > 0$, corresponding to the fundamental mode γ_c and all of the spatial modes corresponding to the first and higher harmonics of $\gamma_c (2\gamma_c, 3\gamma_c, \ldots)$ and the mean $(0\gamma_c)$ decay exponentially as $t \to \infty$. The harmonics and mean and the higher spatial modes in the fundamental have no life of their own when $\gamma = \gamma_c$ and T is near T_c. In the next section we shall show how the picture is modified through nonlinear interactions due to the nonlinear terms that have been neglected.

As we have noted earlier, our primary interest in this review is a consideration of the case when the outer cylinder is at rest $(\mu = 0)$. We shall conclude this section, however, with a few summarizing comments about other values of $\mu < \eta^2$.

For the case of small gap $(\eta \to 1)$ and $\mu > 0$ the result for T_c obtained using an average value for the Couette velocity is very good, not only for $\mu > 0$ but also for $\mu = 0$ and slightly negative. *Chandrasekhar* [Ref. 6.30, p. 313] has given the following formula which includes a correction term in $(1 - \mu)$:

$$T_c = \frac{3416}{1 + \mu} \left[1 - 7.61 \times 10^{-3} \left(\frac{1 - \mu}{1 + \mu} \right)^2 \right], \quad \alpha_c = 3.12 , \quad \text{as} \quad \mu \to 1 \ (\eta = 1) . \tag{6.22}$$

This formula is quite satisfactory for $-0.25 < \mu < 1$.

For negative values of μ only the region near the inner cylinder is unstable according to Rayleigh's criterion, and this suggests that the proper length scale, rather than d, is d_1, the distance from the inner cylinder to the zero of the Couette velocity profile. This idea was explored by *Di Prima* [6.38] with the following improved numerical results given by *Harris* and *Reid* [6.39]:

$$T_c \to 1178.6(1 - \mu)^4 , \quad \alpha \to 2.034(1 - \mu) \quad \text{as} \quad \mu \to -\infty (\eta = 1) . \tag{6.23}$$

Also see *Duty* and *Reid* [6.40]. Formula (6.23) holds with quite reasonable accuracy for $\mu < -1$. Thus for the small gap problem, (6.22) and (6.23) include all values of μ except a small range $-1 < \mu < -0.25$, approximately.

Approximate formulas for $T_c(\mu, \eta)$ for $\eta \neq 1$ have been obtained by a somewhat heuristic argument for $\mu \geq 0$ by *Walowit* et al. [Ref. 6.35, Eq. (30)], and for $\mu > 0$ and $\mu < 0$ (two formulas) by *Coles* [Ref. 6.41, Eqs. (12) and (13)]; also, of course, there are Taylor's original formulas. An important observation made by *Coles* is that for a wide range of parameter values the stability boundary can be described by two dimensionless parameters, $-Aa^2/B$ and B/v, rather than the usual parameters μ, η, and T or $\eta, \Omega_1 a^2/v$, and $\Omega_2 b^2/v$.

The reader is referred to the papers mentioned in this section for details of methods of solving the linear stability problem for axisymmetric disturbances and for numerical results.

6.3 Growth of Taylor Vortices

The Taylor vortex flow was first calculated by *Davey* [6.24]. His calculations are formal and follow the method of an amplitude expansion proposed by *Stuart* [6.25] and *Watson* [6.26]. He calculated the Taylor vortex flow for several values of η, for the value of α_c corresponding to that value of η, and for $R - R_c$ (or $T - T_c$) small. We describe briefly the procedure followed by *Davey*.

The full nonlinear disturbance equations for axisymmetric disturbances with i) w' and p' eliminated, ii) $v' = a\Omega_1 v$ and $u' = (va\Omega_1/2Ad^2)u$, $z = d\zeta$, and $t = d^2\tau/v$, and with the notation of (6.11) are

$$\left(DD^* + \frac{\partial^2}{\partial \zeta^2} - \frac{\partial}{\partial \tau}\right)\left(DD^* + \frac{\partial^2}{\partial \zeta^2}\right)u - TG(x)\frac{\partial^2 v}{\partial \zeta^2} = N_1(u, v) , \tag{6.24}$$

$$\left(DD^* + \frac{\partial^2}{\partial \zeta^2} - \frac{\partial}{\partial \tau}\right)v - u = N_2(u, v) , \tag{6.25}$$

$$u = v = Du = 0 \quad \text{at} \quad x = \pm 1/2 , \tag{6.26}$$

where $N_1(u, v)$ and $N_2(u, v)$ are quadratic polynomials in u and v and their derivatives. These quantities are recorded in [Ref. 6.42, p. 9] with only a slight change from the present notation.

We assume that μ and η are fixed and consider a disturbance with wave number α_c as predicted by linear theory, and finally we assume that $T - T_c$ is small. *Stuart* and *Watson* argued physically as follows. For $T > T_c$ a disturbance of the form[4] $u(x, \zeta, \tau) = u_{11}(x)\exp(\sigma\tau + i\alpha\zeta)$ will grow exponentially in time according to linear theory. The spatial form in x and ζ should be correct to first order, but the exponential growth in time must be modified, so the factor $\exp(\sigma\tau)$ is replaced by an (possibly complex-valued) amplitude function $F(\tau)$ that is to be determined. The quadratic nonlinearity will force terms $O(F^2)$ proportional to the first harmonic and a mean motion. In turn, these terms

---

4 There is a similar expression for v which we omit. Also we recall that Im $\{\sigma\} = 0$.

interact with the basic disturbance to correct the x dependence of the fundamental mode and to generate the second harmonic at $O(F^3)$. The process continues and the velocity field has the form

$$u(x, \zeta, \tau) = F(\tau)u_{11}(x)E_1 + F^2(\tau)u_{22}(x)E_2 + |F(\tau)|^2 u_{20}(x)$$
$$+ F(\tau)|F(\tau)|^2 u_{31}(x)E_1 + F^3(\tau)u_{33}(x)E_3 + \ldots + \text{c.c.} \tag{6.27}$$

with a similar expression for v with $u_{ij}(x)$ replaced by $v_{ij}(x)$. Here $E_m = \exp(im\alpha\zeta)$, (u_{11}, v_{11}) is the eigenfunction of (6.12) and (6.13), and u_{22}, u_{20}, u_{31}, $u_{33}, \ldots$ are the solutions of certain nonhomogeneous boundary value problems. Associated with (6.27) is an amplitude equation, which *Stuart* and *Watson* showed has the form

$$dF/d\tau = \sigma F - a_1 F|F|^2 + O(F^5) . \tag{6.28}$$

The parameter $\sigma(T)$ is the eigenvalue of the boundary value problem (6.12) and (6.13); it gives the growth rate of the disturbance according to linear theory. The parameter a_1, which turns out to be real for the Taylor vortex problem, is determined by a solvability condition on the equations for (u_{31}, v_{31}) so that these functions are nonsingular at (α_c, T_c). It is often referred to as the Landau constant.

Since $\sigma = 0$ at $T = T_c$, it follows that $\sigma = \sigma_0 \varepsilon + O(\varepsilon^2)$, where $\varepsilon = T - T_c$ and $\sigma_0 > 0$. *Davey* [6.24] carried out the necessary calculations to determine a_1, finding that it was real and positive. In this case (6.28) has an equilibrium solution $|F_e|^2 = \sigma/a_1$ which exists for $T > T_c$. The phase in ζ is indeterminate, but this corresponds only to a translation in the axial coordinate. The truncation at cubic order is consistent since $F_e = O(\varepsilon^{1/2})$ and $\sigma = O(\varepsilon)$, so the first two terms on the right-hand side of (6.28) are $O(\varepsilon^{3/2})$, while the first neglected term is $O(\varepsilon^{5/2})$. With this truncation, the terms $O(F)$ and $O(F^2)$ in the velocity field can be calculated so the velocity is known with an error which is $O(\varepsilon^{3/2})$.

The amplitude equation (6.28) and velocity field (6.27) have also been obtained formally by *Eckhaus* [6.43–45] for a different problem and by a generalization of Eckhaus' procedure for the Taylor vortex problem by *Di Prima* [6.46]. The perturbation is first expanded as a Fourier series in ζ; then the coefficients are expressed as series of the spatial eigenfunctions of the corresponding linear problems with unknown time-dependent coefficients. The series have the form of (6.19), but in dimensionless variables, with $A_{nk} \exp(\beta_{nk}t)$ replaced by $A_{nk}(\tau)$. The resulting system of nonlinear ordinary differential equations for the $A_{nk}(\tau)$ is reduced to a single equation of the form (6.28) for $A_{10}(\tau)$ by exploiting the fact the amplification rate σ is small.

The first rigorous mathematical proof of the existence of Taylor vortex flow for $T > T_c$ was given by *Velte* [6.47] using the method of topological degree of *Leray-Schauder*. A constructive proof was given by *Kirchgässner* and *Sorger* [6.48] using the method of *Liapunov* and *Schmidt*. Other papers on the

Table 6.3. Comparison of different calculations of Taylor vortex torques

Authors	η	α	A	B
[6.54]	1	3.13	9.015×10^{-4}	$(-8.3 \pm 0.3) \times 10^{-7}$
[6.55]	1	3.13	9.047×10^{-4}	-8.44×10^{-7}
[6.54][a]	0.95	3.0	7.884×10^{-4}	$(-7.1 \pm 0.3) \times 10^{-7}$
[6.48]	0.95	3.0	7.882×10^{-4}	-6.7×10^{-7}
[6.54]	0.5	3.0	1.421×10^{-4}	$(-3.5 \pm 0.3) \times 10^{-8}$
[6.48]	0.5	3.0	1.422×10^{-4}	-4.12×10^{-8}

[a]   Results for $\eta = 0.95$ were obtained by linear interpolation from those given for $\eta = 0.9490$ and 0.9524.

mathematical bifurcation problem for Taylor vortex flow are *Yudovich* [6.49, 50], *Ivanilov* and *Iakovlev* [6.51], *Kirchgässner* and *Sorger* [6.52], and *Ovchinnikova* and *Yudovich* [6.53].

The quintic term in the amplitude equation (6.28), which allows determination of the Taylor vortex through terms $O(\varepsilon^2)$, has been calculated by *Di Prima* and *Eagles* [6.54]. The calculations of *Kirchgässner* and *Sorger* [6.48] were also carried to this order. The mean torque on the inner cylinder with the outer cylinder at rest is given by

$$\mathscr{G} = \frac{\pi h \varrho v^2 (1+\eta)\eta R}{2(1-\eta)^2} \, |G_1 + G_t| \, , \tag{6.29}$$

where h is the height of the cylinders, $G_1 = -8/(1+\eta)^2$ is the dimensionless torque due to Couette flow, and G_t is the dimensionless Taylor vortex torque. Following the notation of *Di Prima* and *Eagles*, G_t can be expressed in the form

$$G_t/G_1 = A(\eta, \alpha)\,[T - T_c(\eta, \alpha)]/2 + B(\eta, \alpha)\,[T - T_c(\eta, \alpha)]^2/4 \, . \tag{6.30}$$

The coefficients $A(\eta, \alpha)$ and $B(\eta, \alpha)$ should not be confused with the parameters A and B that appear in expression (6.1) for Couette flow. Values of $A(\eta, \alpha)$ and $B(\eta, \alpha)$ obtained by *Di Prima* and *Eagles*, by *Kirchgässner* and *Sorger* [6.48], and by *Reynolds* and *Potter* [6.55] are given in Table 6.3. All of the results are consistent for the values of A; however, there are differences in the values of B. We note that the values of B given by *Di Prima* and *Eagles* agree with those obtained by *Reynolds* and *Potter* for $\eta \to 1$, but do not agree with those obtained by *Kirchgässner* and *Sorger* for $\eta = 0.95$ and $\eta = 0.5$. Also, the value of B given for $\eta \to 1$ has been confirmed in an *independent* calculation ($B = -8.41 \times 10^{-9}$) by *Eagles* et al. [6.56].

A calculation such as we have described can be carried out for any value of α and for values of T (or R) slightly greater than the critical value of T for the value of α. Thus for a fixed value of T slightly greater than T_c (see Fig. 6.4), a Taylor vortex flow with axial wavelength $\lambda = 2\pi d/\alpha$ exists for each value of α in

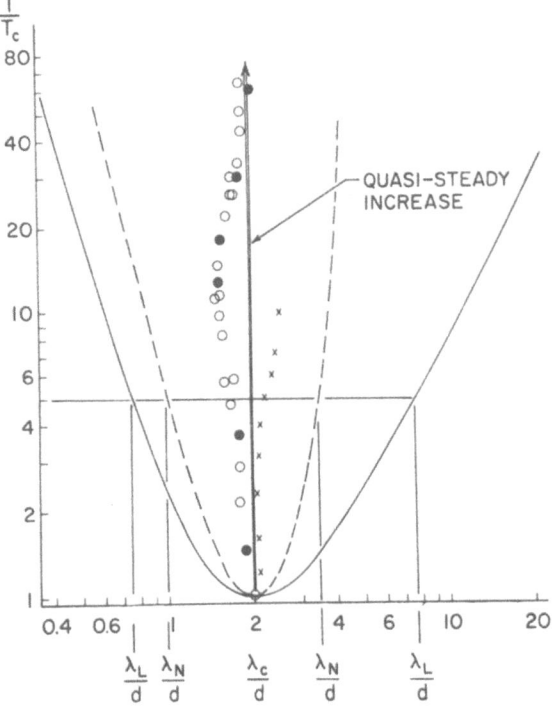

Fig. 6.5. Stability diagram for supercritical Taylor vortex flow with the observed wavelength ($\lambda = 2\pi d/\alpha$) of vortices for $\eta = 0.727$, $\mu = 0$. The solid line shows the neutral curve for the instability of Couette flow as given by *Chandrasekhar* [Ref. 6.30, p. 303], and the dashed line shows the stability boundary for Taylor vortex flow determined by *Kogelman* and *Di Prima* [6.42]. Experimental points: 0 – sudden starts, x – filling experiments (annulus was filled with fluid after inner cylinder was rotating at Ω_1) [Ref. 6.61, Fig. 6]

the interval $\alpha_- < \alpha < \alpha_+$. However, *Kogelman* and *Di Prima* [6.42], following *Eckhaus* [6.45, 57], showed that as $T \to T_c$, Taylor vortex flows with wave numbers outside a band of width $1/\sqrt{3}$ times the width of the neutral curve (centered on α_c) are unstable to axisymmetric disturbances; see Fig. 6.5. This result is consistent with the finding (based on reasonable assumptions) by *Kirchgässner* and *Sorger* [6.52] that branching solutions for $\alpha \neq \alpha_c$ are initially unstable.

In the analysis by *Kogelman* and *Di Prima*, the Taylor vortex flow for each value of α was computed to $O[T - T_c(\alpha)]$ where $T_c(\alpha)$ is the critical value of T corresponding to the given value of α. *Nakaya* [6.58] has considered the same stability problem[5] for $\eta = 0.5$, but with the Taylor vortex flow calculated to

---

5 For $\eta = 0.5$ axisymmetric Taylor vortex flows can exist for values of T much greater than T_c, whereas for $\eta \to 1$ (or near one) Taylor vortex flows become unstable for values of T slightly greater than T_c. This instability to a nonaxisymmetric disturbance leads to the wavy vortex flow shown in Fig. 6.1b. It will be discussed in the next section.

$O\{[T-T_c(\alpha)]^2\}$. He found that the band of wave numbers corresponding to stable Taylor vortex flows is somewhat narrower than the result given by *Kogelman* and *Di Prima*.

Snyder [6.59, 60], and *Burkhalter* and *Koschmieder* [6.61] have shown experimentally that by varying the way in which a final value of $T>T_c$ is reached (varying the initial conditions), Taylor vortex flows of different wave numbers can be obtained at the same value of T (see also [6.62]). However, the wave numbers corresponding to Taylor vortex flows obtained by *Snyder* for $\eta=0.5$ and by *Burkhalter* and *Koschmieder* for $\eta=0.727$ lie within the "$1/\sqrt{3}$" band[6]. The latter results are shown in Fig. 6.5. While the theory and the experimental results are noncontradictory, there is much that remains to be done on the problem of nonuniqueness of the final state at a given value of $T>T_c$. This problem is discussed in *Di Prima* and *Eagles* [6.54] and in several of the references cited in that paper.

In closing this discussion of Taylor vortex flows, we note that the stability analysis of *Kogelman* and *Di Prima* [6.42] and the original work of *Eckhaus* [6.57] is intricate and difficult to understand. *Stuart* and *Di Prima* [6.63] have clarified the analysis by studying a generalization of the amplitude equation (6.28) which allows for slow spatial variations in the axial direction of the disturbance. It is shown that the Eckhaus mechanism of instability of two-dimensional flows, which are periodic in one spatial dimension, and the *Benjamin-Feir* [6.64] instability mechanism of the two-dimensional Stokes water wave are related. In each case sideband perturbations resonate with the first harmonic of the fundamental mode of the periodic flow to mutually reinforce each other, leading to an instability.

Finally, the form of the expansion (6.27) for the velocity field has been confirmed experimentally. Since the equilibrium solution F_e is $O(\varepsilon)$, it follows from (6.27) that the fundamental mode has amplitude $O(\varepsilon^{1/2})$, the first harmonic and mean motion have amplitude $O(\varepsilon)$, the second harmonic and first correction to the fundamental have amplitude $O(\varepsilon^{3/2})$, and so on. *Donnelly* and *Schwarz* [6.65] used an ion conduction technique and found that the ion current increased as $\varepsilon^{1/2}$, as expected. *Snyder* and *Lambert* [6.66] measured the total shear at the inner cylinder using hot thermistor anemometers and also found that the fundamental mode had amplitude $O(\varepsilon^{1/2})$. *Gollub* and *Freilich* [6.67], using the laser Doppler technique for an apparatus with $\eta=0.612$, measured the radial component of the velocity over a large range of Reynolds number. They found that the amplitude of the fundamental mode was $O(\varepsilon^{0.50\pm0.03})$ with correction $O(\varepsilon^{1.42\pm0.05})$, in good agreement with theory. However, the amplitude of the first harmonic was observed to be $O(\varepsilon^{0.67\pm0.05})$ rather than $O(\varepsilon)$.

The growth or decay rate of a disturbance as predicted by (6.28) is $O(\varepsilon)$. *Donnelly* and *Schwarz* found that when R was suddenly increased beyond R_c,

---

6 The wave numbers obtained by *Snyder* and by *Burkhalter* and *Koschmieder* also lie within the narrower band calculated by *Nakaya* [6.58].

the approach to steady state was in time $O(\varepsilon)$, as predicted; on the other hand, a different behavior was observed when R was suddenly decreased. However, the more extensive measurements of *Gollub* and *Freilich* showed that the growth and decay rates were both $O(\varepsilon^{1.0\pm0.03})$, over a wide range in R above and below R_c.

6.4 Wavy Vortex Flow

As was noted in Sect. 6.1, as the speed of the inner cylinder is increased ($\mu=0$), a second critical Reynolds number R_c' (or Taylor number T_c') is reached at which the axisymmetric Taylor vortex flow becomes unstable. This instability leads to a wavy vortex flow as shown in Fig. 6.1b. Experiments show that R_c' depends strongly on the radius ratio η; it is about $1.05R_c$ to $1.10R_c$ for $\eta \simeq 0.95$ and is very much larger, $10R_c$ or greater, for $\eta=0.5$ [6.60, 66, 68]. *Cole* [6.69] has shown that R_c' is also dependent on $h/(b-a)$, where h is the height of the cylinders[7]. In his experiments, critical speeds for the onset of Taylor vortices and the later development of wavy vortices were determined by torque measurements and visual observations. He observed that values of $h/(b-a)$ greater than 40 are required before R_c' is within a few percent of the value calculated for infinite cylinders. *Walden* and *Donnelly* [6.29], and *Walden* [6.70] have also found that the Reynolds numbers for some higher transitions depend on a $h/(b-a)$.

The wavy vortex flow has been extensively studied photographically by *Coles* [6.6] in an apparatus with $\eta=0.874$ and $h/(b-a)=27.9$, which accommodates no more than 32 cells. Let a given flow be denoted by p/m where p is the number of Taylor vortices and m is the number of azimuthal waves. Then *Coles* found at a rising sequence of definite, and repeatable, speeds of the inner cylinder ($\mu=0$) the following sequence of states: 28/0 (Taylor vortices), 28/4 (wavy vortices at about $1.5T_c$ or $1.2R_c$), 24/5, 22/5, 22/6, 22/5, 22/4, 22/0. In all cases for which $m \neq 0$, the boundaries between neighboring cells were wavy. The angular wave speed was about equal to $0.5\Omega_1$ (i.e., the average angular velocity between the cylinders) at the first appearance of the wavy vortices, but decreased to about $0.34\Omega_1$ as the speed of the inner cylinder was increased. *Coles* also observed that in the range of speeds for which wavy vortex flows are possible, different states p/m could be attained at the same final speed, the state depending upon the manner in which the final speed was reached. This nonuniqueness of the flow is discussed further in the next section.

Schwarz et al. [6.71], using an apparatus with $\eta=0.95$ and $h/(b-a)=261$, observed a transition to a nonaxisymmetric state with $m=1$ rather than 4 as observed by *Coles*. This mode, which became evident at about 3 to 8% above T_c, appeared to be a subtle modification of the Taylor vortex mode, having a

---

7 In the same experiments *Cole* also observed that the value of R_c at which axisymmetric Taylor vortex flow occurs is rather insensitive to annulus length.

Table 6.4. Critical values of the Taylor number T, the axial wave number α, and the phase velocity $\text{Im}\{\beta\}/m\Omega_1$, for $\mu=0$ with $\eta\to 1$ for the linear stability of Couette flow as given by Krueger et al. [6.14]

m	α_c	T_c	$\text{Im}\{\beta\}/m\Omega_1$	$T_c(m)/T_c(m=0)$
0	3.127	3390.1	0	1
1	3.131	3402.5	0.5262	1.0037
2	3.143	3440.3	0.5265	1.0148
3	3.163	3504.8	0.5270	1.0338
4	3.190	3598.6	0.5277	1.0615
5	3.225	3725.6	0.5285	1.0990

regular vortex spacing in the axial direction with planes perpendicular to the axis of the cylinders separating neighboring vortices on which the axial component of the velocity vanished. As T was increased, wavy vortex boundaries became evident at $T/T_c \simeq 1.2$.

Mobbs and co-workers [6.72–75], using an apparatus with $\eta=0.91$ and $h/(b-a)=65$, observed waves with $m=1$ appearing in the center of the vortex cell at $T/T_c=1.03$; the vortex boundaries remained straight. The vortex boundaries began to oscillate at $T/T_c=1.08$; this transition was marked by a slight decrease in the slope of the torque-speed curve. At $T/T_c=1.23$ there was a transition in azimuthal wave number from $m=1$ to $m=3$, where again there was a decrease in the slope of the torque-speed curve.

The linear problem for the stability of Couette flow to nonaxisymmetric disturbances has been studied by *Di Prima* [6.76] and by *Krueger* et al. [6.14]. The form of the disturbance in the original physical variables is

$$u'(r, \theta, z, t) = u(r)\exp[\beta t + i(\gamma z + m\theta)] , \qquad (6.31)$$

with similar expressions for the other velocity components and the pressure. In contrast to the Taylor vortex problem, the eigenvalue β will now be complex corresponding to a wave travelling in the azimuthal direction with phase angular velocity $\text{Im}\{\beta\}/m\Omega_1$.

In the papers by *Di Prima* and by *Krueger* et al. the small gap limit $(\eta\to 1)$ is employed. As a consequence it is necessary to scale the azimuthal variable so that the wave number is treated as a continuous variable. The continuous variable can be related to integer values of m in (6.31) by choosing a value of η near one. For the outer cylinder at rest $(\mu=0)$ it was found, as expected, that the critical Taylor number for nonaxisymmetric disturbances was higher than the critical Taylor number for axisymmetric disturbances. However, the increase is very small. Also, the critical Taylor number appears to be a monotone increasing function of the azimuthal wave number. The results obtained by *Krueger* et al. are recorded in Table 6.4. The values of the azimuthal wave number m correspond to $\eta = 0.952$. For μ sufficiently negative ($\mu < -0.78$ when $\eta \to 1$), *Krueger* et al. found the critical disturbance was nonaxisymmetric. In particular, for $\mu = -1$ and $\eta = 0.95$ the critical disturbance has four waves in the azimuthal direction.

The instability of the Taylor vortex to wavy vortex disturbances was studied by *Davey* et al. [6.27] for the case $\mu = 0$ with $\eta \to 1$. We shall briefly summarize that work using the notation of Sect. 6.3 and with reference only to the azimuthal component of velocity. They considered the interaction of four primary modes,

$$v_{c10}(x, \tau) \cos \alpha \zeta = F_c(\tau) f_0(x) \cos \alpha \zeta \ ,$$

$$v_{s10}(x, \tau) \sin \alpha \zeta = F_s(\tau) f_0(x) \sin \alpha \zeta \ ,$$

$$v_{c11}(x, \tau) e^{im\theta} \cos \alpha \zeta = G_c(\tau) g_0(x) e^{im\theta} \cos \alpha \zeta \ ,$$

$$v_{s11}(x, \tau) e^{im\theta} \sin \alpha \zeta = G_s(\tau) g_0(x) e^{im\theta} \sin \alpha \zeta \ .$$

(6.32)

The first two modes are Taylor vortex modes; the latter two (with their complex conjugates) are nonaxisymmetric modes. The functions f_0 and g_0 are solutions of the linear eigenvalue problems for axisymmetric and non-axisymmetric disturbances, respectively. The amplitude functions $F_c(\tau)$ and $F_s(\tau)$ are real, while $G_c(\tau)$ and $G_s(\tau)$ are complex valued. Nonlinear interactions produce first harmonics of the primary modes and a mean motion with magnitude proportional to products of the amplitudes. These terms interact with the primary modes to correct the x dependence of the primary modes and to generate second harmonics at cubic order in the amplitudes.

The four amplitude functions F_c, F_s, G_c, and G_s satisfy a set of four coupled first-order nonlinear autonomous equations which can be consistently truncated at cubic terms in the amplitudes. The coefficients in these equations are determined as part of the formal expansion procedure; they are constants, except for the coefficients of the linear terms which depend on T. Depending upon the values of the coefficients, several different flows are possible. Of special interest to us is the Taylor vortex flow ($G_c = G_s = 0$, with F_s a multiple of F_c), and wavy vortex flow ($F_s = G_c = 0$; F_c and G_s are nonzero). The $\cos \alpha \zeta$ dependence of the Taylor vortex mode and the $\sin \alpha \zeta$ dependence of the nonaxisymmetric mode assures that the wavy vortex mode is indeed a wavy vortex.

Davey et al. [6.27] showed that for a fixed value of α (chosen as α_c) the Taylor vortex flow exists for $T > T_c$, but becomes unstable at T'_c which depends on m and is only slightly greater than T_c. Moreover, for $T > T'_c(m)$ a wavy vortex flow with m waves in the azimuthal direction exists and is stable. The *Davey* et al. calculations were extended to fifth order in the amplitudes by *Eagles* [6.77]. Also, he carried out the calculations with and without the small gap approximation. In Table 6.5 we record the values of $T'_c(m)$ obtained by *Eagles* at which Taylor vortex flow becomes unstable. For the case of small gap, his

Table 6.5. Values of $T'_c(m)$ at which Taylor vortex flow with $\alpha = 3.127$ becomes unstable for $\mu = 0$ and $\delta = 0.05$ according to Eagles [6.77]. The figures in parentheses are percentages above T_c

Transition	Small gap approximation	Full equations
T_c	3390	3506
$T'_c(m=1)$	3670 (8 %)	3892 (11 %)
$T'_c(m=2)$	3676 (8 %)	3902 (11 %)
$T'_c(m=4)$	3710 (9 %)	3692 (12 %)

results are very similar to those obtained by *Davey* et al. Also, we note that since the Taylor number is proportional to the square of the speed of the inner cylinder Ω_1, the critical speed at which Taylor vortex flow becomes unstable when the gap between the cylinders is small is only about 5 % above Ω_{1c}.

Nakaya [6.78] has carried out similar calculations. In his analysis he did not require that the axial wave numbers of the axisymmetric and non-axisymmetric modes be the same. However, he found that the nonaxisymmetric disturbance grows most rapidly when it has the same axial wave number as that of the Taylor vortex flow and differs from the Taylor vortex flow in phase by $\pi/2$. He also calculated the second stability boundary T'_c at which Taylor vortex flow becomes unstable as a function of μ and compared this curve with the first stability boundary at which Couette flow becomes unstable and Taylor vortex flow occurs. For $d/a = 0.05$ ($\delta = 0.0488$) he found that the curves intersect at $\mu = -0.78$. This is consistent with the calculations of *Krueger* et al. [6.14] (mentioned earlier in this section) for the linear stability of Couette flow to nonaxisymmetric disturbance.

While the nonlinear theory shows the instability of the Taylor vortex flow and the occurrence of the wavy vortex flow, it is incomplete in determining the wavy vortex flow that will occur. It is clear from Table 6.5 that the values of $T'_c(m)$ for $m = 1, 2$, and 4 are grouped very closely together with $m = 1$ giving the lowest value of T'_c. The experiments of *Coles* [6.6] suggest that a wavy vortex flow with $m = 4$ is preferred as T is increased slowly to values greater than T'_c. However, since the stability analysis is based on the interaction of a Taylor vortex mode with a nonaxisymmetric mode with a definite azimuthal periodicity, it is not possible to draw a conclusion about the emergent mode when the critical Taylor numbers for the different vortex flows are grouped so closely. Further analysis involving nonlinear interactions with *several* nonaxisymmetric modes of different azimuthal periodicity is required.

Eagles [6.79] has calculated the torque corresponding to wavy vortex flows with $m = 1, 2, 3$, and 4 for $\delta = 0.05$. His results and the experimental data are shown in Fig. 6.6. It is clear that for T (or R) beyond the point at which the Taylor vortex flow becomes unstable, the $m = 4$ wavy vortex flow is in good

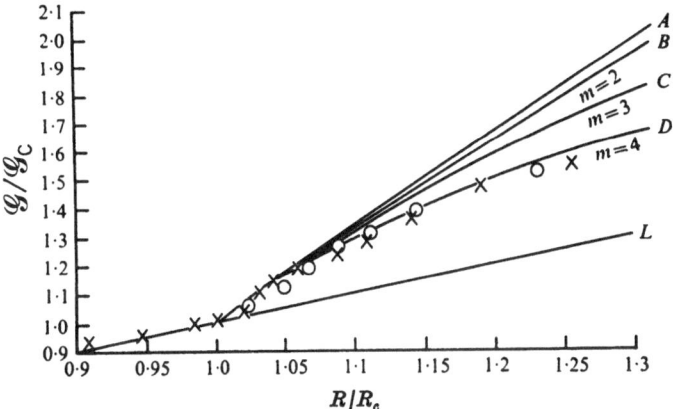

Fig. 6.6. The relative total torque for $\eta=0.95$, $\mu=0$. Curve L shows the laminar Couette torque. Curve A shows the total torque with Taylor vortices. Curves B, C, and D show the total torque with wavy vortices for $m=2, 3$, and 4. Experimental points: $\times - $ [6.81]; $\circ - $ [6.68]. [Ref. 6.79, Fig. 2]

agreement with the torque measurements of *Donnelly* [6.80] (also see [6.81]) and *Debler* et al. [6.68]. This certainly suggests that the $m=4$ nonaxisymmetric mode does indeed dominate the other nonaxisymmetric modes to produce a wavy vortex flow with four waves in the azimuthal direction.

6.5 Higher Instabilities and Turbulence

It should be clear by now that the nonlinear analyses which successfully described the growth of Taylor vortices and the onset of wavy vortex flow are too difficult in practice to be extended to study higher transitions and turbulence. Almost all of our knowledge of the flow for Reynolds numbers beyond the onset of wavy vortex flow comes from a few experiments. We shall summarize the experimental results and then discuss recent numerical studies of simple models which exhibit some of the features of the experimental data.

Experimental studies of the transition to turbulence in the flow between concentric cylinders with the inner cylinder rotating have primarily either used flow visualization techniques or have determined the power spectrum of some property of the flow. The major experiments are summarized in Table 6.6. These studies were all performed for cylinders with radius ratios $\eta \simeq 7/8$. The transition to Taylor vortex flow has been studied for radius ratios as small as $\eta=0.20$ [6.82], but for small radius ratios the transition to wavy vortex flow has not been observed even at fairly large Reynolds number. Therefore, the transition to turbulence for radius ratios significantly different from $\eta=7/8$ is likely to be qualitatively different from that observed in the experiments done thus far.

Table 6.6. Experiments on the transition to turbulence between concentric cylinders with the inner cylinder rotating ($\eta \simeq 7/8$)

Authors	Primary techniques	Radius ratio a/b	R_c for infinite cylinders [6.36]	Aspect ratio $h/(b-a)$	Maximum R/R_c studied	R/R_c for the appearance of noise[a]	R/R_c for the disappearance of azimuthal waves
Schultz-Grunow and Hein [6.83]	Flow visualization	0.840	105.1	50	200	≈ 10	Between 18 and 38[b]
Coles [6.6]	Flow visualization Hot wire anemometry	0.874 0.889	117.7 125.1	27.9 30	15 25	≈ 11 –	– 23
Gollub and Swinney [6.28], Fenstermacher et al. [6.7]	Laser Doppler spectra and flow visualization	0.877	119.1	20.0	45	≈ 12	21.9
Walden and Donelly [6.29]	Ion current spectra	0.875	118.2	18 to 80	67	≈ 11	21–25[c]
Bouabdallah and Cognet [6.92]	Electrochemical current spectra and flow visualization	0.909[d]	137.8	40	70	≈ 12	19.5
Koschmieder [6.5]	Flow visualization	0.896[e]	129.1	123	62	≈ 10	26
Mobbs et al. [6.75], Barcilon et al. [6.93]	Hot film spectra and flow visualization	0.908	137.1	65	320	≈ 5.5	21

[a] In flow visualization experiments this is the Reynolds number at which the flow appeared to develop small-scale structure; in experiments where power spectra were obtained this is the Reynolds number at which a broad spectral component was first observed.

[b] Waves were present at $R/R_c = 18$ and had disappeared at $R/R_c = 38$.

[c] The Reynolds number at which the waves disappeared increased monotically with aspect ratio.

[d] Some measurements were also made with $a/b = 0.820$ and 0.954.

[e] Some measurements were also made with $a/b = 0.727$.

6.5.1 Flow Visualization Experiments

Taylor [6.4] used dye to observe the vortex flow pattern. He found that for large Reynolds number the flow became time dependent, but the dye mixed in the fluid fairly rapidly so it was not possible to observe steady-state time-dependent flows. Some general features of the time-dependent flow were revealed by the beautiful photographic studies of *Schultz-Grunow* and *Hein* [6.83] and *Coles* [6.6], who suspended flat particles (aluminum paint pigment) in the fluid[8]. (This method of flow visualization, which was used to obtain the photographs in Fig. 6.1, had been used earlier, but less effectively, by *Lewis* [6.84].) At higher Reynolds number, $R/R_c \gtrsim 11$, *Schultz-Grunow* and *Hein*, and *Coles* found that the wavy vortices began to appear noisy, as shown in Fig. 6.1c. At yet higher Reynolds number the azimuthal waves disappeared (Fig. 6.1d), while the axial periodicity persisted even at the highest Reynolds number studied, $R/R_c \simeq 200$. *Townsend* [6.85] has recently observed that well-defined Taylor vortices exist even for $R/R_c \gtrsim 1000$.

As mentioned in the previous section, *Coles* discovered that a flow state, characterized by the axial and azimuthal wave numbers, is *not* a unique function of the Reynolds number and boundary conditions[9]. The different spatial states were achieved by approaching the final Reynolds number with different acceleration rates and by rotating and then stopping the outer cylinder. The number of axial vortices ranged from 18 to 32, corresponding to axial wavelengths from 1.74d to 3.10d, and the number of azimuthal waves ranged from 3 to 7. Subsequently *Snyder* [6.59, 60] and *Burkhalter* and *Koschmieder* [6.87] found that the time-independent Taylor vortex flow is also not unique; in their apparatus several different axial states were stable at some Reynolds numbers. Recently, *Benjamin* [6.88, 89] has observed different spatial states in Taylor vortex flow even in an annulus so short that only three or four vortices could be accommodated.

In retrospect it is not surprising that there exist multiple stable states. Uniqueness of the solutions of the Navier-Stokes equations has been proved only for very small Reynolds number [6.8], and nonlinear systems often exhibit multiple stable solutions; nevertheless, we know of no other physical system for which the existence of multiple stable states has been so vividly demonstrated.

Coles measured the frequency ω_1 of the azimuthal waves and found that ω_1/m was a universal function of Reynolds number, independent of the azimuthal mode number m. The frequency ω_1/m decreased from $0.5\Omega_1$ at the onset of wavy vortex flow to $0.34\Omega_1$ at "the present limit of observation $[R/R_c \simeq 22]$, where the noise level is so high that a dominant frequency for the

---

8 *Coles* produced an excellent movie, "Transition in Circular Couette Flow", which is available for $5.00 rental (in 1980) from the Engineering Societies Library, 354 East 47th Street, New York, New York 10017. The catalog number is C-2.

9 The nonuniqueness of the flow between concentric rotating cylinders was apparently first noted by *Pai* [6.86], whose hot wire anemometry studies of turbulent Taylor vortices revealed in 1943 two distinct types of flow at the same Reynolds number, depending in the starting conditions.

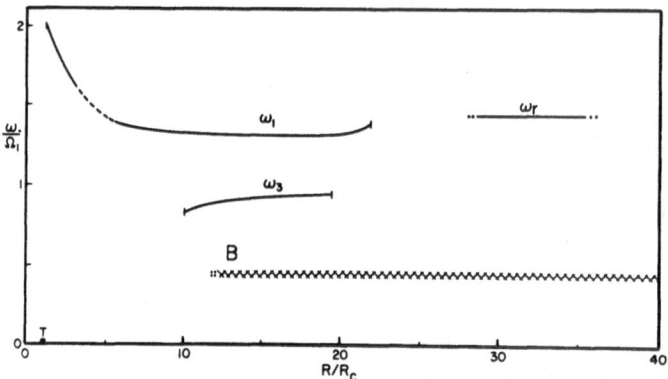

Fig. 6.7. Frequencies observed in the flow between concentric cylinders with the inner cylinder rotating; $\eta = 0.88$, $h/(b-a) = 25$. The curves describe a system with 4 azimuthal waves in the range where the azimuthal waves exist, $R'_c < R < 23 R_c$. "T" indicates the region where there are stable time-independent Taylor vortices, $R_c < R < R'_c$. The graph is based on data obtained from several independent experiments with slightly different boundary conditions; see the discussion in the text

large scale motions can no longer be isolated by conventional filtering techniques" [Ref. 6.6, p. 403]. Recent measurements of the velocity power spectrum, to be described later, agree well with Coles' results for ω_1, except that they show a small variation (1 to 2 %) in ω_1/m for different modes m [6.90]; see also [6.91]. In addition, the spectra show that the amplitude of the ω_1 mode in the spectra goes to zero at a well-defined Reynolds number [6.7], which is about $R/R_c = 22$ but which varies slightly with $h/(b-a)$ [6.29]; see Table 6.6.

The Reynolds number dependence of ω_1 for a state with 4 waves is shown in Fig. 6.7. The curve for ω_1 is dashed in the range $2.3 < R/R_c < 4.1$ because *Coles* found that the 4 wave state was not stable there. States with 5, 6, and 7 waves were found to be stable in that region, and recently *Bouabdallah* and *Cognet* [6.92] and *Mobbs* et al. [6.75] have found stable states in this region with 8 to 10 waves.

With increasing Reynolds number *Coles* observed that the flow pattern gradually becomes more complex, and he concluded that "the discrete spectrum changes gradually and reversibly to a continuous one by broadening of the initially sharp spectral lines" [Ref. 6.6, p. 385]. While this conclusion is not supported by the power spectra obtained later, his observation of the "first appearance of randomness" [Ref. 6.6, Fig. 19e] at $R/R_c \simeq 11.4$ is in good agreement with the spectroscopic studies.

Recently *Barcilon* et al. [6.93] have reported flow visualization studies extending up to $R/R_c = 320$ in a system with $\eta = 0.908$ and $h/(b-a) = 65$. The "herring-bone" like streaks observed in the photographs for $R/R_c \gtrsim 20$ led them to conjecture that Görtler vortices and Taylor vortices coexist on two widely separated length scales. The smaller length scale, corresponding to the characteristic distance between streaks, agrees quantitatively with the predicted

Görtler vortex spacing; however, high resolution stop-action photographs such as Fig. 6.1d[10] do not show the particle alignment expected if Görtler vortices were present, while such photographs clearly show that the particles align with the Taylor vortex pattern even at much larger Reynolds number.

Although there are stable flows of different axial wavelengths at each Reynolds number, there is a tendency for the wavelength to increase with increasing R [Ref. 6.6, p. 417]. *Koschmieder* [6.5] (see also [6.61]) has found that the wavelength increases with increasing R only up to $R/R_c \simeq 10$, and then remains constant or decreases with further increase in R, depending on the way that the final R is approached[11].

6.5.2 Studies of the Flow Spectrum

Different dynamical regimes of a flow can be distinguished by examining high-resolution power spectra of a time-dependent property of the flow. Transitions that are obvious in the power spectra, such as the broadening of a spectral line or the appearance of a new characteristic frequency in the flow, can be undetected in a direct inspection of the time records or flow photographs. Thus power spectra have become a major tool for the the study of the transition from laminar to turbulent flow.

The power spectrum of a time series is the squared modulus of its Fourier transform. Discrete Fourier transforms of digital records can easily be performed using the Cooley-Tukey algorithm on a modern minicomputer (see, e.g., [Ref. 6.95, Chap. 6]). The resolution $\Delta\omega/\omega_{max}$ can be made quite high ($\sim 10^{-4}$) by obtaining long data records with many samples. (If no averaging is employed to reduce noise, $\Delta\omega \simeq 2/t$ and $\omega_{max} = n/2t$, where n is the number of samples and t is the duration of the record [Ref. 6.95, Sects. 8.3, 8.4]). Thus changes in the dynamics of a flow can be detected with high sensitivity.

Power spectra have long been used in studies of strong turbulence, but the first use to study the transition to turbulence was in an analysis of heat flux measurements in Rayleigh-Bénard convection [6.96]. Subsequently, velocity power spectra were obtained for the flow between concentric cylinders [6.7, 28, 97]. The fluid velocity was determined from measurements of the Doppler shift of scattered laser light (see, e.g., [6.98, 99]). The Doppler shifts were typically $\sim 10^5$ Hz while the characteristic frequencies of the fluid were ~ 0.1 to 10 Hz, so measurements of the Doppler shift in short time intervals yielded essentially the instantaneous fluid velocity. In the laser Doppler technique any velocity component can be selected by appropriate choice of the

---

10   The exposure time of 50 µs gives a streak length no longer than the particle size ($\simeq 25$ µm).

11   For η appreciably smaller than 7/8 the flow behavior can be quite different from that described in this section. For example, *Burkhalter* and *Koschmieder* [6.87] found in a system with $\eta = 0.727$ that the vortices persist with unvarying wavelength to quite large R when R is increased quasi-statically, and *Snyder* [6.94] observed a variety of different waveforms in flow visualization studies of systems with η varying from 0.20 to 0.96.

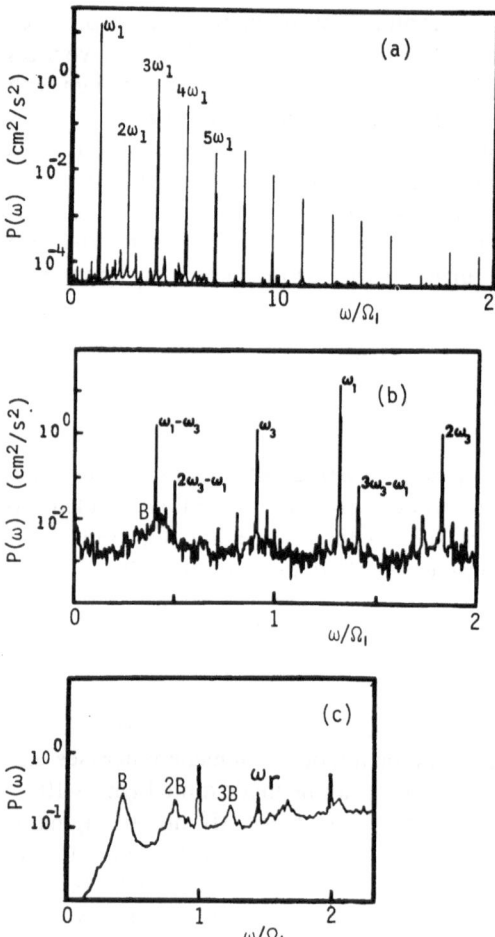

Fig. 6.8a–c. Power spectra for time-dependent Taylor vortex flow. (a) $R/R_c = 5.6$: periodic flow [Ref. 6.7, Fig. 7a]. All components are harmonically related to ω_1; the first 5 harmonics are labeled. (b) $R/R_c = 15.1$: weakly turbulent flow with two discrete fundamental frequencies, ω_1 and ω_3, and a broad component B [Ref. 6.7, Fig. 5d]. (c) $R/R_c = 28$: weakly turbulent flow with the broad component B and a discrete component ω_r, ([Ref. 6.29, Fig. 4d]; the components at $\omega/\Omega_1 = 1$ and 2 are instrumental artifacts). In (a) and (b) the aspect ratio is 20; in (c) it is 80

light scattering geometry; in these experiments the radial component of the velocity was determined.

For Reynolds numbers in the range from the onset of wavy vortex flow up to $R/R_c = 10.1$ the power spectrum contained only a single frequency component and its harmonics, as shown in Fig. 6.8a; hence this flow is strictly periodic. Note that the amplitude of the fundamental is more than 5 orders of magnitude above the instrumental noise level. The same *frequencies* were observed for different positions of the scattering volume within the annulus, although the *amplitudes* of the spectral components varied with the position of the scattering volume.

As the Reynolds number was increased above $R/R_c = 10.1$, a second fundamental frequency of the steady-state flow appeared in the power spectrum[12,13]. This component was labeled ω_3 since a transient called ω_2 was observed at lower Reynolds numbers. Within the experimental resolution the amplitude of the component ω_3 grew continuously from zero as the Reynolds number was increased above $R/R_c = 10.1$; therefore, this transition appears to be continuous (a supercritical bifurcation). Consistent with this observation was the absence of hysteresis – the transition occurred at the same Reynolds number with increasing and decreasing Reynolds number.

The frequency ratio ω_3/ω_1 increased monotonically with increasing Reynolds number (see Fig. 6.7). Hence the frequencies ω_1 and ω_3 appear to be incommensurate; such a system is termed quasi-periodic because it has approximate but not exact recurrence.

As the Reynolds number was increased above $R/R_c \simeq 12$, there appeared in the spectrum a weak broad component, labeled B in Fig. 6.8b. The flow could no longer be described by a small number of well-defined characteristic frequencies; the flow should be described as choatic or turbulent. However, it should be noted that the noise component B at this point contains less than 1 % of the total spectral energy, while essentially all of the remainder resides in the sharp peaks.

It is important to emphasize the essential qualitative change in the behavior of the system when the broad component appears in the spectrum. When only discrete frequencies are present, the behavior of the system is in principle known *for all times*, but when even a small amount of noise or randomness appears, as evidenced by the broad spectral component, we can no longer hope to make accurate predictions of future velocity values.

As the Reynolds number was increased further, the amplitudes of the components at ω_3 and ω_1 decreased, and these components disappeared at $R/R_c = 19.3$ and 21.9, respectively, leaving a spectrum with only the broad component B on top of a background spectral continuum. The transitions marking the disappearance of ω_3 and ω_1 showed no hysteresis within the Reynolds number resolution. No other transitions were observed [6.7].

The ω_3 mode, first observed in velocity power spectra [6.28], has been identified in recent flow visualization experiments by *Gorman* and *Swinney* [6.100]. The amplitude of the waves of the vortex outflow boundaries was found to oscillate at the same frequency as the component called ω_3 in the velocity power spectra. The ω_3 mode was observed over a range of Reynolds number for all azimuthal wave numbers ($m = 3, 4, 5, 6$) and annulus heights ($16 \leq h/d \leq 44$) that were investigated.

Walden and *Donnelly* [6.29] and *Walden* [6.70] have studied the transitions in the flow between concentric cylinders by an entirely different measurement

---

12 The transition Reynolds numbers given here are those observed for a flow with 17 axial vortices and, where the azimuthal waves exist, 4 azimuthal waves [6.7].

13 The "doubly periodic" flow described by *Coles* [6.5] had two *spatial* frequencies, axial and azimuthal; the *temporal* frequencies were not determined. The measurements described here show that the spatially doubly periodic flow is characterized by one temporal frequency for $R/R_c < 10.1$ and two temporal frequencies for $10.1 < R/R_c < 19.3$.

technique. Their outer cylinder (radius 2.54 cm) had 1 mm diameter ion collectors embedded in the wall. The ion current ($\approx 10^{-13}$ A) between a collector and the gold-plated inner cylinder was recorded as a function of time in a computer and the records were Fourier-transformed to obtain ion current power spectra. The ion current provides a local probe (volume $\approx 5 \times 10^{-5}$ cm^3) of the radial component of the velocity in a narrow boundary layer at the outer cylinder.

The ion current power spectra that *Walden* and *Donnelly* obtained [Ref. 6.29, Fig. 2] for a fluid height to gap ratio of 20 contained the same frequency components ω_1, ω_3, and B that were observed by *Fenstermacher* et al. [6.7] in the velocity power spectra for a system with height to gap ratio of 20; also, the Reynolds numbers for the transitions agreed within the experimental uncertainty.

Additional studies were performed by *Walden* and *Donnelly* for $h/(b-a)$ ranging from 18 to 80. The Reynolds number at which the spectral component at ω_1 disappeared was found to increase monotonically from $R/R_c = 22$ to 26 as $h/(b-a)$ was increased. Moreover, for $h/(b-a)$ greater than 25 a new spectral component was observed in the power spectrum in the range $28 \lesssim R/R_c \lesssim 36$. This component is labeled ω_r in Figs. 6.7 and 6.8c. This component has not been identified with any feature in the flow photographs. In the region where ω_r is present the flow is turbulent, as is clear from the presence of B, the broad background continuum in the power spectra, and the noisy appearance of the flow in the photographs. The frequency ω_r may correspond to a periodic large scale structure of the type often observed in turbulent flows (see, e.g., [6.101]).

Bouabdallah and *Cognet* [6.92] have obtained power spectra of the time-dependent flow between concentric cylinders using another measurement technique. The gradient of the velocity in the radial direction (at the inner cylinder wall) was determined from measurements of the current in an electrolytic reaction, the reduction of ferricynide ion on a nickel electrode [6.102]. The cylinder contained an array of electrodes in both the vertical and azimuthal directions. The power spectrum of the current was determined with an electronic spectrum analyzer rather than by Fourier-transforming a digital record. The spectral resolution was limited to 256 points (compared with several thousand in the previously cited studies), but the spectra appear generally consistent with those of *Fenstermacher* et al. [6.7], and *Walden* and *Donnelly* [6.29]. For example, the frequencies ω_1, ω_3, and B were apparently all present, and B appeared at $R/R_c \simeq 12$ and ω_1 and ω_3 disappeared at $R/R_c \simeq 20$; however, the component which is presumably ω_3 appeared at $R/R_c \simeq 5$ rather than 10 as found by *Fenstermacher* et al.

An interesting new result was found by *Bouabdallah* and *Cognet* [6.92] from a study of the axial dependence of the spectrum. They observed that in the range where the flow first becomes turbulent, $R/R_c \gtrsim 12$, the spectral density of the background continuum is greater and extends to higher frequencies at axial positions corresponding to velocity maxima than at other axial positions. Thus turbulence appears to originate at the vortex boundaries corresponding to fluid outflow. Only at much greater Reynolds number did the spectrum become essentially independent of the axial position.

Recently *Mobbs* et al. [6.75] have studied wavy vortex flow using hot film probes. Power spectra of the digitized probe signals show that as the Reynolds number at which a change in azimuthal wave number from m to $m \pm 1$ is approached, there is a range of R for which the modes m and $m \pm 1$ (and sometimes $m \pm 2$ and $m \pm 3$) coexist; within this range the strongest peak in the power spectrum often corresponds to $m = 1$. *Mobbs* et al. also found that in the region $5.5 \lesssim R/R_c \lesssim 8.4$ the power spectra indicate that there are several wave number changes, and "the flow appears visually to be chaotic and could easily be described as turbulent".

All the measurements of power spectra that we have described were made on annuli with $\eta \simeq 7/8$. Recently *Zhuravel* et al. [6.103] have used the laser Doppler velocimetry technique to study transitions in an annulus with $\eta = 0.636$ (see also [6.104]). Their velocity spectra show: i) at $R/R_c = 13$, a spectral component appears with frequency $\omega_1/\Omega_1 = 1.9$; ii) at $R/R_c = 16$, a second component appears with frequency $\omega_2/\Omega_1 = 0.58$; iii) at $R/R_c = 17$, a third component appears with frequency $\omega_3/\Omega_1 = 0.36$; iv) as R is increased further, ω_3 disappears, but at $R/R_c = 17.4$ another component appears, $\omega_4/\Omega_1 = 0.95$; v) the frequency component corresponding to the waves, ω_1, apparently begins to broaden even before ω_2 appears, so the flow is chaotic before the components ω_2, ω_3, and ω_4 appear. For $R/R_c \gtrsim 22$ all spectral components are broad and for $R/R_c \gtrsim 110$ there is a broad spectral continuum with no peaks.

6.5.3 Summary of the Experiments

A remarkably clear and consistent picture of the transition to turbulence has emerged from the experiments we have described (for annuli with $\eta \simeq 7/8$), as summarized in Table 6.6 and Fig. 6.7. In wavy vortex flow there are many distinct stable spatial states p/m with different numbers of axial vortices and azimuthal waves, depending on the height and radius ratio of the annulus and the initial conditions of the velocity field. As the Reynolds number is varied, a state p/m can become unstable, resulting in a transition to a state p'/m', where usually $\Delta p = \pm 2$ or ± 4 and $\Delta m = 0$; or $\Delta p = 0$ and $\Delta m = \pm 1$ or ± 2. If the direction of change in Reynolds number is now reversed, the new state p'/m' will generally remain stable far into the region where the system was formerly in state p/m. Thus transitions between different spatial states exhibit large hysteresis. However, the frequency ω_1/m is the same within a percent or two for all spatial states.

Experiments in which power spectra of time-dependent flow properties have been obtained show that the flow in the region $R_c' < R < 10R_c$ is strictly periodic. The time dependence is entirely characterized by a single fundamental frequency and its harmonics. The second fundamental frequency of the steady state flow appears at $R/R_c = 10$ and disappears at $R/R_c = 19.3$, while both the spectral and flow visualization experiments show that ω_1 (the azimuthal wave frequency) persists up to $R/R_c = 22$ [14]. For a given axial state these transitions are nonhysteretic within the experimental resolution. For $h/(b-a)$ greater than 25, another frequency component appears in the range $28 \lesssim R/R_c \lesssim 36$.

The photographic and spectral measurements all show a qualitative change in the flow behavior at $R/R_c \simeq 12$, where the photographs first begin to show noise (small scale structure; see Fig. 6.1c), and the power spectra show the appearance of a broad spectral line. Although the system at this point is certainly to a large degree ordered in space and time, the experimental indicators of disorder all mark $R/R_c \simeq 12$ as the Reynolds number corresponding to the onset of chaotic or turbulent flow. Some spatial and temporal order remains even at the highest Reynolds number studied, where the photographs still show the axial vortices, and the spectra still contain a well-defined feature, the broad component B, in addition to the background continuum.

6.5.4 Model Systems

Some insight into the transition to chaos can be gained from numerical analyses of model systems with a small number of modes. The classic prototype of such studies is the investigation by *Lorenz* [6.105] of a three-variable model of Rayleigh-Bénard convection, described in Chap. 4. Several other nonlinear model systems are discussed in Chap. 9.

Most of the sets of coupled nonlinear equations with three or more dependent variables that have been investigated exhibit chaotic behavior for some range of the parameter space. Therefore, the goal in constructing and studying model systems is not simply to find chaotic behavior but to discover a model that exhibits the qualitative and even quantitative features of the experiments. Ideally such a model should show which modes are important and should predict features of the transitional behavior which are as yet unobserved in the experiments.

Model equations are usually constructed by expanding the velocity and pressure fields in suitable basis functions, such as Fourier series. The expansion is then truncated, with the hope that the terms retained are the physically important ones, at least for Reynolds numbers that are not too large. The sums representing the velocity and pressure are substituted into the Navier-Stokes equations, and the resulting set of coupled nonlinear ordinary differential

---

14 These Reynolds numbers correspond to a system with $\eta = 7/8$ and $h/(b-a) = 20$, and a flow with 17 vortices and 4 azimuthal waves; the transition Reynolds numbers for annuli with this radius ratio but other heights or spatial states have not been studied in detail and may be quite different.

equations for the time-dependent amplitudes of the different modes is integrated numerically for different Reynolds numbers.

Sherman and *McLaughlin* [6.106] have considered a model set of five amplitude equations with quadratic coupling. They assumed values for the growth rates and frequencies of the individual modes and for the coupling coefficients. The frequencies were chosen to have roughly the same ratios as those measured by *Swinney* et al. [6.107]. *Sherman* and *McLaughlin* found that as the growth rates of the modes are increased, there is a transition from a spectrum with sharp frequency components to one with broad components, similar to the component called *B* observed in the experiments; however, their equations are, of course, technically inapplicable to the experiments.

Yahata [6.108, 109] has considered a set of time-dependent amplitude equations derived using the Galerkin method. The velocity field was Fourier-analyzed in the axial and azimuthal directions, and each Fourier coefficient in this double Fourier series was expanded in a complete set of functions in the radial variable. The Fourier series were severely truncated: in the axial direction only the fundamental mode $\cos \gamma z$ was retained (where the value of the wave number γ was taken to be that given by experiments, $2\pi/\gamma = 2.5d$); in the azimuthal direction only the $m=0$ and $m=4$ modes, corresponding to the mean motion and 4 azimuthal waves, were retained. After truncation of the radial functions the resultant model consisted of 32 coupled ordinary differential equations for the mode amplitudes.

Yahata calculated power spectra at four Reynolds numbers for the radial velocity mode with the largest growth rate. Assuming $\eta = 0.875$, as in the experiments in Table 6.6, he found: i) At $R/R_c = 7.97$ the spectrum contained a single intense sharp peak at $\omega/\Omega_1 \simeq 1.5$; this presumably corresponds to the azimuthal wave frequency ω_1. ii) At $R/R_c = 15.94$ the spectrum contained a second sharp intense component, whose frequency $\omega/\Omega_1 \simeq 1.1$ was related to the first frequency in approximately the same ratio as the ω_3 and ω_1 frequencies observed in the experiments. iii) The spectrum at $R/R_c = 22.31$ still contained the component at $\omega/\Omega_1 \simeq 1.5$, but the component at $\omega/\Omega_1 \simeq 1.1$ had disappeared, and other sharp components, much weaker than the component at $\omega/\Omega_1 \simeq 1.5$, had appeared. This behavior is similar to the experimental observation of the disappearance of ω_3 with ω_1 remaining sharp. iv) Finally, at $R/R_c = 23.91$ the component at $\omega/\Omega_1 \simeq 1.5$ was broad. At $R/R_c = 22.31$ it was found that phase space orbits corresponding to almost identical initial conditions remained close together as time evolved, while at $R/R_c = 23.91$ orbits initially close in phase separated exponentially in time. The "sensitive dependence on initial conditions" observed at $R/R_c = 23.91$ is a hallmark of chaotic behavior (see Chap. 4 and [6.110]).

The Yahata model reproduces well some aspects of the experimental data; however, the sensitive dependence on initial conditions in a model should correspond to the observation of noise in experiments. The experiments in Table 6.6 indicate that noise appears at $R/R_c \simeq 12$, well below the Reynolds number at which the waves disappear, $R/R_c \simeq 22$.

6.6 Finite Annulus Length Effects

We noted at the beginning of Sect. 6.4 that Cole [6.69] has shown experimentally that the value of R_c at which axisymmetric Taylor vortex flow occurs is rather insensitive to annulus length, but that the value of R'_c at which wavy vortex flow occurs is dependent on the height to gap ratio, h/d. While certain features of the flow in a annulus of moderate length, such as torque, can be correctly predicted (within some reasonable accuracy) by the infinite length problem, we must recognize that there are essential differences in the two problems.

Couette flow is not a solution of the finite length problem. Indeed, the presence of, say, fixed end plates forces a solution with all three velocity components as functions of the radial and axial variables (we restrict our attention to the axisymmetric problem). For cylinders of moderate or large length compared to the gap we can expect the azimuthal velocity to be nearly that given by Couette flow except for boundary layers at the ends, but also there will be a slow circulation which has some axial structure in planes containing the axis of the cylinders. As the Reynolds number is increased this circulatory motion slowly develops with the possibility of smooth transitions in the cellular structure (i.e., more or fewer cells may develop) until R is near the R_c for infinitely long cylinders; then there is a rapid, but smooth, development of the classical Taylor vortex flow over most of the length of the cylinders. This picture is consistent with the observation of "ghost" or "shadow" Taylor vortices at values of R below R_c [6.66, 87, 111–113].

Benjamin [6.88] has given a very interesting, but qualitative, discussion of the possibility of multiple solutions, their stability and associated bifurcations and transitions for flows in bounded domains with special reference to the flow between rotating concentric cylinders of finite length. In particular, he called attention to the possibility that end effects could introduce a quadratic term and a nonhomogeneous term in an amplitude equation such as (6.28). We leave aside, for the moment, the question of how the true problem can be formulated in terms of an amplitude equation. However, we note that such terms in an amplitude equation can introduce hysteresis effects and change bifurcations to smooth transitions; see, for example, Benjamin [6.88], Stuart [6.114], and other papers in the volume in which [6.114] appears.

Benjamin [6.89] has also described a series of experiments in cylinders of very short length with fixed end plates; the observed phenomena are consistent with his qualitative discussion. For his experiments $a = 36.9$ mm and $b = 60$ mm, so $a/b = 0.615$. For all values of h/d considered in the experiments and for R (or T) sufficiently small there is a unique primary flow (with axial structure) which develops smoothly with increasing R. For $h/d < 3.6$ the primary flow has two cells, and for $h/d > 3.72$ the primary flow has four cells. However, there are curves in the $(h/d, R)$ plane such that if $h/d < 3.6$ and R is to the right of the curve, a secondary flow with four cells is possible. Similarly, if $h/d > 3.72$ and R is to the right of the curve, a secondary flow with four cells is possible. It is important to note that these secondary flows cannot be obtained by slowly increasing R for a fixed value of h/d; rather, both h and R must be varied to achieve a secondary flow in regions of the $(h/d, R)$ plane where these flows exist. For $3.6 < h/d < 3.72$ there exists an $R_c(h/d)$ such that at this value of R the primary flow (which may be a rudimentary four-cell flow or have traits of both a two-cell and four-cell structure) loses its stability with an abrupt transition to a flow with a two-cell structure. When R is then decreased, the flow does not jump back to the primary flow until a value of R slightly less than R_c is reached; thus there is a hysteresis effect. Also, for this range of values of h/d and for R sufficiently large, a stable four-cell secondary flow is possible. These results are summarized by *Benjamin* [Ref. 6.89, Fig. 5].

A direct numerical analysis of the finite length problem with fixed end plates has been given by *Alziary de Roquefort* and *Grillaud* [6.115]. They employed a finite difference procedure with implicit fractional steps to integrate the time-dependent equations. Their calculations are for the parameter values $\eta = 0.933$ and $h/d = 10^{15 \cdot}$ The grid for the computations was 129 points in the axial direction and 33 points in the radial direction. Computations were carried out for several values of R, with the solution for one value of R used as the initial conditions for the next value of R.

Their computed results (Fig. 6.9) show what appear to be smooth transitions with increasing R, from two cells at $R/R_c = 0.0039$ (where R_c corresponds to the infinite cylinder) through four cells and eight cells to, at $R/R_c = 1.17$, ten cells of equal size with the appearance of a classical Taylor vortex flow except for small end effects. They noted that the cellular vortices appear in the vicinity of the midplane, and that for R/R_c between 0.97 and 1.17 the vortex intensity increased rapidly. Finally by varying the initial conditions they were able to obtain at the same value of $R > R_c$ several steady-state solutions differing in the number of cells. However, all such solutions were among the allowable states predicted by *Kogelman* and *Di Prima* [6.42]; see Fig. 6.5. We note that the results of these numerical calculations are consistent with the earlier discussion of this section: axial structure for R small, smooth development of the flow with rapid increase of vortex activity for R near R_c, and the existence of multiple states for R sufficiently large.

Only very recent theoretical work has dealt with problems in which end boundary conditions are imposed. *Blennerhassett* and *Hall* [1.116] have considered the small-gap linear stability problem for finite length cylinders with the inner cylinder rotating and with the end plates assumed to be sufficiently elastic to adjust to circular Couette flow as an exact solution. Solutions that are even in ζ (u, v, p even and w odd) or odd in ζ (u, v, p odd and w even) can be treated separately and lead to critical Taylor numbers $T_{ce}(L)$ and $T_{co}(L)$, respectively, where $L = h/2d$. The curves of T_{ce} and T_{co} against L lie very close to each other and interlace [Ref. 6.116, Fig. 1]. For $L > 6$, the values of T_{ce} and T_{co} are essentially the same and are nearly equal to the value of T_c for $L \to \infty$. One result that may be relevant to the experiments of *Benjamin* [6.89] is that the

---

15 Because of this value of h/d the results are not applicable to the two-cell and four-cell interactions described by *Benjamin*; however, they are very interesting in their own right.

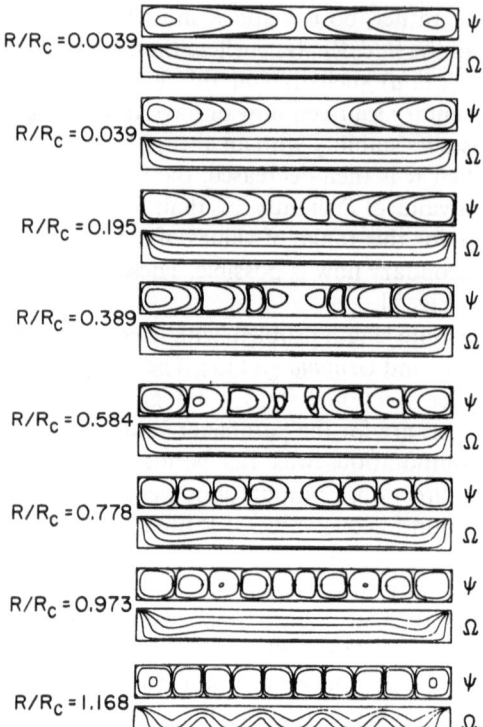

R/R$_c$ =0.0039

R/R$_c$ =0.039

R/R$_c$ = 0.195

R/R$_c$ = 0.389

R/R$_c$ = 0.584

R/R$_c$ = 0.778

R/R$_c$ = 0.973

R/R$_c$ = 1.168

Fig. 6.9. Contour plots of the stream function ψ and angular velocity Ω for an annulus with $\eta = 0.933$ and $h/d = 10$ [6.115]

instability changes from a two-cell motion to a four-cell motion at $L \simeq 1.3$. In those experiments, conducted in an annulus with $\eta = 0.615$, the primary flow changed from two cells to four cells at $L \simeq 1.85$.

Hall [6.117] has extended the analysis of the idealized problem to take of nonlinear effects. He derived *two* coupled amplitude equations for the amplitudes $A(\tau)$ and $B(\tau)$ of the even and odd eigenfunctions, respectively. In the formal analysis of *Stuart* and *Di Prima* [6.118] these equations have the form

$$dA/d\tau = \sigma_A A + a_{20}A^2 + a_{02}B^2 + a_{30}A^3 + a_{12}AB^2 + \dots, \qquad (6.33)$$

$$dB/d\tau = \sigma_B B + a_{11}AB + a_{03}B^3 + a_{21}A^2B + \dots. \qquad (6.34)$$

The parameters σ_A and σ_B are the growth rates of the even and odd modes, respectively, as given by linear theory. The coefficients $a_{20}, \dots, a_{21}$ are determined by integrability conditions associated with nonhomogeneous boundary value problems for certain functions in the expansion of the velocity field.

Of particular interest is the appearance of the quadratic terms in (6.33) and (6.34) which are not present in the amplitude equation (6.28) for the case $L \to \infty$. Of course the coefficients depend on L, and we can expect $a_{20} \to 0$, $a_{02} \to 0$, and $a_{11} \to 0$ as $L \to \infty$. However, for a finite value of L these coefficients will in general be nonzero and can produce transcritical (two-sided) bifurcations (see

Fig. 3.7c). The many different possibilities that can arise from (6.33) and (6.34) for different ranges of L and T, including mixed solutions ($A \neq 0$, $B \neq 0$), have been discussed by *Hall* [6.117]. If the "perfect" boundary conditions at $\zeta = \pm L$ have small imperfections, then nonhomogenous terms are introduced into the amplitude equations. The nonhomogeneous terms change the bifurcation phenomena to a smooth transition[16]. Thus there is a smoothly developing flow with cellular structure even when T is small (a primary flow), which develops rapidly in the neighborhood of the bifurcation point (which may be transcritical) corresponding to the perfect boundary conditions.

Schaeffer [6.119] has considered an alternative idealized problem for which the boundary conditions at the end plates are

$$(1-\alpha)\partial u_r/\partial n + \alpha u_r = 0 \ ,$$

$$(1-\alpha)\partial u_\theta/\partial n + \alpha u_\theta = 0 \ , \tag{6.35}$$

$$u_z = 0 \ .$$

Here α is a parameter with $0 \leq \alpha \leq 1$ and $\partial/\partial n$ is the outward normal derivative. If $\alpha = 1$, the boundary conditions correspond to fixed plates. If $\alpha = 0$, the boundary conditions define an artificial problem for which Couette flow is an exact solution and for which there is a bifurcation to a Taylor vortex flow at a certain $T_c(L, \eta)$. On the other hand, if $\alpha \neq 0$, then Couette flow is not an exact solution, and we can anticipate the smooth development of a flow with axial structure (even for small T) for some range of values of T. It is hoped that for α small, but nonzero, the results which are obtained by a perturbation analysis will approximate the true problem ($\alpha = 1$).

For $\alpha = 0$, the linear stability problem can be solved for $2m$ cell disturbances to give curves of T vs L for each value of m (we assume that η is fixed). The neutral curves for $2m$ and $2m + 2$ cell disturbances will intersect at (T_m^*, L_m^*). In the neighborhood of such a point the interaction of $2m$ and $2m + 2$ cell disturbances can be studied using the full nonlinear equations and perturbing from $\alpha = 0$.

Schaeffer [6.119] has given a discussion of such interactions by analyzing the bifurcations of a 2×2 system of nonlinear (cubic) algebraic equations depending on a bifurcation parameter and three auxiliary parameters. While he did not calculate the coefficients in the equation, he did relate the analysis to phenomena such as are observed in *Benjamin's* [6.89] experiments. On the other hand, since only cubic nonlinearities are allowed, he could not discuss the two cell-four cell problem in which quadratic nonlinearities are generated.

The two cell-four cell problem with the boundary conditions (6.35) has been considered by *Hall* [6.122] using the method of amplitude equations and perturbation theory for small α. For the case $\eta \to 1$, he found $L_1^* = 1.45$ and $T_1^* = 4010$ with the two-cell disturbance being critical for $L < L_1^*$ and the four-

---

16 For a somewhat similar problem of convection in a box, *Hall* and *Walton* [6.120] have shown the "imperfections" in the assumption of perfect insulating end walls lead to a smoothly developing cellular motion rather than a bifurcation; also see *Daniels* [6.121].

cell disturbances being critical for $L > L_1^*$. ($T_c = 3390$ for $L \to \infty$.) *Hall* studied the evolution of the two-cell eigenfunction with amplitude $A(\tau)$ and the four-cell eigenfunction with amplitude $B(\tau)$ for $L - L_1^*$ and $T - T_1^*$ each $O(\alpha^{1/2})$. The amplitude equation for A has a quadratic term AB arising from a two-cell and four-cell interaction, and the amplitude equation for B has a quadratic term A^2 arising from a two-cell interaction with itself. Moreover, each amplitude equation has a nonhomogeneous term which results from satisfying the boundary conditions (6.35) for $\alpha \neq 0$. The analysis of these equations shows the instability of a primary mode, which could have two cells or four cells, the occurrence of a secondary flow (which was not computable in the model), and finally for larger T the existence of a stable four-cell secondary motion. These results are in qualitative agreement with the observations of *Benjamin* [6.89]. However, the model is not able to predict the hysteresis effect found by *Benjamin*.

It is clear that significant strides have been made on the problem of Taylor vortex flows in cylinders of finite length; it is equally clear that there is more to be done.

Acknowledgements. R. C. *DiPrima* acknowledges the partial support of the Army Research Office and the Fluid Dynamics Branch of the Office of Naval Research. *H. L. Swinney* acknowledges the support of the National Science Foundation (Grant CME 79-09585) and The Trull Foundation.

References

6.1   A. Mallock: Determination of the viscosity of water. Proc. R. Soc. London A **45**, 126–132 (1888)

6.2   A. Mallock: Experiments on fluid viscosity. Philos. Trans. R. Soc. London A **187**, 41–56 (1896)

6.3   M. M. Couette: Études sur le frottement des liquides. Ann. Chim. Phys. **6**, Ser. 21, 433–510 (1890)

6.4   G. I. Taylor: Stability of a viscous liquid contained between two rotating cylinders. Philos. Trans. R. Soc. London A **223**, 289–343 (1923)

6.5   E. L. Koschmieder: Turbulent Taylor vortex flow. J. Fluid Mech. **93**, 515–527 (1979)

6.6   D. Coles: Transition in circular Couette flow. J. Fluid Mech. **21**, 385–425 (1965)

6.7   P. R. Fenstermacher, H. L. Swinney, J. P. Gollub: Dynamical instabilities and the transition to chaotic Taylor vortex flow. J. Fluid Mech. **94**, 103–129 (1979)

6.8   J. Serrin: On the stability of viscous fluid motions. Arch. Ration. Mech. Anal. **3**, 1–13 (1959)

6.9   D. D. Joseph, B. R. Munson: Global stability of spiral flow. J. Fluid Mech. **43**, 545–575 (1970)

6.10 D. D. Joseph: *Stability of Fluid Motions I*, Springer Tracts in Natural Philosophy, Vol. 27 (Springer, Berlin, Heidelberg, New York 1976)

6.11 W. L. Hung: "Stability of Couette Flow by the Method of Energy"; M.S. Thesis, University of Minnesota (1978)

6.12 G. P. Neitzel: Private communication (1978)

6.13 D. D. Joseph, W. Hung: Contributions to the nonlinear theory of stability of viscous flow in pipes and between rotating cylinders. Arch. Ration. Mech. Anal. **44**, 1–22 (1971)

6.14 E. R. Krueger, A. Gross, R. C. DiPrima: On the relative importance of Taylor-vortex and non-axisymmetric modes in flow between rotating cylinders. J. Fluid Mech. **24**, 521–538 (1966)

6.15 Lord Rayleigh: On the dynamics of revolving fluids. Proc. R. Soc. London A **93**, 148–154 (1916)

6.16 T. von Kármán: Some aspects of the turbulence problem. Proc. 4th Int. Cong. Appl. Mech. (Cambridge, 1934) pp. 54–91

6.17   C.C.Lin: *The Theory of Hydrodynamic Stability* (Cambridge University Press, Cambridge 1955)

6.18   J.L.Synge: The stability of heterogeneous liquids. Trans. R. Soc. Canada **27**, 1–18 (1933)

6.19   J.L.Synge: On the stability of a viscous liquid between two rotating coaxial cylinders. Proc. R. Soc. London A **167**, 250–256 (1938)

6.20   F.Schultz-Grunow: Zur Stabilität der Couette Strömung. Z. Angew. Math. Mech. **39**, 101–110 (1959)

6.21   F.Schultz-Grunow: On the stability of Couette flow. NATO Advisory Group for Aeronautical Research and Development, Report 265 (1960)

6.22   F.Schultz-Grunow: Stabilität einer rotierenden Flüssigkeit. Z. Angew. Math. Mech. **43**, 411–415 (1963)

6.23   R.C.DiPrima, R.N.Grannick: "A Non-linear Investigation of the Stability of Flow Between Counter-rotating Cylinders", in *Instability of Continuous Systems*, ed. by H.Leipholz (Springer, Berlin, Heidelberg, New York 1971) pp. 55–60

6.24   A.Davey: The growth of Taylor vortices in flow between rotating cylinders. J. Fluid Mech. **14**, 336–368 (1962)

6.25   J.T.Stuart: On the non-linear mechanics of wave disturbances in stable and unstable parallel flows. Part 1. The basic behaviour in plane Poiseuille flow. J. Fluid Mech. **9**, 353–370 (1960)

6.26   J.Watson: On the non-linear mechanics of wave disturbances in stable and unstable parallel flows. Part 2. The development of a solution for plane Poiseuille flow and for plane Couette flow. J. Fluid Mech. **9**, 371–389 (1960)

6.27   A.Davey, R.C.DiPrima, J.T.Stuart: On the instability of Taylor vortices. J. Fluid Mech. **31**, 17–52 (1968)

6.28   J.P.Gollub, H.L.Swinney: Onset of turbulence in a rotating fluid. Phys. Rev. Lett. **35**, 927–930 (1975)

6.29   R.W.Walden, R.J.Donnelly: Reemergent order of chaotic circular Couette flow. Phys. Rev. Lett. **42**, 301–304 (1979)

6.30   S.Chandrasekhar: *Hxdrodynamic and Hydromagnetic Stability* (Oxford University Press, Oxford 1961) p. 303

6.31   C.S.Yih: Spectral theory of Taylor vortices. Part II: Proof of nonoscillation. Arch. Ration. Mech. Anal. **47**, 288–300 (1972)

6.32   K.N.Astill, K.C.Chung: A numerical study of instability of the flow between rotating cylinders. Am. Soc. Mech. Eng. Pub. 76-FE-27 (1976)

6.33   K.Chung: "Stability Study of a Viscous Flow Between Rotating Coaxial Cylinders"; Ph.D. Thesis, Tufts University (1976)

6.34   E.M.Sparrow, W.D.Munro, V.K.Jonsson: Instability of the flow between rotating cylinders: the wide gap problem. J. Fluid Mech. **20**, 35–46 (1974)

6.35   J.Walowit, S.Tsao, R.C.DiPrima: Stability of flow between arbitrarily spaced concentric cylindrical surfaces including the effect of a radial temperature gradient. Trans. Am. Soc. Mech. Eng., J. Appl. Mech. **31**, 585–593 (1964)

6.36   P.H.Roberts: The solution of the characteristic value problems. (Appendix to [6.65]). Proc. R. Soc. London A **283**, 550–556 (1965)

6.37   R.C.DiPrima, G.J.Habetler: A completeness theorem for non-selfadjoint eigenvalue problems in hydrodynamic stability. Arch. Ration. Mech. Anal. **34**, 218–227 (1969)

6.38   R.C.DiPrima: Application of the Galerkin method to the calculation of the stability of curved flows. Quart. Appl. Math. **13**, 55–62 (1955)

6.39   D.L.Harris, W.H.Reid: On the stability of viscous flow between rotating cylinders. Part 2. Numerical analysis. J. Fluid Mech. **20**, 95–101 (1964)

6.40   R.L.Duty, W.H.Reid: On the stability of viscous flow between rotating cylinders. Part 1. Asymptotic analysis. J. Fluid Mech. **20**, 81–94 (1964)

6.41   D.Coles: A note on Taylor instability in circular Couette flow. Trans. Am. Soc. Mech. Eng., J. Appl. Mech. **89**, 527–534 (1967)

6.42   S.Kogelman, R.C.DiPrima: Stability of spatially periodic supercritical flows in hydrodynamics. Phys. Fluids **13**, 1–11 (1970)

6.43   W.Eckhaus: Problèmes non linéaires dans las theorie de la stabilité. J. Mécanique **1**, 49–77 (1962)

6.44   W.Eckhaus: Problèmes non linéaires de stabilité dans un espace à deux dimensions. Premiere partie: Solutions périodiques. J. Mécanique **1**, 413–438 (1962)

6.45   W.Eckhaus: *Studies in Non-Linear Stability Theory*, Springer Tracts in Natural Philosophy, Vol. 6 (Springer, Berlin, Heidelberg, New York 1965)

6.46   R.C.Di Prima: "Vector Eigenfunction Expansions for the Growth of Taylor Vortices in the Flow Between Rotating Cylinders", in *Nonlinear Partial Differential Equations*, ed. by W. F. Ames (Academic Press, New York 1967) pp. 19–42

6.47   W.Velte: Stabilität und Verzweigung stationärer Lösungen der Navier-Stokesschen Gleichungen beim Taylor-Problem. Arch. Ration. Mech. Anal. **22**, 1–14 (1966)

6.48   K.Kirchgässner, P.Sorger: Branching analysis for the Taylor problem. Quart. J. Mech. Appl. Math. **22**, 183–190 (1969)

6.49   V.I.Yudovich: Secondary flows and fluid instability between rotating cylinders. J. Appl. Math. Mech. **30**, 822–833 (1966)

6.50   V.I.Yudovich: The bifurcation of a rotating flow of a liquid. Sov. Phys. Dokl. **11**, 566–568 (1966)

6.51   I.P.Ivanilov, G.N.Iakovlev: The bifurcation of fluid flow between rotating cylinders. J. Appl. Math. Mech. **30**, 910–916 (1966)

6.52   K.Kirchgässner, P.Sorger: "Stability Analysis of Branching Solutions of the Navier-Stokes Equations", in *Proceedings of the Twelfth International Congress of Applied Mechanics, 1968*, ed. by M. Hetényi, W. G. Vincenti (Springer, Berlin, Heidelberg, New York 1969)

6.53   S.N.Ovchinnikova, V.I. Yudovich: Stability and bifurcation of Couette flow in the case of a narrow gap between rotating cylinders. J. Appl. Math. Mech. **34**, 1025–1030 (1974)

6.54   R.C.Di Prima, P.M.Eagles: Amplification rates and torques for Taylor-vortex flows between rotating cylinders. Phys. Fluids **20**, 171–175 (1977)

6.55   W.C.Reynolds, M.C.Potter: A finite amplitude state-selection theory for Taylor-vortex flow. Unpublished report, Dept. Mech. Eng. Stanford Univ. (1967); also see Bull. Am. Phys. Soc. **12**, 834 (1967);
       J.T.Stuart: In *Annual Reviews of Fluid Mechanics*, Vol. 3, (Annual Reviews, Palo Alto, CA, 1971) pp. 347–370

6.56   P.M.Eagles, J.T.Stuart, R.C.DiPrima: The effects of eccentricity on torque and load in Taylor-vortex flow. J. Fluid Mech. **87**, 209–231 (1978)

6.57   W.Eckhaus: Problemes non-linéaries de stabilité dans un espace à deux dimensions. Deuxieme partie: Stabilité des solutions périodiques. J. Mécanique **2**, 153–172 (1963)

6.58   C.Nakaya: Domain of stable periodic vortex flows in a viscous fluid between concentric circular cylinders. J. Phys. Soc. Jpn. **36**, 1164–1173 (1974)

6.59   H.A.Snyder: Wavenumber selection at finite amplitude in rotating Couette flow. J. Fluid Mech. **35**, 273–298 (1969)

6.60   H.A.Snyder: Change in waveform and mean flow associated with wavelength variations in rotating Couette flow. Part 1. J. Fluid Mech. **35**, 337–352 (1969)

6.61   J.E.Burkhalter, E.L.Koschmieder: Steady supercritical Taylor vortices after sudden starts. Phys. Fluids **17**, 1929–1935 (1974)

6.62   E.L.Koschmieder: Stability of supercritical Bénard convection and Taylor vortex flow. Adv. Chem. Phys. **32**, 109–133 (1975)

6.63   J.T.Stuart, R.C.Di Prima: The Eckhaus and Benjamin-Feir resonance mechanisms. Proc. R. Soc. London A **362**, 27–41 (1978)

6.64   T.B.Benjamin, J.E.Feir: The disintegration of wave trains on deep water. Part I. Theory. J. Fluid Mech. **27**, 417–430 (1967)

6.65   R.J.Donnelly, K.W.Schwarz: Experiments on the stability of viscous flow between rotating cylinders. VI. Finite-amplitude experiments. Proc. R. Soc. London A **283**, 531–546 (1965)

6.66   H.A.Snyder, R.B.Lambert: Harmonic generation in Taylor vortices between rotating cylinders. J. Fluid Mech. **26**, 545–562 (1966)

6.67   J.P.Gollub, M.H.Freilich: Optical heterodyne test of perturbation expansions for the Taylor instability. Phys. Fluids **19**, 618–626 (1976)

6.68   W.Debler, E.Füner, B.Schaaf: "Torque and Flow Patterns in Supercritical Circular Couette Flow", in *Proceedings of the Twelfth International Congress of Applied Mechanics, 1968*, ed. by M. Hetényi, W. G. Vincenti (Springer, Berlin, Heidelberg, New York 1969)

6.69   J.A.Cole: Taylor-vortex instability and annulus-length effects. J. Fluid Mech. **75**, 1–15 (1976)

6.70   R.W.Walden: "Transition to Turbulence in Couette Flow Between Concentric Cylinders"; Ph.D. Thesis, University of Oregon (1978)

6.71   K.W.Schwarz, B.E.Springett, R.J.Donnelly: Modes of instability in spiral flow between rotating cylinders. J. Fluid Mech. **20**, 281–289 (1964)

6.72   P.Castle, F.R.Mobbs: Hydrodynamic stability of the flow between eccentric rotating cylinders: visual observations and torque measurements. Proc. Inst. Mech. Eng. (London) **182**, 41–52 (1967–68)

6.73   P.Castle, F.R.Mobbs, P.H.Markho: Visual observations and torque measurements in the Taylor vortex regime between eccentric rotating cylinders. J. Lub. Tech., Trans. Am. Soc. Mech. Eng. **93**, 121–129 (1971)

6.74   P.H.Markho, C.D.Jones, F.R.Mobbs: Wavy modes of instability in the flow between eccentric rotating cylinders. J. Mech. Eng. Sci. **19**, 76–80 (1977)

6.75   F.R.Mobbs, S.Preston, M.S.Ozogan: An experimental investigation of Taylor vortex waves. Taylor Vortex Flow Working Party, Leeds (1979)

6.76   R.C.DiPrima: Stability of nonrotationally symmetric disturbances for viscous flow between rotating cylinders. Phys. Fluids **4**, 751–755 (1961)

6.77   P.M.Eagles: On stability of Taylor vortices by fifth-order amplitude expansions. J. Fluid Mech. **49**, 529–550 (1971)

6.78   C.Nakaya: The second stability boundary for circular Couette flow. J. Phys. Soc. Jpn. **38**, 576–585 (1975)

6.79   P.M.Eagles: On the torque of wavy vortices. J. Fluid Mech. **62**, 1–9 (1974)

6.80   R.J.Donnelly: Experiments on the stability of viscous flow between rotating cylinders. I. Torque measurements. Proc. R. Soc. London A **246**, 312–325 (1958)

6.81   R.J.Donnelly, N.J.Simon: An empirical torque relation for supercritical flow between rotating cylinders. J. Fluid Mech. **7**, 401–418 (1960)

6.82   H.A.Snyder: Stability of rotating Couette flow. II. Comparison with numerical results. Phys. Fluids **11**, 1599–1605 (1968)

6.83   F.Schultz-Grunow, H.Hein: Beitrag zur Couetteströmung. Z. Flugwiss. **4**, 28–30 (1956)

6.84   J.W.Lewis: An experimental study of the motion of a viscous liquid contained between coaxial cylinders. Proc. R. Soc. London A **117**, 388–407 (1928)

6.85   A.Townsend: Private communication (1979)

6.86   S.I.Pai: Turbulent flow between rotating cylinders. National Advisory Committee for Aeronautics Technical Note No. 892 (1943)

6.87   J.E.Burkhalter, E.L.Koschmieder: Steady supercritical Taylor vortex flow. J. Fluid Mech. **58**, 547–560 (1973)

6.88   T.B.Benjamin: Bifurcation phenomena in steady flow of a viscous fluid. I. Theory. Proc. R. Soc. London A **359**, 1–26 (1978)

6.89   T.B.Benjamin: Bifurcation phenomena in steady flows of a viscous fluid. II. Experiments. Proc. R. Soc. London A **359**, 27–43 (1978)

6.90   H.L.Swinney, P.R.Fenstermacher, J.P.Gollub: Transition to turbulence in circular Couette flow, Turbulent Shear Flow Symposium (Pennsylvania State Univ., 1977) pp. 17.1–17.6

6.91   G.Cognet: Utilization de la polargraphie pour l'étude de l'écoulement de Couette. J. Mécanique **10**, 65–90 (1971)

6.92   A.Bouabdallah, G.Cognet: "Laminar-turbulent transition in Taylor-Couette flow", in *Laminar-Turbulent Transition*, ed. by R. Eppler and H. Fasel (Springer, Berlin, Heidelberg, New York 1980) pp. 368–377

6.93   A.Barcilon, J.Brindley, M.Leesen, F.R.Mobbs: Marginal instability in Taylor-Couette flows at very high Taylor number. J. Fluid Mech. **94**, 453–463 (1979)

6.94   H.A.Snyder: Waveforms in rotating Couette flow. Int. J. Non-Linear Mech. **5**, 659–685 (1970)

6.95   R.K.Otnes, L.Enochson: *Applied Time Series Analysis* (John Wiley, New York 1978)

6.96   G.Ahlers: Low temperature studies of the Rayleigh-Bénard instability and turbulence. Phys. Rev. Lett. **33**, 1185–1188 (1974)

6.97   P.R.Fenstermacher: "Laser Doppler Velocimetry Study of the Onset of Chaos in Taylor Vortex Flow"; Ph.D. Thesis, City College of the City University of New York (1979)

6.98   T.S.Durrani, C.A.Greated: *Laser Systems in Flow Measurement* (Plenum, New York 1977)

6.99   F.Durst, A.Melling, J.H.Whitelaw: *Principles and Practice of Laser-Doppler Anemometry* (Academic Press, London 1976)

6.100 M.A.Gorman, H.L.Swinney: Visual observation of a second characteristic mode in wavy vortex flow. Phys. Rev. Lett. **43**, 1871–1875 (1979)

6.101 A.Roshko: Structure of turbulent shear flows: a new look. Am. Inst. Aero. Astron. J. **14**, 1349–1357 (1976)

6.102 L.P.Reiss, T.J.Hanratty: Measurement of instantaneous rates of mass transfer to a small sink on a wall. Am. Inst. Chem. Eng. J. **8**, 245–247 (1962)

6.103 Z.B.Krugljak, E.A.Kuznetsov, V.S.L'vov, Yu.E.Nesterikhin, A.A.Predtechensky, V.S. Sobolev, E.N.Utkin, F.A.Zhuravel: "Laminar-turbulent transition in circular Couette flow", in *Laminar-Turbulent Transition*, ed. by R. Eppler and H. Fasel (Springer, Berlin, Heidelberg, New York 1980), pp. 378–387

6.104 E.A.Kuznetsov, V.S.L'vov, A.A.Predtechenskii, V.S.Sobolev, E.N.Utkin, JETP Lett. **30**, 207–210 (1979)

6.105 E.N.Lorenz: Deterministic nonperiodic flow. J. Atmos. Sci. **20**, 130–141 (1963)

6.106 J.Sherman, J.B.McLaughlin: Power spectra of nonlinearly coupled waves. Commun. Math. Phys. **58**, 9–17 (1978)

6.107 H.L.Swinney, P.R.Fenstermacher, J.P.Gollub: "Transition to Turbulence in a Fluid Flow", in *Synergetics, a Workshop*, ed. by H. Haken (Springer, Berlin, Heidelberg, New York 1977) pp. 60–69

6.108 H.Yahata: Temporal development of the Taylor vortices in a rotating fluid. Prog. Theor. Phys. Suppl. **64**, 176–185 (1978)

6.109 H.Yahata: Temporal development of the Taylor vortices in a rotating fluid II. Prog. Theor. Phys. **61**, 791–800 (1979)

6.110 D.Ruelle: Sensitive dependence on initial condition and turbulent behavior of dynamical systems. Ann. N.Y. Acad. Sci. **316**, 408–416 (1979)

6.111 J.A.Cole: Taylor vortices with short rotating cylinders. J. Fluids Eng. **96**, 69–70 (1974)

6.112 J.A.Cole: Taylor vortex behavior in annular clearances of limited length. *Proc. of the Fifth Australasian Conf. on Hydraulics and Fluid Mechanics*, pp. 514–521 (1974)

6.113 P.A.Jackson, B.Robati, F.R.Mobbs: "Secondary Flows Between Eccentric Rotating Cylinders at Subcritical Taylor Numbers", in *Superlaminar Flow in Bearings*, Proc. of the Second Leeds-Lyon Symposium on Tribology (Institute of Mechanical Engineers, London 1975) pp. 9–14

6.114 J.T.Stuart: "Bifurcation Theory in Non-linear Hydrodynamic Stability", in *Applications of Bifurcation Theory*, ed. by P. H. Rabinowitz (Academic Press, New York 1977) pp. 127–147

6.115 T.Alziary de Roquefort, G.Grillaud: Computation of Taylor vortex flow by a transient implicit method. Comput. Fluids **6**, 259–269 (1978)

6.116 P.J.Blennerhasset, P.Hall: Centrifugal instabilities of circumferential flow in finite cylinders: linear theory. Proc. R. Soc. London A **365**, 191–207 (1979)

6.117 P.Hall: Centrifugal instabilities of circumferential flows in finite cylinders: nonlinear theory. Proc. R. Soc. London A **372**, 317–356 (1980)

6.118 J.T.Stuart, R.C.Di Prima: On the mathematics of Taylor-vortex flows in cylinders of finite length. Proc. R. Soc. London A **372**, 357–365 (1980)

6.119 D.G.Schaeffer: Qualitative analysis of a model for boundary effects in the Taylor problem. Math. Proc. Camb. Philos. Soc. **87**, 307–337 (1980)

6.120 P.Hall, I.C.Walton: The smooth transition to a convective regime in a two dimensional box. Proc. R. Soc. London A **358**, 199–221 (1977)

6.121 P.G.Daniels: The effect of distant side walls on the transition to finite amplitude Bénard convection. Proc. R. Soc. London A **358**, 173–197 (1977)

6.122 P.Hall: Centrifugal instabilities in finite containers: a periodic model. J. Fluid Mech. **99**, 575—596 (1980)

7. Shear Flow Instabilities and Transition

S. A. Maslowe

With 10 Figures

The classical linear theory remains an important part of the subject of stability of parallel flows despite a general recognition that it is not sufficient, by itself, to predict transition. Thus, this chapter begins by developing the traditional normal-mode approach and then outlines recent improvements in techniques for solving the Orr-Sommerfeld and related equations. Following that outline, some of the nonlinear theories are presented that appear promising at the present time. Several important experiments are then reviewed and the relationship discussed between the observations of transition to turbulence and the theory.

It, of course, can be said at the outset that no single theory seems capable of describing the entire transition process. Nonetheless, a fair degree of success has been achieved in modelling isolated segments of this process, and the relevant analyses are presented with that objective in mind. The flows receiving primary attention here are the boundary layer and free shear layer. These are the most important flows in practice and it will be seen that they make for an interesting contrast.

7.1 Overview

From a practical point of view, the overriding goal of hydrodynamic stability theory is to predict the transition from laminar to turbulent flow. Drag forces and heat-transfer rates are so much higher in turbulent boundary layers, for example, that the ability to control transition would open significant possibilities to the design engineer. Usually, one wishes to delay transition (e.g., to reduce the heat-transfer to a space vehicle reentering the earth's atmosphere). However, there are other important applications where turbulence is desirable in order to prevent boundary layer separation or to promote rapid mixing in a chemical process. An example, perhaps noted by the reader while travelling by jet, is the use of vortex generators to induce transition on the upper surface of the wings on Boeing 707-class aircraft. These rectangular arrays of metal plates, by promoting turbulence, improve the control characteristics of the plane by preventing early separation of the boundary layer and possible stalling.

The mathematical side of the subject, especially in terms of quantity of papers, has been dominated by the normal mode approach to linear stability. In that approach, one supposes that there is a mean parallel flow $\bar{u}(y)$ in the x direction. A perturbation is then added to this basic flow that is proportional to $\exp[i\alpha(x-ct)]$, where the governing equations have been linearized; α the wave number is real, while $c = c_r + ic_i$ is complex. The quantity c_r is the phase speed of the wavelike perturbation, while αc_i is termed the amplification factor. Instability corresponds to $c_i > 0$ and the object is to compute c_i as a function of α and, possibly, other parameters. Thus, linear *instability* implies that a specified *infinitesimal* perturbation will grow exponentially.

Typical of the output from a linear stability calculation are the results shown in Fig. 7.1 obtained by solving numerically the Orr-Sommerfeld equation in the case of a Blasius boundary layer. As the boundary layer grows in thickness with increasing x, the Reynolds number increases. At some "critical Reynolds number", about 520 here, instability can occur. It was the hope of some early investigators that turbulence would follow shortly thereafter, i.e., that there was a direct link between linear instability and transition. Certainly, the prediction of an exponential rate of growth for the instability suggests that a major change may take place in the basic state. However, the exponential behavior is itself a consequence of the linearization, and the prediction is only valid so long as the perturbation remains small. Ultimately, nonlinear effects become significant and should they be stabilizing, as they generally are at some point, a new equilibrium state may be possible. One or more such states might occur before turbulence results.

How much can one really say then about transition when $c_i > 0$? The answer depends very much upon the magnitude of αc_i, the particular flow being studied, and whom you ask. When αc_i is large there is more reason to anticipate that linear instability will lead to a dramatic change in the flow. However, that quantity is often rather small, particularly in cases where the flow is bounded. Nonetheless, there do exist empirical methods that employ linear theory and these can be of use in applications until more substantial schemes become available. It should be recognized though that the value of such methods is limited. As an illustration of that point, it is well known that boundary layer transition can be induced by free-stream disturbances or wall roughness without the occurrence of linear instability.

To summarize then, an accurate prediction and description of transition is somewhat beyond the capability of a linear theory. Because of imperfections in the system (wall roughness, structural vibrations, etc.) and the inherent importance of nonlinear effects during the transition process, a successful predictive scheme would require, as a minimum, not only a critical value of the Reynolds number, but also some nonlinear dependence on an amplitude parameter (possibly characteristic of one of the aforementioned imperfections).

Despite these limitations, linear stability theory remains an important area of study both in its own right and in the context of transition. First, there is little doubt that it correctly describes the onset and early evolution of infinitesimal perturbations. This has been verified in numerous experiments where the environment was suitably controlled. Secondly, linear theory does

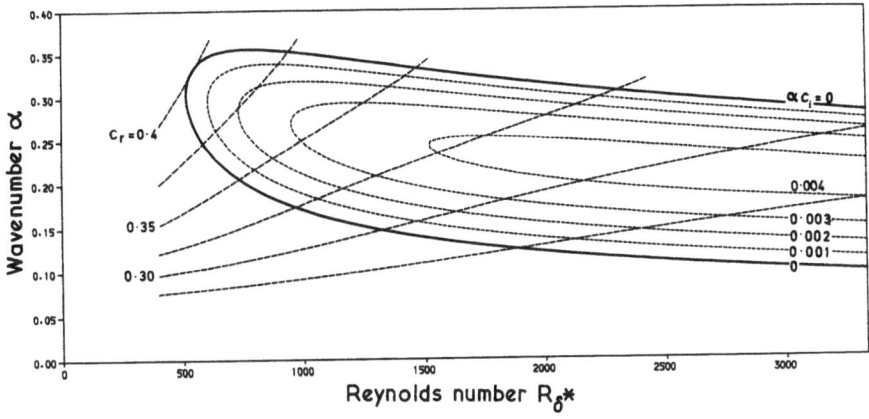

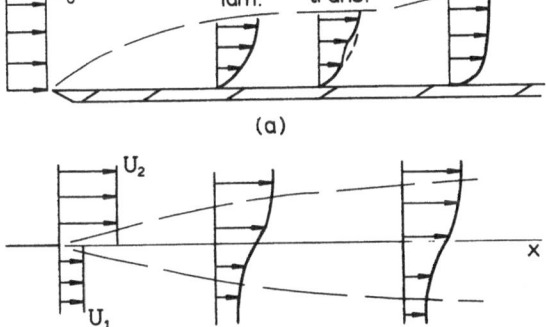

Fig. 7.1. Stability diagram for Blasius boundary layer: $R_{\delta*}$ is Reynolds number based on displacement thickness [7.1]

Fig. 7.2a, b. Quasi-parallel flows undergoing transition. (a) Transitional boundary layer on a flat plate; (b) mixing layer between two uniform streams

seem to give a qualitatively correct indication of the relative stability of the flows of interest. To take an extreme example, flow through a circular pipe is stable at all Reynolds numbers according to the linear result. While turbulence is often observed in actual fact, it is nevertheless true that pipe flow is much more stable than, say, a boundary layer or channel (Poiseuille) flow. A third use of the linear theory is that it can serve as a starting point in a perturbation scheme for weakly nonlinear studies. This is most clearly illustrated by the Stuart-Watson theory (Sect. 7.4.1), where αc_i is the first term in a series expansion of the amplitude evolution equation.

Before commencing the mathematical development of the theory, a few remarks will be made comparing the transition of boundary layers and free shear layers. These important flows have both been the object of detailed experimental studies and some of the more significant experiments are described in Sect. 7.5. The mean velocity profile of a boundary layer undergoing transition is sketched in Fig. 7.2a. After having reached a fully turbulent state, the (time-averaged) velocity profile is seen to be much fuller than was the laminar profile. The eddying motion in the turbulent boundary layer is substantial so it is clear that a major change has occurred in the character of the flow.

The free shear layer (Fig. 7.2b) results from the mixing of two streams and, once turbulent, also contains large-amplitude eddies. However, there is a tendency toward the formation of organized two-dimensional structures that are embedded in the smaller-scale random turbulence. Such organized structures in a turbulent boundary layer seem to be three-dimensional if they occur at all. Also, in contrast to the case of the boundary layer, there is little change in the *mean* velocity profile between the laminar and turbulent flows. Yet another interesting contrast has to do with the final stage of breakdown into turbulence. That stage in the case of a boundary layer is marked by very high-frequency fluctuations that spread out and merge rapidly. Such high-frequency "bursts" have not yet been observed in free shear layers.

These and other differences emphasize the point that transition in each of the two flows compared above has its own character; it follows that no single set of model equations can describe both. The most profitable approach will in all likelihood continue to be a direct attack on the full equations applying to the specific flow in question. Singular perturbation techniques such as the methods of matched asymptotic expansions and multiple scaling, so useful in other areas of fluid mechanics, prove to be equally valuable in stability investigations. The reader not familiar with these techniques may wish to consult a text such as *Nayfeh* [7.2] or *Kevorkian* and *Cole* [7.3]. However, with the exception of Sects. 7.2.5 and 7.4, most of the chapter should be accessible without such familiarity.

7.2 Linear Stability via the Normal-Mode Approach

Our basic starting point is the vorticity equation for incompressible flow [7.31], namely,

$$\nabla^2\psi_t + \psi_y\nabla^2\psi_x - \psi_x\nabla^2\psi_y = \mathrm{Re}^{-1}\nabla^2(\nabla^2\psi) \,, \tag{7.1}$$

where all quantities have been nondimensionalized with respect to characteristic length and velocity scales L and U. The stream function ψ is related to the horizontal velocity components u and v by $u = \psi_y$ and $v = -\psi_x$, while the Reynolds number $\mathrm{Re} = UL/v$, v being the kinematic viscosity. We consider a parallel shear flow $\bar{u}(y)$ and study its stability by superimposing a small wave like perturbation.

While it is well-known that in order for a parallel flow to satisfy (7.1) exactly $\bar{u}(y)$ must be parabolic, the approximation is a reasonable one for large Re because of the following considerations. First, any $\bar{u}(y)$ is a solution in the inviscid limit $\mathrm{Re} \to \infty$. Transition usually occurs at large values of Re and the flow is therefore nearly parallel; that is observed to be the case. Now, if instability results and αc_i is not small, then the perturbation will grow rapidly compared to the basic flow, so our analysis is valid, at least locally. However, it is clear that there are two situations when the parallel flow approximation needs to be reexamined: 1) when it turns out that the critical value of Re is not large and/or 2) when αc_i is small. We assume for the moment that such is not the case, but note that these situations can occur and nonparallel effects may then be significant.

7.2.1 The Orr-Sommerfeld Equation

To begin, we separate the stream function into a mean and a fluctuating part by writing

$$\psi(x, y, t) = \bar{\psi}(y) + \varepsilon \hat{\psi}(x, y, t) , \tag{7.2}$$

where $\varepsilon \ll 1$ is an amplitude parameter. Assuming that $\bar{\psi}(y)$ is a solution of (7.1), the equation for $\hat{\psi}$ is

$$\nabla^2 \hat{\psi}_t + \bar{u} \nabla^2 \hat{\psi}_x - \bar{u}'' \hat{\psi}_x + \varepsilon(\hat{\psi}_y \nabla^2 \hat{\psi}_x - \hat{\psi}_x \nabla^2 \hat{\psi}_y) = \nabla^2(\nabla^2 \hat{\psi})/\mathrm{Re} . \tag{7.3}$$

In the linear normal mode approach, the terms multiplied by ε are neglected and one considers perturbations of the form

$$\hat{\psi} = \phi(y) \exp\left[i\alpha(x - ct)\right] , \tag{7.4}$$

where, as discussed above, c is complex. The result of linearizing and employing (7.4) in (7.3) is the Orr-Sommerfeld (OS) equation

$$(\bar{u} - c)(\phi'' - \alpha^2 \phi) - \bar{u}'' \phi = \frac{1}{i\alpha \mathrm{Re}} (\phi^{IV} - 2\alpha^2 \phi'' + \alpha^4 \phi) . \tag{7.5}$$

Assuming for the moment that there are solid boundaries at $y = (y_1, y_2)$, the boundary conditions (BCs) are that

$$\phi(y_1) = \phi'(y_1) = \phi(y_2) = \phi'(y_2) = 0 , \tag{7.6}$$

corresponding to the vanishing of the normal and tangential velocity components. The consideration of (7.5) subject to the homogeneous BCs (7.6) leads to an eigenvalue problem whose solution will be denoted $f(\alpha, c, \mathrm{Re}) = 0$. There is a single arbitrary complex constant in the solution of (7.5) which we may view as the initial perturbation amplitude. Because the choice of this constant is free in an eigenvalue problem its definition is a matter of convenience, so we will leave it unspecified for the moment.

It should be remarked at this point that (7.4) can be readily generalized to include oblique waves. However, a transformation due to Squire (see, e.g., the monograph by *Lin* [7.4]) shows that the growth rate of an unstable three-dimensional wave is equal to that of a two-dimensional wave at a smaller Re. Hence, the first observed instability will generally be two-dimensional, although three-dimensional effects are ultimately significant.

At this point, we leave the OS equation in order to consider its inviscid limit on the basis that Re is large in most applications. The equation resulting from (7.5) when the rhs is neglected is known as the Rayleigh equation, and for many flows its solution provides the major part of the story. It should be noted, however, that the order of the equation is now reduced. Hence, this is a possible singular perturbation problem, i.e., there are situations in which the viscous terms play a subtle, but important, role no matter how large the Reynolds number.

7.2.2 The Rayleigh Equation

The inviscid limit of the problem posed by (7.5) and (7.6) corresponds to solving the Rayleigh equation

$$\phi'' - \alpha^2\phi - [\bar{u}''/(\bar{u}-c)]\phi = 0 \tag{7.7}$$

with the BCs $\phi(y_1) = \phi(y_2) = 0$. The latter require the normal velocity to vanish at a solid boundary. As it turns out, this theory is most important for unbounded flows, where the BCs must be modified. Consider, for example, the free shear layer $\bar{u} = \tanh y$. As $y \to \infty$, $\bar{u}'' \to 0$ rapidly so the asymptotic solution of (7.7) has the form

$$\phi \sim A_1 \exp(-\alpha y) + A_2 \exp(\alpha y) . \tag{7.8}$$

The desired boundary condition is clearly $A_2 = 0$.

We now turn to the statement and proof of some important theorems involving (7.7). The first of these is Rayleigh's inflection point theorem (1880) which may be stated as follows: *A necessary condition for instability is that $\bar{u}''$ change sign at some point in the flow*. This means that $\bar{u}''$ must vanish somewhere in (y_1, y_2). In more physical terms, the vorticity $\bar{u}'$ has at least one extremum in the unstable case. Not only is the mathematical result useful, but its physical significance is far-reaching, as will be seen.

We begin the proof by assuming that there is *instability* so that $c_i > 0$ and then multiplying (7.7) by ϕ^*, the complex conjugate of ϕ. After multiplying the numerator and denominator of the $\bar{u}''$ term by $\bar{u} - c^*$, integrating by parts, and imposing the BCs, we obtain

$$-\int_{y_1}^{y_2} (|\phi'|^2 + \alpha^2|\phi|^2)dy = \int_{y_1}^{y_2} \frac{\bar{u}''(\bar{u}-c^*)}{|\bar{u}-c|^2}|\phi|^2 dy . \tag{7.9}$$

Examining now the imaginary part of (7.9) leads to

$$c_i \int_{y_1}^{y_2} \bar{u}'' \frac{|\phi|^2}{|\bar{u}-c|^2} \, dy = 0 \; , \tag{7.10}$$

from which the theorem follows.

On an inviscid basis we conclude therefore that flows such as the Blasius boundary layer and Poiseuille flow, $\bar{u} = 1 - y^2$, are stable. The fact that they are known to be unstable observationally means that viscosity somehow plays a destabilizing role as will be discussed in Sect. 7.2.5.

Note that Rayleigh's theorem says nothing about Couette flow where $\bar{u}'' = 0$ for all y nor does it distinguish between a vorticity maximum and a minimum. Our next theorem, due to Fjørtoft, resolves this dilemma. By further consideration of (7.9), including its real part as well as (7.10), Fjørtoft (1950) showed that an additional necessary condition for instability is that $\bar{u}''(\bar{u}-\bar{u}_s) < 0$ for some value of y, where $\bar{u}_s$ is evaluated at the inflection point at y_s. That result in conjunction with Rayleigh's theorem is equivalent to the statement that a necessary condition for instability is that the absolute value of the vorticity must have a relative *maximum* at y_s. Thus, Couette flow is stable. Another interesting example is $\bar{u} = \sinh y$. Here, there is actually an inflection point at $y = 0$, but $\bar{u}'$ is a minimum there so the flow is stable. For further discussion and a proof of Fjørtoft's theorem, the reader is referred to the survey article by *Drazin* and *Howard* [7.5].

There is an illuminating physical interpretation of the foregoing result based upon the vorticity distribution. *Lin* [Ref. 7.4, Sect. 4.4], by considering the variation of vorticity in the y direction, showed that a fluid particle displaced vertically in either direction is forced generally to return to its starting point. However, at a vorticity extremum this restoring mechanism is not present. *Gill* [7.6] has shown further that by taking into account variations in the x direction it becomes clear why the extremum in vorticity must be a maximum to produce instability. It should be noted that such *local* arguments do not ensure that instability will take place because they ignore the possible stabilizing influence of boundaries. However, the fact of the matter is that a vorticity maximum usually does lead to a strong instability and so this is a most important mechanism.

Returning now to the mathematical side of the inviscid theory, another useful result pertaining to (7.7) is the "semicircle theorem" proved by *Howard* [7.7]. The derivation of this result illustrates an ingenious technique that has since been employed successfully by numerous other investigators (including the author) to obtain a wide variety of results. Howard's procedure is a generalization of Rayleigh's method of proof that allows one to consider singular neutral modes. It was first used in the context of density-stratified shear flows.

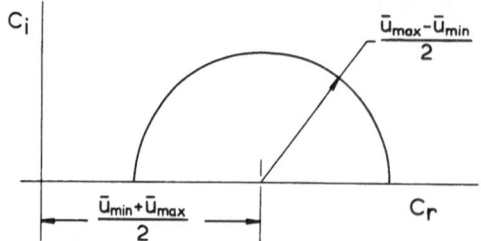

Fig. 7.3. Unstable eigenvalues of Rayleigh's equation are bounded above by Howard's semicircle, as indicated here in the complex c plane

We again begin by' taking $c_i > 0$ and make the change of variable $H = W^{n-1}\phi$, where $W = \bar{u} - c$. The Rayleigh equation becomes

$$\frac{d}{dy}\left(W^{2(1-n)}\frac{dH}{dy}\right) - [\alpha^2 W^{2(1-n)} + nW^{1-2n}\bar{u}'' + W^{-2n}n(1-n)\bar{u}'^2]H = 0 .$$

$$(7.11)$$

Now the idea is to multiply (7.11) by H^* and integrate from y_1 to y_2, employing the BCs to eliminate terms, as was done in deriving (7.9). Then one simply examines separately the real and imaginary parts of the resulting equation looking for conditions analogous to (7.10). At the same time, different values of n are chosen in order to obtain as many such conditions as required; it, of course, is natural to choose values of n for which the equation simplifies due to the vanishing of terms. There is no telling in advance the nature of the outcome. The semicircle theorem itself follows from the choice $n = 0$; consideration of both the real and imaginary parts of various integrals, followed by some manipulation leads to the result

$$[c_r - \tfrac{1}{2}(\bar{u}_{max} + \bar{u}_{min})]^2 + c_i^2 \leqq [\tfrac{1}{2}(\bar{u}_{max} - \bar{u}_{min})]^2 .$$

$$(7.12)$$

Equation (7.12) states that all unstable modes lie in the semicircle in Fig. 7.3.

Besides giving an upper bound on the growth rate for an unstable mode, the "real part" of the semicircle theorem illustrates the importance of critical points where $\bar{u} = c$. Because $\bar{u}_{min} < c_r < \bar{u}_{max}$, all neutral modes that lie on stability boundaries (i.e., they are the limits of unstable modes as $c_i \to 0$) have critical points. This does not necessarily mean that such neutral modes are singular, however, because it often happens that $\bar{u} - c = 0$ at an inflection point of $\bar{u}$. We shall return to this point after first discussing the Reynolds stress.

There remains one peculiarity to note about the Rayleigh equation before moving on. It can readily be verified that if $(\phi; \alpha, c)$ is a solution of (7.7), then so is $(\phi^*; \alpha, c^*)$. Hence, to every damped mode with $c_i < 0$, there corresponds an unstable mode having the same value of c_i with the sign reversed. In general, the solution with $c_i > 0$ is the more meaningful one; the damped mode may or may not have some significance, but the growing mode does have a clear relationship to the solution of the OS problem. As $Re \to \infty$, the unstable (and neutral) solutions of (7.5) do approach those of the Rayleigh equation as shown in [7.4, Chap. 8]. That is not usually the case for damped modes; an exception though is the long-wave solution found by *Tatsumi* et al. [7.8] for the tanh y shear layer. There, at least one family of damped waves in the inviscid theory corresponds to the large Re limit of (7.5). When there is no direct relationship between damped solutions of (7.7) and the inviscid limit of (7.5), one might conclude from the discussion in *Lin* that such solutions are meaningless. However, as pointed out to the writer by Professor *Lin*, these may still be valid limits of the inviscid *nonlinear* equations. In any case, we take the point of view for the moment that singular neutral modes obtained from Rayleigh's equation are the limits of *unstable* modes as $c_i \rightarrow 0$.

7.2.3 The Reynolds Stress

A mathematically and physically important quantity in the subject is the Reynolds stress defined by

$$\tau = -\varrho \overline{u'v'} ,$$

where the primes denote fluctuating quantities and the bar signifies an average over x. The Reynolds stress plays an important role in turbulence theories, where the velocity fluctuations are often large and are not periodic in x. However, for the purpose at hand, we take u' and v' to be small so that they can be evaluated from linearized theory. The averaging will be over one wavelength. Finally, we restrict attention to incompressible flows, where ϱ can be scaled out, and this leads to

$$\tau = -\frac{\alpha}{2\pi} \int_0^{2\pi/\alpha} u'v' dx . \tag{7.13}$$

The stream function, which will be employed to evaluate u' and v', can be normalized by writing

$$\psi' = \{\phi \exp[i\alpha(x-ct)] + \phi^* \exp[-i\alpha(x-ct)]\}/2$$

(which is real as required to evaluate the nonlinear integrand). Substituting into (7.13) now, and differentiating with respect to y, we obtain

$$\frac{d\tau}{dy} = \tfrac{1}{4} i\alpha(\phi\phi''^* - \phi''\phi^*)e^{2\alpha c_i t} . \tag{7.14}$$

The quantity in parentheses can be simplified by using the Rayleigh equation and its complex conjugate. Multiplying (7.7) by ϕ^* and subtracting the product of ϕ and the complex conjugate of (7.7) leads, finally, to

$$\frac{d\tau}{dy} = \tfrac{1}{2}\alpha c_i \frac{\bar{u}''}{|\bar{u}-c|^2}|\phi|^2 e^{2\alpha c_i t} , \tag{7.15}$$

with BCs $\tau(y_1) = \tau(y_2) = 0$.

Some important conclusions about neutral waves can be drawn from (7.15). Clearly, as $c_i \to 0$, the rhs vanishes except, possibly, at points where $\bar{u}=c$. At such a point the otherwise constant Reynolds stress can be discontinuous. The magnitude of the discontinuity is given by

$$\tau(y_c^+) - \tau(y_c^-) = \frac{\alpha\pi}{2}\left(\frac{\bar{u}_c''}{|\bar{u}_c'|}\right)|\phi_c|^2 . \tag{7.16}$$

This result was originally derived by Tollmien using the method of Frobenius, i.e., power series solutions obtained by expanding about $y=y_c$. The relevant expansions will be given in Sect. 7.2.5. However, a simpler approach is to integrate (7.15) with $c_i > 0$ from $y_c - \delta$ to $y_c + \delta$ and then take the limit as $c_i \to 0$. Thus, we obtain

$$\tau(y_c^-) - \tau(y_c^+) = \frac{\alpha}{2}\int_{\bar{u}(y_c-\delta)}^{\bar{u}(y_c+\delta)} \frac{\bar{u}''}{\bar{u}'}|\phi|^2 \frac{c_i d\bar{u}}{(\bar{u}-c_r)^2 + c_i^2}$$

$$= \frac{\alpha}{2}\frac{\bar{u}_c''}{\bar{u}_c'}|\phi_c|^2 \left[\tan^{-1}\left(\frac{\bar{u}-c_r}{c_i}\right)\right]_{\bar{u}(y_c-\delta)}^{\bar{u}(y_c+\delta)}$$

from which the result (7.16) follows. An alternative derivation could be based upon the Plemelj formulas from complex variable theory (see e.g., [Ref. 7.9, p. 413]). The contour of integration would be as indicated in Fig. 7.4 with $\bar{u}$ viewed by analytic continuation as a function of $\zeta = y + iz$, say. Since the singularity occurs at $\zeta = y_c + ic_i$, as $c_i \to 0$ the path of integration must be indented as shown.

In most cases, the possibility $\phi_c = 0$ can be ruled out. That fact permits us to draw some useful conclusions about neutral modes from (7.16). For example, in the case of a monotonic velocity profile, $\tau = 0$ on either side of y_c so there can be no jump in τ. It follows that $\bar{u}_c'' = 0$ and this condition determines c for a neutral mode. On the other hand, for a jet-type profile it is possible that the jumps across the two critical points will cancel, thus permitting τ to be a nonzero constant between these points.

Two important examples for which regular neutral mode solutions are known in closed form are the following:

a) Free shear layer: $\bar{u} = \tanh y$

$$c = 0, \quad \alpha = 1, \quad \phi = \operatorname{sech} y ;$$

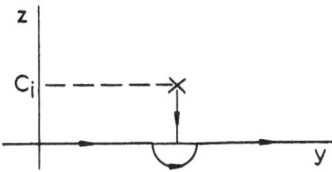

Fig. 7.4. Contour of integration for solving Rayleigh's equation when $c_i \to 0$ in the case $\bar{u}'_c > 0$

b) Bickley jet: $\bar{u} = \mathrm{sech}^2 y$
  i) odd (varicose) mode; $c = 2/3$, $\alpha = 1$, $\phi = \mathrm{sech}\, y \tanh y$;
  ii) even (sinuous) mode; $c = 2/3$, $\alpha = 2$, $\phi = \mathrm{sech}^2 y$.

Note that even though the second example involves a nonmonotonic profile, $\tau = 0$ everywhere because both inflection points occur at the same value of $\bar{u}$. That would not be the case for an asymmetric jet; unfortunately, though, there are no such closed-form solutions to illustrate the point. For further examples of neutral solutions, the reader is referred to the very useful collection in [7.5].

The Reynolds stress is also significant in the energy approach to stability theory. Although that approach is not the subject of this chapter, it is nevertheless worth writing down the energy transfer equation because of some physical insights that it provides. We begin by writing $u = \bar{u}(y, t) + u'(x, y, t)$ and $v = v'(x, y, t)$, where the fluctuating quantities have zero mean and are presumed periodic in x. These representations are substituted into the Navier-Stokes equations and averages are taken by integrating over one wavelength in x. Equations result for both the mean and fluctuating quantities. After forming a suitable linear combination of these equations and then integrating in the y direction between the boundaries, one obtains, without approximation, the energy equation

$$\frac{\partial}{\partial t} \int\int \tfrac{1}{2}(u'^2 + v'^2)dxdy = \int\int (-u'v')\frac{\partial \bar{u}}{\partial y} dxdy$$

$$- \frac{1}{\mathrm{Re}} \int\int \left(\frac{\partial v'}{\partial x} - \frac{\partial u'}{\partial y}\right)^2 dxdy . \qquad (7.17)$$

Before discussing the interpretation of (7.17), it should be remarked that its detailed derivation is not trivial and the form above, due to *Stuart* [7.10], differs somewhat from that discussed by a number of earlier investigators. The mean flow here is not the original laminar flow, but rather the x-averaged flow which is modified by the action of the time-dependent Reynolds stresses. For a more detailed derivation and a discussion of the reason for this choice of $\bar{u}$ the reader is referred to [7.10].

Returning now to (7.17), the lhs represents the change with time of the perturbation kinetic energy E, say. The dissipative term multiplied by Re^{-1} always causes E to decrease so instability can result only if the first term on the rhs is positive. That term represents the exchange of energy between the mean flow and perturbation through the action of the Reynolds stresses and it can be of either sign. It is easy to conclude incorrectly from (7.17) that viscosity is always stabilizing. However, as discussed by *Prandtl* [7.11] in 1935, viscosity can cause a certain phase relationship to exist between u' and v' across a thin horizontal layer such that the mean flow will transfer energy to the perturbation via the Reynolds stresses. It is this very subtle mechanism, known as Tollmien-Schlichting instability, which leads to linear instability in noninflectional velocity profiles. The mathematical aspects will be developed shortly in Sect. 7.2.5.

7.2.4 Broken-Line Profiles

It is often a useful simplification in stability studies to employ discontinuous models as approximations to continuous flows, one advantage being that the dispersion relation is obtainable in closed form. A certain amount of insight can be gained into mechanisms of instability and estimates of growth rates readily computed. An early example is the Kelvin-Helmholtz flow investigated during the late 1800's. This two-layer model consists of the velocity and density profiles

$$\bar{u} = \begin{cases} U_1 & y>0 \\ U_2 & y<0, \end{cases} \qquad \bar{\varrho} = \begin{cases} \varrho_1 & y>0 \\ \varrho_2 & y<0. \end{cases}$$

The objective was to study the mechanism by which wind generates waves on the surface of water. Suppose that the equation of the interface between the fluids (whose exact location is unknown) is given by $y=\eta(x,t)$. There are two conditions which must be imposed at the interface. First, a kinematic condition stating essentially that fluid particles on either side of the interface follow the same streamline trajectory and, second, that the pressure is continuous across the interface. A good source for the mathematical statement of these nonlinear conditions is the text by *Stoker* [7.12].

The solution procedure generally followed is to expand all the unknown quantities in Taylor series about $y=0$. Once the perturbation stream function $\hat{\psi}$ has been found in each region, the interface conditions are imposed. That is a simple matter in the linearized Kelvin-Helmholtz problem where $\hat{\psi}$ satisfies Laplace's equation; the required solution is

$$\hat{\psi} = \{\exp[i\alpha(x-ct)] + *\} \exp(-S\alpha y),$$

where $S=\text{sgn}(y)$. Imposing simultaneously the interface conditions leads to a matrix eigenvalue problem whose solution yields the dispersion relation

$$c = \frac{\varrho_1 U_1 + \varrho_2 U_2}{\varrho_1 + \varrho_2} \pm \left[\frac{g}{\alpha}\left(\frac{\varrho_2 - \varrho_1}{\varrho_2 + \varrho_1}\right) - \frac{\varrho_1 \varrho_2}{(\varrho_1 + \varrho_2)^2}(U_1 - U_2)^2 \right]^{1/2}. \tag{7.18}$$

Clearly, $c_i \neq 0$ if $(U_1 - U_2)^2 > g(\varrho_2^2 - \varrho_1^2)/(\alpha \varrho_1 \varrho_2)$, where g is the gravitational constant.

There is an interesting relationship between the Kelvin-Helmholtz flow and any *continuous* stratified mixing layer whose velocity and density profiles approach constant values as $y \to \pm \infty$. *Drazin* and *Howard* [7.13] have shown that (7.18) is a valid long-wave approximation to any such flow, i.e., in an expansion in powers of α and Richardson number (see Chap. 8), (7.18) is the lowest-order approximation. Thus, on the scale of a long wave, any mixing-layer profile "looks like a vortex sheet". Without stratification (i.e., $\varrho_1 = \varrho_2$), the two-layer model is called the Helmholtz flow and there is instability for all α.

Broken-line profiles have been used widely, especially in geophysical fluid dynamics. They provide good estimates for αc_i when α is small, say less than 1/3. However, there are also some dangers in the use of discontinuous models and not all fluid dynamicists seem to be aware of these. They can on occasion yield spurious modes of instability (see [7.14] for discussion), but more often do the reverse – fail to reveal unstable modes that become degenerate in the long-wave limit. The compressible mixing layer provides a good illustration of the latter circumstance. It is known that a compressible vortex sheet is stable for all Mach numbers greater than $2^{1/2}$. However, *Blumen* et al. [7.15] found numerically that a second mode of instability exists in the continuous case and is present at all finite Mach numbers. As $\alpha \to 0$, $c_i \sim \alpha$ for this second mode and in the vortex-sheet ($\alpha = 0$) limit it consequently disappears.

Some other peculiarities, as well as the positive side of the coin, can be illustrated with the three-layer approximation to the homogeneous $\tanh y$ shear layer, namely,

$$\bar{u} = \begin{cases} y & -1 \leq y \leq 1 \\ \text{sgn}(y) & |y| > 1 . \end{cases} \tag{7.19}$$

As more layers are added, one expects the results to become a better approximation to the corresponding continuous flow, and that is generally the case. For the three-layer model (7.19), the dispersion relation turns out to be

$$c^2 = (1 + 4\alpha^2 - 4\alpha - e^{-4\alpha})/4\alpha^2 . \tag{7.20}$$

The neutral wave number according to (7.20) is $\alpha_n \approx 0.64$ which is not very close to the result $\alpha_n = 1$ for the $\tanh y$ shear layer. (However, it is a significant improvement compared with the Helmholtz flow where there is instability for all α.) Since there is instability in this example, of primary interest is the wave number of the fastest growing wave and the corresponding value of αc_i. Agreement here is quite good between the numerical computations of *Michalke* [7.16] for the $\tanh y$ profile ($\alpha c_i = 0.19$ at $\alpha \approx 0.44$) and the result obtained by *Miles* and *Howard* [7.17] from (7.20), viz., $\alpha c_i \approx 0.20$ at $\alpha = 0.40$.

However, this model also illustrates some peculiarities that are characteristic of broken-line profiles. First of all, (7.20) yields a continuous spectrum of travelling neutral waves for $\alpha > \alpha_n$, whereas neither the solutions of the OS equation nor those of the Rayleigh equation in the continuous case exhibit such behavior. (Interestingly, as pointed out in [7.45], these modes may have their counterparts in the nonlinear critical layer theory.) A second possibly anomalous feature has to do with the behavior of αc_i as the stability boundary is approached.

Suppose we consider the complex frequency $\omega = \alpha c$ and assume that there is instability, i.e., $\omega_i > 0$. From (7.20), the "group velocity" is given by

$$\frac{d\omega}{d\alpha} = i \frac{1 - 2\alpha - \exp(-4\alpha)}{[\exp(-4\alpha) + 4\alpha - 1 - 4\alpha^2]^{1/2}} . \tag{7.21}$$

As $\alpha \uparrow 0.64$ the denominator of (7.21) vanishes so that $\omega'(\alpha)$ becomes infinite; this behavior can be seen in the numerical results of [7.17]. This physically implausible result, while not of any real concern in a linear stability study, has important consequences in weakly nonlinear developments, because there one usually perturbs away from the stability boundary (see Sect. 7.4.1). To summarize, discontinuous models can serve as very useful approximations, but they need to be used with care. One should not jump to major conclusions solely on the basis of a study of discontinuous flows, but rather view such results as suggesting profitable areas of further study.

7.2.5 Asymptotic Solution of the Orr-Sommerfeld Equation

The asymptotic theory of the OS equation (7.5) has been discussed at great length elsewhere (see, e.g., the authoritative article of *Reid* [7.18]) so here only an outline will be presented. Although the particular scheme that will be employed does not yield the most accurate results possible, they are satisfactory for most purposes and entail a minimum of complexity. Thus, we shall follow what is basically the "heuristic approach" in [7.4, Sect. 8.5]. This procedure has been further refined by *Graebel* [7.19] using the method of matched asymptotic expansions and those refinements are incorporated herein. The principal advantages of the latter method are that higher approximations can be obtained in a systematic way and that it provides some valuable physical insights.

To begin with, we set $\delta = (\alpha \,\mathrm{Re})^{-1}$ and, assuming that $\delta \ll 1$, let us try to find an "outer expansion" of (7.5) in a straightforward power series of the form

$$\phi \sim \sum_{n=0}^{\infty} \delta^n \phi^{(n)} .$$

The zeroth-order term $\phi^{(0)}$ satisfies the Rayleigh equation (7.7); however, solutions of (7.7) fail to describe the flow either in the neighborhood of a critical point or in the region adjacent to a solid boundary. Within such layers one must conclude that the viscous terms are significant even at very large values of Re.

Near the critical point y_c the general solution of (7.7) can be obtained by the method of Frobenius and, after first expanding $\bar{u} - c$ and $\bar{u}''$ in Taylor series about y_c, we obtain

$$\phi^{(0)} = A\phi_A + B\phi_B ,\tag{7.22}$$

where

$$\phi_A = (y - y_c) + (\bar{u}_c''/\bar{u}_c')(y - y_c)^2/2 + \dots$$

and $$\tag{7.23}$$

$$\phi_B = 1 + \dots + (\bar{u}_c''/\bar{u}_c')\phi_A \log(y - y_c) + \dots .$$

[Note that (7.23) is valid for $c_i \neq 0$, as well as for $c_i = 0$; the critical point is then located off the real axis.] The logarithmic singularity appearing in ϕ_B is significant and leads to two difficulties in the case of a neutral or nearly neutral mode. First, the horizontal perturbation velocity is proportional to ϕ' and this quantity becomes unbounded as $y \to y_c$. (In reality, viscosity mitigates that effect, but strong gradients in u' are indicated nonetheless so this region is signalled as being of physical significance.)

The second difficulty concerns the eigenvalue problem. In the usual approach to turning-point problems, one tries to find a solution continuous in the complex plane by following a path passing above or below the singular point. If ϕ_B is written in the form (7.23) for $y > y_c$, then the $\log(y - y_c)$ term must be written as $\log|y - y_c| \pm i\pi$ when $y < y_c$, the sign depending on whether the contour is indented above or below (as in Fig. 7.4). Within the context of the inviscid normal mode theory, it cannot be decided which is the proper branch for the logarithm. However, by considering the inviscid limit of (7.5), it will be shown that the minus sign is the correct choice (for $\bar{u}_c' > 0$) and one says that there is a "$-\pi$ phase change" as y_c is crossed from above. (An alternative formulation is to always write the log term as $\log|y - y_c|$ and then allow the constants A and B to be different on opposite sides of y_c.)

To investigate the critical layer, (7.5) must be rescaled in such a manner that the previously neglected viscous terms enter into the primary balance. This is accomplished by introducing the "inner variables"

$$\eta = (y - y_c)/\delta^{1/3} \quad \text{and} \quad \chi(\eta) = \phi(y) ,$$

in terms of which, (7.5) becomes

$$\chi^{IV} - i\bar{u}_c'\eta\chi'' = \delta^{1/3}i\bar{u}_c''(\eta^2\chi''/2 - \chi) + O(\delta^{2/3}) .\tag{7.24}$$

Similarly, a viscous wall layer must be introduced in order to satisfy the no-slip condition at $y = 0$; its thickness turns out to be $O(\delta^{1/2})$. Hence, the picture in the

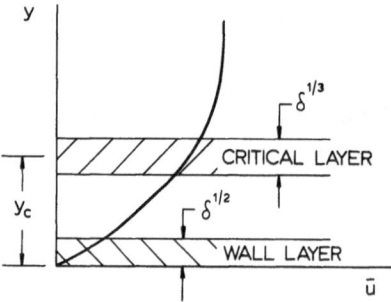

Fig. 7.5. Boundary layer profile with layers indicated where viscosity is most significant

case where the critical layer and wall layers are distinct corresponds to the sketch in Fig. 7.5.

Of course, a difficulty in doing asymptotics in an eigenvalue problem is that one does not know *a priori* that c_r, and therefore y_c, will be large enough so that the critical and wall layers are distinct. It often turns out, in fact, that c_r is so small that a single layer beginning at $y=0$ and having a thickness of $O(\delta^{1/3})$ includes both. That case, insofar as the eigenvalue problem is concerned, is simpler because there is just one viscous layer and an inviscid outer region.

We shall assume here that the critical layer is separate from the wall layer and attempt to find a solution of (7.24) in the form

$$\chi \sim \sum_{n=0}^{\infty} \delta^{n/3} \chi_n .$$

(7.25)

The first term χ_0 satisfies (7.24) with the rhs equal to zero. This is just Airy's equation for χ_0''; four independent solutions are

$$\chi_{01}=\eta, \quad \chi_{02}=1, \quad \chi_{03}=\int_{\infty}^{\eta} d\eta \int_{\infty}^{\eta} d\eta \eta^{1/2} H_{1/3}^{(1)} [\tfrac{2}{3}(i\alpha_0\eta)^{3/2}]$$

and

$$\chi_{04}=\int_{-\infty}^{\eta} d\eta \int_{-\infty}^{\eta} d\eta \eta^{1/2} H_{1/3}^{(2)} [\tfrac{2}{3}(i\alpha_0\eta)^{3/2}] ,$$

where $\alpha_0=(\bar{u}_c')^{1/3}$ and the $H_{1/3}^{(i)}$ are Hankel functions. Writing the general solution as

$$\chi_0=\sum_{i=1}^{4} C_i^{(0)} \chi_{0i} ,$$

(7.26)

the constants $C_i^{(0)}$ are determined by letting $|\eta|\to\infty$ and matching to the outer solution. This leads to the results $C_1^{(0)}=A\delta^{1/3}$, $C_2^{(0)}=B=1$, and $C_4^{(0)}=0$. The constant B has been set equal to unity as it is the arbitrary constant in the eigenvalue problem, while $C_4^{(0)} = 0$ because χ_{04} exhibits exponential growth and therefore cannot be matched to the outer expansion; $C_3^{(0)}$ is still undetermined at this point, but is fixed eventually by the no-slip condition at the wall.

With the solution now determined above and in the critical layer, the proper branch for the logarithm when $y < y_c$ is found by matching (7.22, 23) to the expansion of χ valid as $\eta \to -\infty$. To settle the issue one must consider the second term of the inner expansion, which satisfies

$$\chi_1^{IV} - i\bar{u}_c'' \eta \chi_1'' = i\bar{u}_c''(\eta^2 \chi_0''/2 - \chi_0). \tag{7.27}$$

The solution of (7.27) can be found by variation of parameters and it develops that the particular integral involving χ_{02} is the required part of the general solution. Using the asymptotic expansion of $H_{1/3}^{(1)}$, it is found that

$$\chi_{02} + \delta^{1/3}\chi_{12} \sim 1 + \delta^{1/3}(\bar{u}_c''/\bar{u}_c')\eta \log \eta + \dots, \quad \text{as} \quad \eta \to \infty \tag{7.28}$$

and this clearly matches to ϕ_B for $y > y_c$. [Note also that if (7.28) is expressed in outer variables a $\log \delta$ term arises that cannot be matched to the outer expansion; this difficulty is eliminated by modifying the inner expansion to be $\chi \sim 1 + \delta^{1/3} \log \delta \chi_1 + \delta^{1/3}\chi_2 + \dots$ in place of (7.25).] Finally, expanding $H_{1/3}^{(1)}$ for large negative values of η shows that $\phi^{(0)}$ can be matched to χ only if the log term in (7.28) is written with a $-\pi$ phase shift for $y < y_c$.

The success of this procedure turns out to depend upon the sign of c_i. Let us first consider $c_i > 0$ and then outline what happens as this quantity passes through zero to negative values. First, we note that there is always a small circular region centered at the critical point (which is generally located off the real axis; cf. Fig. 7.4) in which viscosity is important (see [Ref. 7.4, Fig. 8.2]). As $c_i \to 0$, this region approaches the real axis. When the viscous region does include part of the real axis a critical layer is required and the matching is influenced by Stokes' phenomenon, i.e., the required expansions of the Hankel function are valid only in certain sectors of the complex plane. For c_i small and positive the sectors in which the expansions are valid include the real axis to the extent required, i.e., there is an overlap domain in which both inner and outer solutions are valid so matching can be accomplished. This is also true in the case of a neutral mode and it is found that there is a smooth transition from the unstable case as $c_i \downarrow 0$.

However, the damped waves, at least for bounded flows, do not conform to the above picture and must be studied by the comparison equation method. The reason is that the effect of viscosity is no longer confined to a critical layer whose thickness approaches zero as $Re \to \infty$. Thus, a solution uniformly valid in y must be found (see, e.g., [7.20]) and this solution is not related to one associated with Rayleigh's equation. It is believed by some that these damped modes are relevant to turbulence, but whether or not that speculation has merit, it can be more safely said that they are of some significance in nonlinear studies of flows such as pipe flow that do not exhibit linear instability.

At the start of this section, it was noted that more accurate asymptotic procedures have been formulated. In particular, *Lakin* et al. [7.21] have employed the generalized Airy functions introduced earlier by *Reid* and obtained a very close approximation to the neutral curve for Poiseuille flow. Their results seem to have an accuracy comparable with some of those found by computer. This seems to have been achieved by concentrating upon the approximation of the eigenvalue relation describing the neutral curve rather than upon the eigenfunction as in previous efforts.

7.2.6 Numerical Solution of the Rayleigh and Orr-Sommerfeld Equations

It is evident from the discussion above that the Rayleigh equation is of most interest in the context of describing the instability of inflectional velocity profiles. Often, it is desired to compute curves of constant αc_i in order to find the fastest growing wave in a particular flow. This wave will, of course, be the most conspicuous in an experiment. It is not difficult to solve (7.7) numerically; a fourth-order Runge-Kutta scheme is usually adequate and is easy to program. The procedure is to obtain two linearly independent solutions by integrating from one boundary y_1, say, to the other, y_2. By a suitable choice of the constants multiplying these two solutions a linear combination satisfying the condition $\phi(y_1)=0$ is determined. However, for an arbitrary choice of the pair (α, c) the BC $\phi(y_2)=0$ will generally not be satisfied. An iteration procedure must then be employed to solve the eigenvalue problem relating α and c. For example, α can be kept fixed while c_r and c_i are perturbed by an algorithm such as Newton's method until $\phi_r(y_2)=\phi_i(y_2)=0$.

The procedure just described is an example of the initial-value or "shooting" method. The first three chapters of the text by *Betchov* and *Criminale* [7.22] provide a useful introduction to such methods in the context of hydrodynamic stability. Many computational results are presented there, but the reader is cautioned to be aware of printing errors in Chap. III.

When singular neutral modes are to be computed (e.g., in the case of an asymmetric jet), some analytical work is required to take account of the transition across the critical points. The power series solutions (7.23) can be employed on either side of y_c with the $-\pi$ phase shift imposed. However, some accuracy will be lost in the region near the critical point due to steep gradients in ϕ there. This can be avoided by writing ϕ as the sum of a regular and a singular part and solving only for the regular part on the computer. It should be added that these considerations apply not just for neutral modes, but whenever c_i is small.

When the flow is unbounded, the integration is initiated at some large finite value of y, say $y_1 = -3$. The BC at $-\infty$ is satisfied if $A_1 = 0$ in (7.8). A possible procedure for solving the eigenvalue problem would be to then integrate (7.7) from $y_1 = -3$ to $y_2 = 3$ and attempt to find values of c such that

$$\phi'(3) + \alpha\phi(3) = 0 \; ,$$

i.e., the solution behaving as $\phi \sim \exp(-\alpha y)$ is desired. However, there is a danger of numerical instability at large positive values of y. To see how this instability arises, suppose that the values of α and c corresponding to a true eigensolution are employed in (7.7) and we integrate the equation out to large y. At some point, the numerical solution for ϕ will begin to behave like $\exp(\alpha y)$; even if the solution were exact up to some particular value of y, say y_i, the inevitable truncation error in marching from y_i to $y_i + \Delta y$ will be interpreted by the computer as being due to a small component of the growing solution, which we may write as $\varepsilon \exp(\alpha y)$. Despite the smallness of ε, it does not take long for this solution to dominate because the true solution is exponentially decaying. One way to avoid the difficulty is to integrate inward from both sides and match the two solutions at $y = 0$ (e.g., by making their Wronskian vanish).

A final note about solving (7.7) is that sometimes a coordinate transformation can be employed to great advantage. The study described in [7.16] of the mixing layer $\bar{u} = \tanh y$ provides a nice example. In that work, the independent variable $z = \tanh y$ was used in conjunction with the Riccati transformation $\Phi = \phi'/\phi$. The result is a first-order nonlinear ODE for $\Phi(z)$. As it turns out, $\Phi(z)$ is a slowly varying function that is better suited to numerical computation than $\phi(y)$. While such a successful transformation is not often feasible, the potential benefits are great enough so that the attempt to find a suitable transformation is deserving of some effort.

The OS equation is considerably more difficult to solve than its inviscid counterpart due to the small-scale viscous structure present at large values of Re. *Thomas* [7.23], who obtained the first numerical solution of (7.5), used the finite-difference (or matrix) method. Because of the necessity to invert a large, possibly ill-conditioned, matrix in that procedure, it was not used extensively in the 1960's. Instead, adaptations of the shooting method were most widely employed. Recently, however, due to advances in matrix inversion techniques, the finite-difference method has returned to popularity. Both methods are discussed briefly below beginning with the initial-value approach.

In order to see the source of the numerical instability, let us consider the case of a boundary layer and suppose that we integrate inward from the free stream $y \gg 1$ to $y = 0$. Two linearly independent solutions are desired which, for large values of y, behave as

$$\phi_1 \sim \exp(-\alpha y) \quad \text{and} \quad \phi_2 \sim \exp\{-\alpha[1 + i(1-c)\text{Re}/\alpha]^{1/2}y\} \; ,$$

where it has been assumed that $\bar{u} \to 1$ and $\bar{u}'' \to 0$ for large y. There is no difficulty in computing ϕ_2; however, in attempting to compute ϕ_1, it will develop after several steps that the solution begins to look like ϕ_2. The reason is that the always-present round-off and truncation errors introduce a small amount of the viscous solution into the computation of ϕ_1 and this dominates before long because ϕ_2 grows as $\exp(\alpha \, \text{Re})^{1/2}$. The two solutions are no longer linearly independent and the cure for this ailment is to impose an "orthogonality condition" which essentially subtracts out the spurious part of the result at each step. A second aspect of the difficulty has to do with the actual magnitudes of the two solutions, i.e., there is an arbitrary constant multiplying each ϕ_i and it may be necessary to decrease that constant as the integration proceeds in order to maintain accuracy. The procedure in which one alters the constants as well as maintaining linear independence is known as orthonormalization and the reader is referred to [7.24] for a description of a relatively simple program incorporating these techniques.

The finite-difference method avoids the difficulties described above because the homogeneous BCs at both end points are imposed at the same time the solution is being determined in the interior. This prevents the unbounded growth described above, but difficulties in the form of spurious modes may be encountered on occasion. Nonetheless, the procedure is straightforward and efficient library subroutines are readily available for the matrix inversion. The computations by *Thomas* [7.23] for Poiseuille flow must certainly be regarded as an impressive feat. Using single-precision arithmetic he found $\text{Re}_{\text{crit}} = 5780$, a result very close to the "official" value of 5772.22 due to *Orszag* [7.25].

The author [7.26] has found a simple version of this method having second-order accuracy to be adequate in a study of the weakly nonlinear stability of a tanh y shear layer. When greater accuracy is required this can be accomplished efficiently by using Richardson extrapolation [7.27]. Alternatively, a fourth-order difference scheme used in conjunction with a new iteration method was reported by *Osborne* [7.28] to be very fast. Finally, *Gary* and *Helgason* [7.29] have employed the QR algorithm in a procedure that is quite effective when a good initial guess is not available for the eigenvalues. A second advantage is that all the higher modes are obtained at the same time as the fundamental for a given α and Re. The disadvantage of the QR algorithm is that it requires more computation time, because the relevant matrix is no longer banded.

A third technique (which is also a matrix method) that has been used on occasion to solve (7.5) is to expand ϕ in a series of orthogonal polynomials. Chebyshev polynomials have the greatest capability to yield results of high accuracy and solutions in [7.25] were obtained for values of Re as large as 50,000. There are some rather involved manipulations required, so this method should be used only if great accuracy is required and/or high values of Re are to be considered.

7.3 The Linear Initial-Value Problem

Up to this point, we have restricted attention to perturbations consisting of individual normal modes. In certain situations, however, that approach is not entirely adequate and consideration of more general disturbances is desirable as discussed below.

7.3.1 Inviscid Theory

One situation, already encountered, where difficulties arise in the normal mode approach is when the velocity profile under consideration has no inflection point. In that case, (7.7) simply has no solution, stable or unstable, that satisfies the BCs. How then can one, without including viscosity, describe the evolution with time of a specified initial perturbation? It develops that we can do so by employing certain solutions belonging to a continuous spectrum which were excluded by the normal mode approach.

It is easiest to see how these pseudo-modes, as they were termed in [7.30], arise by considering the Couette flow $\bar{u} = y$, $-1 \leq y \leq 1$. The Rayleigh equation for that case takes the form

$$(y-c)(\phi'' - \alpha^2 \phi) = 0, \qquad \phi(-1) = \phi(1) = 0 \tag{7.29}$$

and, if we divide through by $(y-c)$, there are no solutions vanishing at both boundaries. However, suppose singular solutions are permitted by rewriting (7.29) as

$$\phi'' - \alpha^2 \phi = \delta(y-c) . \tag{7.30}$$

The solution of (7.30) satisfying the BCs is simply the Green's function associated with (7.29). Hence, ϕ is a generalized function having a discontinuity in its first derivative at $y = c$. A continuum of such modes exists filling out the range of $\bar{u}(y)$ (the existence of the continuous spectrum was, in fact, recognized by Rayleigh). By summing these pseudo-modes the singularities are smoothed out and the initial-value problem corresponding to the IC $\hat{\psi}(x, y, 0) = F(x, y)$ can be solved. The result is obtained in the form of an integral too complex to be evaluated in closed form; however, the stability of the flow can be assessed using asymptotic methods and it is found that $\hat{\psi}$ decays algebraically as $t \to \infty$, so the flow is stable. (For an interesting and more complete discussion of the Couette flow problem, including Orr's solution obtained in 1907, the reader is referred to the text of [7.31].)

The first investigation of the initial-value problem during the present era was reported by *Eliassen* et al. [7.32] who considered Couette flow with both stable and unstable density stratification. Subsequently, *Case* [7.33] (who was evidently unaware of the work in [7.32] and repeated much of it) employed more systematic procedures, namely transform methods, and drew attention to this sort of approach. Both a Fourier transform in x and a Laplace transform in t are required, in general. However, the essential features are associated with the Laplace transform inversion so let us write $\hat{\psi} = \exp(i\alpha x)\Phi(y, t)$ and substitute into the linearized, inviscid equivalent of (7.3). (To deal with an arbitrary initial condition a sum over α would be employed.) We obtain

$$\frac{\partial}{\partial t}(\Phi_{yy} - \alpha^2 \Phi) + \bar{u} i\alpha(\Phi_{yy} - \alpha^2 \Phi) - i\alpha \bar{u}'' \Phi = 0 \tag{7.31}$$

which will be reduced to an ODE by taking a Laplace transform in time.

If we denote the transform of Φ by $\tilde{\Phi}(y, p)$, then according to the inversion formula

$$\Phi(y, t) = \frac{1}{2\pi i} \int_{\gamma - i\infty}^{\gamma + i\infty} \tilde{\Phi}(y, p) e^{pt} dp ,$$ (7.32)

where the path of integration is a line parallel to the imaginary axis in the complex p plane and to the right of all singularities. After transforming (7.31), it is found that the quantity $\tilde{\Phi}$ satisfies

$$(\bar{u} - ip/\alpha)(\tilde{\Phi}'' - \alpha^2 \tilde{\Phi}) - \bar{u}'' \tilde{\Phi} = -(i/\alpha)F(y) ,$$ (7.33)

where $F(y) = (\Phi_{yy} - \alpha^2 \Phi)$ evaluated at $t = 0$. Equation (7.33) is, of course, just the analog of the Rayleigh equation. Once (7.33) has been solved for $\tilde{\Phi}$, the techniques of complex variable theory can be used to evaluate (7.32) provided that $\tilde{\Phi} \to 0$ on the appropriate semicircle in the left side of the complex p plane. The solution will then consist of a sum of residues at the poles of $\tilde{\Phi}$ plus the contribution of integrals along any branch cuts. Clearly, the usual normal modes will be recovered from poles that occur in the homogeneous solution of (7.33), instability corresponding to the real part of p positive.

The continuous spectrum, on the other hand, is associated with branch points occurring at singularities of (7.33) where $p = -i\alpha\bar{u}(y)$. Thus, there are a continuum of pseudo-modes analogous to those found by a slightly different, but equivalent, procedure in the context of (7.29). The asymptotic evaluation of the integrals arising from such branch cuts is a difficult matter and even in the simplest case – that of Couette flow – there is disagreement as to the rate of decay of $\hat{\psi}$. For an initial disturbance occupying a finite domain in x, it was found in [7.32] that $\hat{\psi} \sim O(t^{-2})$, whereas it was incorrectly concluded in [7.33] and [7.31] that $\hat{\psi} \sim O(t^{-1})$. Engevik [7.34] has, in fact, shown that the latter result obtains only when the integral over α of the initial vorticity is infinite. This difference is significant because the horizontal perturbation velocity is given by $\hat{\psi}_y$ and since $\hat{\psi}$ typically contains terms behaving like $t^{-2} \exp(iyt)$, $\hat{u} \sim O(t^{-1})$; if $\hat{\psi}$ had been $O(t^{-1})$, $\hat{u}$ would not have decayed and one could not say that the flow is stable. The result $\hat{\psi} \sim O(t^{-2})$ seems to be typical of homogeneous parallel flows as shown by the analysis in the Appendix of [7.35].

As a procedure for evaluating the stability of a parallel flow, the initial-value method must be regarded as being primarily of academic interest. The new feature introduced, viz., the continuous spectrum, seems to generally decay and in those rare cases when it can grow (see [7.36] for an example) there are present exponentially growing modes, as well, which will dominate. There are, however, certain problems in geophysical fluid dynamics that are posed most naturally as initial-value problems and these have revived interest in the technique. Although such problems, often involving forced waves, are not, strictly speaking, stability problems similar issues are raised. In particular, critical point singularities often occur and are of great physical significance; some examples will be discussed in Sect. 7.4.3.

In the Laplace transform inversion integral (7.32), the path is dictated by convergence requirements and the issue raised in Sect. 7.2.3 about which way to indent the contour is clear cut. If the path of integration is deformed so that it proceeds along the imaginary p axis, then it must be indented to the right of any singularities on that axis. Thus, a phase change occurs that in the case of a neutral mode agrees exactly with the result obtained by taking the inviscid limit of the OS equation. However, it is not true, as is often stated, that the two procedures are equivalent. Including viscosity allows one to determine the structure of a neutral mode in the critical layer. The initial-value method, on the other hand, breaks down for large times in the neighborhood of y_c and so is not capable of yielding the ultimate steady-state solution, if there is one. This point will be taken up again in Sect. 7.4.3 where recent attempts to resolve that difficulty by adding nonlinear effects will be reviewed.

7.3.2 The Initial-Value Problem at Finite Reynolds Number

For bounded flows (e.g., of the Poiseuille type) the OS equation has a complete set of normal modes and one can therefore describe an arbitrary initial perturbation in terms of them. However, in flows of the boundary layer type, where the domain is semi-infinite, there exists in addition to the normal modes a continuous spectrum consisting of improper eigenfunctions that are bounded at infinity, but do not decay to zero there [7.37]. It is shown in [7.38] how these improper modes can be integrated and then reinterpreted in terms of discrete spectra of the heat-conduction type.

It is natural to enquire as to the relationship of the viscous and inviscid initial-value problems when $\mathrm{Re} \to \infty$. *Lin* [7.39] has discussed this matter by using some simple parabolic PDEs to illustrate certain points. He shows that a pseudo-mode of the "inviscid problem" can sometimes be represented by a sum of normal modes of the viscous problem (assuming the latter are complete). However, if the initial condition depends upon viscosity, then there is no correspondence. Similarly, it is shown in [7.39] that viscous modes may occur which have no limit as $\mathrm{Re} \to \infty$. Although these examples are instructive, they are less applicable to unbounded flows and it seems that only through careful numerical study can these complex issues be resolved.

7.3.3 Wave Packets

It is an exceptional situation when a perturbation consists of only a single wavelength. The utility of the normal mode approach is due to the observed dominance of the most unstable wave and the fact that single frequencies are sometimes forced in an experiment. In reality, there is a finite bandwidth of waves present and it is desirable to generalize the theory to take account of that. The concept of a wave packet is a familiar one in mathematical physics and it is well-known that the envelope of the packet propagates with the group velocity $\omega'(\alpha)$. The method of stationary phase can be used to evaluate the long-time behavior of a Fourier integral and proves to be a useful way to bring out the role of ω' in dealing with a system of dispersive waves (see, e.g., [7.40]). It develops that at a certain point x, far from the disturbance source, most of the solution at time t will be due to the packet of waves whose group velocity $\omega' = x/t$.

These ideas, long familiar in the theory of linear dispersive waves, have been extended recently to include applications in hydrodynamic stability. The method of multiple scales proves to be a useful asymptotic procedure in this endeavor and was first used by *Benney* and *Newell* [7.41] to study the potential sideband instability mechanism in a system of conservative waves. Subsequently, the procedure was utilized in weakly nonlinear studies of thermal convection [7.42, 43] and of Poiseuille flow [7.44]. A general formulation directed specifically toward parallel shear flows at large Reynolds numbers was described by *Benney* and *Maslowe* [7.45]. Their analysis is briefly outlined here and will later be taken up in the context of weakly nonlinear theories.

Suppose that we wish to consider slow modulations of a nearly neutral wave over distances and times that are long compared to the basic wavelength and period. To that end, let us introduce the slow space and time variables $X = \mu x$ and $\tau = \mu' t$, where $\mu, \mu' \ll 1$. Moreover, we replace the constant amplitude of the eigenvalue problem by the quantity $A(X, \tau)$. Thus, the perturbation stream function is now written

$$\hat{\psi} = A(X, \tau)\phi(y)\exp[i\alpha(x - ct)] \, ,$$

and we wish to derive the amplitude evolution equation satisfied by A. The method of multiple-scales [7.2, 3] will be employed wherein x and t derivatives in the basic PDEs are transformed according to

$$\frac{\partial}{\partial t} \to \frac{\partial}{\partial t} + \mu'\frac{\partial}{\partial \tau} \quad \text{and} \quad \frac{\partial}{\partial x} \to \frac{\partial}{\partial x} + \mu\frac{\partial}{\partial X} \, .$$

In general, we may take $\mu' = \mu$ although some important exceptions will be noted later. Finally, to generalize slightly the results in [7.45], let us suppose that there is some parameter R in the problem (other than the wave number) such that instability occurs for R slightly greater than R_n, its neutral value. Under these circumstances, it is shown in [7.45] that A satisfies the PDE

$$\mu\left(\frac{\partial A}{\partial \tau} + \frac{\partial \omega}{\partial \alpha}\frac{\partial A}{\partial X}\right) - \tfrac{1}{2}i\mu^2\frac{\partial^2\omega}{\partial\alpha^2}\frac{\partial^2 A}{\partial X^2} + \ldots = \frac{\partial\omega}{\partial R}(R - R_n)A \, . \tag{7.34}$$

To arrive at this result, $\hat{\psi}$ was expanded in powers of $i\mu$. The quantities $\partial^n\omega/\partial\alpha^n$ are given by expressions involving integrals which may be singular in the inviscid theories. The specific manner of computing these coefficients will vary from one problem to another so such details should not be stressed at this point. What is significant is the form of the amplitude equation (7.34), which seems to be quite general. We mention finally, that in some circumstances it may be advantageous to assume that the dispersion relation is given implicitly, i.e., in the form $F(\omega, \alpha, R) = 0$; *Weissman* [7.46] has recently given a useful formulation applicable to that case.

7.4 Nonlinear Theories

The theories presented here are all based upon the *finite-amplitude* approach; i.e., one tries to formulate an analysis that will exhibit some of the nonlinear features of transition while, at the same time, exploiting the smallness of an amplitude parameter ε in order to simplify the mathematics. First, the now well-established weakly nonlinear theory is presented. That is followed by an outline of the more recent and "more nonlinear", but still finite-amplitude, nonlinear critical layer approach for neutral modes. Finally, current research is described having as its goal the description of time-dependent perturbations evolving toward a nonlinear critical layer state.

7.4.1 Weakly Nonlinear Theory

One way to approach this topic which appeals to the writer is to begin by citing some of the deficiencies of linearized theory that one might hope to remedy by a nonlinear analysis (i.e., as opposed to trying to directly relate the nonlinear theory to transition). That point of view was adopted in an early survey article by *Stuart* [7.47] and it still seems appropriate. Certainly the most obvious flaw of a linear stability analysis is that the outcome is independent of the initial perturbation amplitude. That quantity is an arbitrary constant in the eigenvalue problem. A second defect is that unsteady perturbations amplify or decay exponentially forever – behavior that is not in accord with observation. These and other defects are remedied in the weakly nonlinear theory, where the amplitude $A_1(t)$, say, is found to satisfy an equation of the form

$$\frac{1}{A_1} \frac{dA_1}{dt} = a_0 + a_2 |A_1|^2 + O(|A_1|^4) . \tag{7.35}$$

In a frame of reference moving with the wave speed c, the quantity a_0 can be identified with αc_i, the amplification factor of linearized theory. It is, however, the Landau constant a_2, that is of central interest here. Supposing, for the sake of simplicity, that a_2 is real, the following possibilities arise: i) $a_2 < 0$ means that a linearly unstable perturbation ($a_0 > 0$) will evolve toward a steady finite-amplitude state having an equilibrium amplitude $|A_e|^2 = -(a_0/a_2)$. This is termed the *supercritical* case. ii) With $a_2 > 0$, modes that would be damped

$(a_0 < 0)$ in the linearized theory can now amplify if their initial amplitude satisfies the condition $|A(0)|^2 > -(a_0/a_2)$. Such destablilization by finite perturbations is known as *subcritical instability*.

It is convenient, particularly in the subcritical case, to identify ε^2 with the initial perturbation kinetic energy. Multiple scaling will again be employed with $\tau = \alpha\varepsilon^2 t$ the slow time scale describing the evolution of the $O(1)$ wave amplitude A, where $\varepsilon A = A_1$ in (7.35). We introduce a coordinate system moving at the wave speed by writing

$$\psi = \int_{y_c}^{y} [u(y, \tau) - c]\,dy + \varepsilon\hat{\psi}(\theta, y, \tau) , \tag{7.36}$$

where $\theta = \alpha x$ and note that $\partial/\partial t = 0$ in this system. Following *Stuart* [7.48], u is separated into a mean component $\bar{u}(y)$ presumed to satisfy the steady Navier-Stokes equations and a time-dependent part that is forced by the self-interaction of the $O(\varepsilon)$ perturbation. Thus, we write

$$u(y, \tau) = \bar{u}(y) + \varepsilon^2 AA^* f(y) , \tag{7.37}$$

where the need to break up u in this way will become apparent shortly. Substituting now into (7.1) leads to the equation for $\hat{\psi}$, viz.,

$$\varepsilon^2 \nabla^2 \hat{\psi}_\tau + (u - c)\nabla^2 \hat{\psi}_\theta - u_{yy}\hat{\psi}_\theta + \varepsilon(\hat{\psi}_y\nabla^2\hat{\psi}_\theta - \hat{\psi}_\theta\nabla^2\hat{\psi}_y)$$

$$= (\alpha\,\mathrm{Re})^{-1}[\nabla^2(\nabla^2\hat{\psi}) + \varepsilon f'''|A|^2] + O(\varepsilon^3) , \tag{7.38}$$

where $\nabla^2 = \alpha^2\partial^2/\partial\theta^2 + \partial^2/\partial y^2$. The solution of (7.38) is to be found by expanding $\hat{\psi}$ as follows:

$$\varepsilon\hat{\psi} \sim \varepsilon[\phi_1(\tau, y)e^{i\theta} + \phi_1^* e^{-i\theta}] + \varepsilon^2[\phi_2(\tau, y)e^{2i\theta} + *]$$

$$+ \varepsilon^3[\phi_{31}(\tau, y)e^{i\theta} + * + \phi_{33}(\tau, y)e^{3i\theta} + *] + O(\varepsilon^4) . \tag{7.39}$$

In addition to expanding $\hat{\psi}$, it is usual to also expand certain parameters about their neutral values, the specific nature of this expansion depending upon the properties of the linear neutral curve for the flow in question. The case of Poiseuille flow is favorable as an illustration because $\bar{u} = 1 - y^2$ is an exact solution of (7.1) and there is a critical value of Re. Although it is permissible to expand about any point on the neutral curve, the critical point Re_c (see Fig. 7.6) is a natural choice. Hence, we write

$$\mathrm{Re} = \mathrm{Re}_c + \varepsilon^2\,\mathrm{Re}_2 + \dots . \tag{7.40}$$

Had we perturbed about some point other than Re_c, α would have been expanded as well.

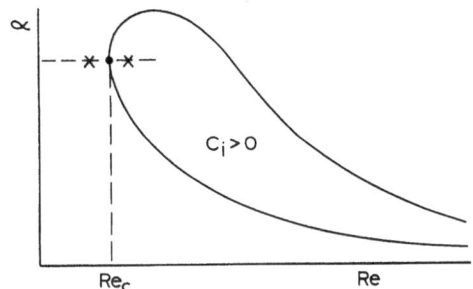

Fig. 7.6. Sketch of neutral curve for Poiseuille flow and location of critical point $Re_c = 5772$ and $\alpha_c = 1.02$

A few words about the scaling would perhaps be appropriate at this point. Anticipating that a_2 will be $O(1)$, we chose τ to be $O(\varepsilon^2)$ in order to balance the nonlinear term in (7.35). Now, a_0 is αc_i and expanding this quantity about the neutral curve as

$$\alpha c_i \approx [\partial(\alpha c_i)/\partial \, Re]_{Re_c}(Re - Re_c) + \dots \tag{7.41}$$

makes it clear why we took $(Re - Re_c) \sim O(\varepsilon^2)$ in (7.40). By so doing, a balance is achieved among the three terms displayed in (7.35).

Returning now to the expansion of $\hat{\psi}$, we begin by equating terms multiplied by equal powers of ε. At lowest order, the variables are separated by writing $\phi_1(\tau, y) = A(\tau)\Phi_1(y)$ and it is found that Φ_1 satisfies the OS equation which is to be solved numerically. At $O(\varepsilon^2)$, the $\exp(2i\theta)$ terms can be separated by taking $\phi_2 = A^2\Phi_2$, where Φ_2 satisfies

$$(\bar{u} - c)(\Phi_2'' - 4\alpha^2\Phi_2) - \bar{u}''\Phi_2 + (i/2\alpha \, Re)(\Phi_2^{iv} - 8\alpha^2\Phi_2'' + 16\alpha^4\Phi_2)$$

$$= (\Phi_1\Phi_1''' - \Phi_1'\Phi_1'')/2 \, . \tag{7.42}$$

Again, (7.42) is to be solved numerically subject to homogeneous BCs on Φ_2 and Φ_2' at both boundaries.

There is also a mean flow distortion produced at this order by the interaction of $\exp(i\theta)$ and $\exp(-i\theta)$ terms in (7.39). Equating such terms [see (7.37, 38)] leads to an ODE for f'' which can be integrated twice to obtain

$$f(y) = i\alpha \, Re \int_{y_1}^{y} (\phi_1^*\phi_1' - \phi_1\phi_1'^*)d\tilde{y} \, . \tag{7.43}$$

Comparing with (7.14) shows clearly the role of the Reynolds stress in distorting the laminar profile to one having less kinetic energy. The largest distortion for noninflectional profiles occurs across the critical layer because there the Reynolds stress is largest.

It is at $O(\varepsilon^3)$ that the slow time dependence first enters the analysis and the amplitude equation will be seen to emerge as a result. Writing

$\phi_{31} = A^2 A^* \Phi_{31}(y)$ and considering all terms multiplied by $\exp(i\theta)$ leads to the equation

$$A^2 A^* \mathscr{L} \Phi_{31} = i \frac{dA}{d\tau}(\Phi_1'' - \alpha^2 \Phi_1) + \frac{i \, \mathrm{Re}_2}{\alpha \, \mathrm{Re}_c^2} A \left(\frac{d^2}{dy^2} - \alpha^2\right)^2 \Phi_1 - A^2 A^* G(y),$$

where $\hspace{10cm}$ (7.44)

$$G(y) \equiv 2\Phi_1'^*(\Phi_2'' - 3\alpha^2 \Phi_2) + \Phi_1^*(\Phi_2''' - 3\alpha^2 \Phi_2') - (2\Phi_2 \Phi_1''' {}^* + \Phi_2' \Phi_1'' {}^*)$$
$$+ f(\Phi_1'' - \alpha^2 \Phi_1) - f'' \Phi_1$$

and $\mathscr{L}$ denotes the OS operator. A necessary and sufficient condition for the existence of a solution to (7.44) is that the rhs be orthogonal to the function satisfying the adjoint problem (see, e.g., [7.49]). The latter consists of solving the equation

$$(\bar{u} - c)(\chi'' - \alpha^2 \chi) + 2\bar{u}' \chi' + (i/\alpha \, \mathrm{Re})(\chi^{iv} - 2\alpha^2 \chi'' + \alpha^4 \chi) = 0 \qquad (7.45)$$

subject to appropriate homogeneous BCs. Imposing the orthogonality condition leads to the following result:

$$i \frac{dA}{d\tau} \int_{y_1}^{y_2} \chi(\Phi_1'' - \alpha^2 \Phi_1) dy + \frac{i \, \mathrm{Re}_2}{\alpha \, \mathrm{Re}_c^2} A \int_{y_1}^{y_2} \chi \left(\frac{d^2}{dy^2} - \alpha^2\right)^2 \Phi_1 dy$$

$$- A^2 A^* \int_{y_1}^{y_2} \chi G dy = 0 \qquad (7.46)$$

which is clearly equivalent to (7.35). Comparing with the latter shows that a_2 is given by the relation

$$a_2 = \varepsilon^2 \int_{y_1}^{y_2} \chi G dy \bigg/ \left[\int_{y_1}^{y_2} \chi(\Phi_1'' - \alpha^2 \Phi_1) dy\right]. \qquad (7.47)$$

The numerical computation required in order to arrive at a value for a_2 is considerable and it was not until 1967 that *Reynolds* and *Potter* [7.50] obtained the result $a_{2r} = 31$. Thus, subcritical instability is predicted in accord with observation. However, experimental values of the transition Reynolds number are typically in the range 1000 to 2500. That is too far away from the linear Re_c to enable quantitative application of the theory. Nonetheless, it was recently demonstrated by *Nishioka* et al. [7.51] that by reducing the background turbulence sufficiently transition could be delayed to the point where comparison with the present theory is possible. Their experiments, which largely support the concepts discussed above, are reviewed in Sect. 7.5.3.

The first actual computation of a_2 for a shear flow was that due to *Schade* [7.52] who considered the inviscid limit of a tanh y shear layer. For that case a solution was obtained in closed form with the result $a_2 = -16/3\pi$. Again, the sign of a_2 seems to be in agreement with experiment, where supercritical equilibrium states are observed to occur. That result, plus the amenability of the tanh y profile to theoretical analysis, has encouraged a number of additional studies (see, e.g. [7.26, 53, 54]). However, the problem remains far from being resolved due to a serious difficulty; the wave number of the fastest growing wave is 0.43 which is not at all close to the inviscid neutral value of $\alpha = 1$ about which one usually perturbs. The situation is only slightly improved at finite values of Re where, for example, $\alpha_n = 0.874$ at Re $= 40$, a value typical of many experiments. Yet another difficulty is that $\bar{u} = \tanh y$ is not a solution of (7.1) and it really seems here that one must come to grips with the nonparallel aspect of the basic flow. A crude attempt in that direction [7.26] succeeded in at least reproducing the asymmetry of the mean flow distortion that has been noted in experiments. Similar difficulties occur in the case of the boundary layer as discussed in [7.55]. Although calculations have been made showing that a_{2r} changes sign along the neutral curve – from negative to positive as one moves from the upper to the lower branch – there seems to be little justification for the parallel flow assumption here.

Finally, we note a class of flows for which even greater obstacles exist in developing the weakly nonlinear approach – those flows for which linear instability does not occur. In such cases, there is no neutral curve to perturb away from, so it is not clear that the amplitude equation converges for the values of A_1 that are of interest [i.e., in the usual situation we can let $A_1 \to 0$ and the form of (7.35) suggests that the series will converge for A_1 sufficiently small]. Nonetheless, it is possible to formulate a finite-amplitude analysis and that has been done for pipe flow [7.56a, b] and Couette flow [7.57]. The Landau constant was positive in each case – an encouraging result, because turbulence can occur in practice. However, more terms of the series would need to be computed before saying very much with confidence.

Extension of the theory to include wave packets is fairly straight-forward following the ideas presented in Sect. 7.3.3. Equation (7.34) can be generalized by rewriting it as

$$\mu \left(\frac{\partial A}{\partial \tau} + \frac{\partial \omega}{\partial \alpha} \frac{\partial A}{\partial X} \right) - \tfrac{1}{2} i \mu^2 \frac{\partial^2 \omega}{\partial \alpha^2} \frac{\partial^2 A}{\partial X^2} + \cdots$$

$$= a_0 A + i \varepsilon^2 \gamma A^2 A^* . \tag{7.48}$$

The case where $\partial \omega / \partial \alpha$ is real is of particular interest. When perturbing away from a point such as Re$_c$ in Fig. 7.6 the latter case corresponds to a packet centered about the most unstable wave of linearized theory. It is then appropriate to introduce a rescaled coordinate system in which the observer is moving at the group velocity by means of the transformation $\xi = X - (\partial \omega / \partial \alpha) \tau$ and $T = \mu \tau$. In terms of these variables, (7.48) takes the form

$$\frac{\partial A}{\partial T} - \frac{i}{2} \frac{\partial^2 \omega}{\partial \alpha^2} \frac{\partial^2 A}{\partial \xi^2} = \left(\frac{a_0}{\varepsilon^2} \right) A + i \gamma A^2 A^* , \tag{7.49}$$

where we have set $\mu = \varepsilon$. For the supercritical case, *Stewartson* and *Stuart* [7.44] have shown how (7.49) can be related directly to the linear initial-value problem, i.e., the solution of the latter for $t \gg 1$ provides the initial condition (IC) for $T \to 0$ of (7.49).

When $\partial^2 \omega / \partial \alpha^2$ and γ are real, with $a_0 = 0$, then (7.49) reduces to the nonlinear Schrödinger equation. The latter equation, which arises in many branches of nonlinear physics, can be solved exactly by means of the inverse-scattering method. If $\gamma \partial^2 \omega / \partial \alpha^2$ is positive, then the wave system is susceptible to a sideband instability mechanism and this will lead to the formation of envelope solitons. In systems with dissipation, however (such as those with which we are concerned here), the constants γ and $\partial^2 \omega / \partial \alpha^2$ will generally be complex. A review of the properties of (7.49), particularly in the supercritical case, is given in [7.58]. When the Landau constant is positive numerical methods are required. *Hocking* and *Stewartson* [7.59] have found in a numerical and analytical study that for certain values of the coefficients of (7.49) unbounded growth of the solution can occur at $\xi = 0$ due to nonlinear focusing. Although that result is of interest it is not clear what physical significance can be attached to it because it coincides with a breakdown of the theory as A increases.

7.4.2 The Nonlinear Critical Layer

Because the theory described in this section is relatively recent, it remains to be seen just how much of an impact it will have. However, it is already clear that it represents an important advance in the subject and has attracted a great deal of attention. Even in those cases, and there are many, where the analysis is not directly applicable it seems that the ideas are often relevant and provide some significant insights. To motivate the approach let us refer to (7.38) and note first of all that the Rayleigh equation results from setting $\varepsilon = (\alpha \, \mathrm{Re})^{-1} = 0$. Suppose now that we focus attention upon the case of a neutral mode (i.e., we ignore the first term on the lhs). Up until this point the difficulties associated with the singularity at y_c were resolved by restoring the neglected viscous terms. It was noted some time ago, however, by *Lin* and *Benney* [7.60] that a possibly more relevant alternative was to retain the nonlinear terms in (7.38) while setting the rhs to zero. That idea was exploited several years later by the use of matched asymptotic expansions in independent work by *Benney* and *Bergeron* [7.61], and *Davis* [7.62]. We follow the approach of [7.61] as that work was more oriented toward stability; *Davis*, on the other hand, was interested in critical layer effects on the generation of water waves by wind.

Returning now to (7.38), we adopt the point of view for the moment that there are two small parameters in the problem and seek an outer solution having the form

$$\hat{\psi} \sim \psi^{(0,0)} + \varepsilon \psi^{(1,0)} + (\alpha \, \mathrm{Re})^{-1} \psi^{(0,1)} + \varepsilon (\alpha \, \mathrm{Re})^{-1} \psi^{(1,1)} + \dots . \qquad (7.50)$$

Clearly, $\psi^{(0,0)}$ satisfies the Rayleigh equation which we know from (7.23) has a weak singularity; it develops, however, that $\psi^{(1,0)}$ and $\psi^{(0,1)}$ are more singular, behaving as $(y-y_c)^{-1}$, and $(y-y_c)^{-2}$, respectively. Thus, the expansion (7.50) is not uniformly valid and a boundary layer analysis is required in the neighborhood of y_c. A balance is desired between the linear and nonlinear inertial terms of (7.38) and the intent, at least initially, is to neglect viscosity. This balance can be achieved by introducing a critical layer whose thickness is $O(\varepsilon^{1/2})$.

Setting $u(y,\tau)=\bar{u}(y)$ in (7.36) and expanding $(\bar{u}-c)$ in a Taylor series about y_c, it can be seen that in the moving coordinate system the mean flow is $O(\varepsilon)$ at the edge of the critical layer, i.e., the mean flow and fundamental perturbation are the same order of magnitude. Appropriate inner variables Y and Ψ are therefore defined by

$$y-y_c=\varepsilon^{1/2}Y \quad \text{and} \quad \psi(\theta,y)=\varepsilon\bar{u}_c'\Psi(\theta,Y) .$$

Employing these variables in (7.38) yields the governing equation in the critical layer, namely,

$$\Psi_Y\Psi_{YY\theta}-\Psi_\theta\Psi_{YYY}=\lambda\Psi_{YYYY}+O(\varepsilon) , \tag{7.51}$$

where $\lambda\equiv 1/\alpha\,\mathrm{Re}\,\varepsilon^{3/2}$. The parameter λ plays an important role in the theory and it should be noted that $\lambda^{1/3}=(\alpha\,\mathrm{Re})^{-1/3}/\varepsilon^{1/2}$, i.e., the ratio of the viscous to the nonlinear critical layer thicknesses. Here, we assume that $\lambda\ll 1$. It is significant to note at this point that had we employed the viscous scaling the nonlinear terms would be multiplied by $\lambda^{-2/3}$. Hence, the viscous theory *and* the weakly nonlinear theory require $\lambda\gg 1$, as well as $\varepsilon\ll 1$. The former requirement places a severe limitation on those theories that was not previously recognized.

Apropos of the eigenvalue problem associated with (7.7), there is again a question of the phase change across y_c. Here, it is most convenient to use real variables so, for $y>y_c$, we write

$$\psi_+^{(0,0)}=(A_+\phi_A+B_+\phi_B)\cos\theta , \tag{7.52}$$

where ϕ_A and ϕ_B are given by (7.23). Moreover, we take B_+ as the arbitrary constant and set it equal to $\bar{u}_c'$. The solution in the critical layer itself is to be matched to (7.52) as $Y\to\infty$. The quantity Ψ can then be expanded for large *negative* Y and the matter is decided by seeing what linear combination of Frobenius solutions can be matched to Ψ. If there is a phase change, then it will turn out that $A_-\neq A_+$ and $\sin\theta$ terms will be generated for $y<y_c$.

Now, we are in a position to examine the behavior of the expansion for ψ as $y \downarrow y_c$; from (7.50), this is given by

$$
\psi_+ \sim \frac{\bar{u}_c'}{2}(y-y_c)^2 + \frac{\bar{u}_c''}{6}(y-y_c)^3 + \dots
$$

$$
+ \varepsilon \left\{ \bar{u}_c' \left[1 + \dots + \frac{\bar{u}_c''}{\bar{u}_c'}(y-y_c)\log|y-y_c| + \dots \right] + A_+(y-y_c) \right\} \cos\theta
$$

$$
- \varepsilon^2 \frac{\bar{u}_c''}{8(y-y_c)}\cos 2\theta + \frac{\varepsilon}{\alpha\,\mathrm{Re}}\frac{\bar{u}_c''}{3(y-y_c)^2}\sin\theta + \dots \,. \tag{7.53}
$$

Writing (7.53) in terms of the inner variables, it is clear that the expansion for Ψ has the form

$$
\Psi \sim \Psi^{(0)} + \varepsilon^{1/2}\log\varepsilon\,\Psi^{(1)} + \varepsilon^{1/2}\Psi^{(2)} + \dots + \lambda\tilde{\Psi}^{(0)} + \dots \tag{7.54}
$$

and that $\Psi^{(0)} \sim Y^2/2 + \cos\theta$ as $Y \to \infty$. Note that we must determine $\Psi^{(2)}$ in order to find out if there is a phase change because the first term in the power series ϕ_A does not enter into the matching of Ψ until $O(\varepsilon^{1/2})$.

The equation governing the zeroth-order term $\Psi^{(0)}$ is (7.51) with the rhs set to zero and its solution can be written simply as

$$
\Psi^{(0)}_{YY} = F'(\Psi^{(0)}) \,, \tag{7.55}
$$

where F' is arbitrary aside from the matching condition that $F' \to 1$ as $\Psi^{(0)} \to \infty$. A further integration can be accomplished by employing a von Mises transformation, i.e., we take $\Psi^{(0)}$ and θ as independent variables. Integrating with respect to $\Psi^{(0)}$ yields the result

$$
\tfrac{1}{2}(\Psi^{(0)}_Y)^2 = F(\Psi^{(0)}) + G(\theta) \,. \tag{7.56}
$$

By matching to the outer expansion as $\Psi^{(0)} \to \infty$ it is readily ascertained that $G = -\cos\theta$; however F remains arbitrary. In order to determine F uniquely, a *viscous secularity condition* was employed in [7.61]. The $O(\lambda)$ term, namely, $\tilde{\Psi}^{(0)}$, in (7.54) satisfies a certain linear, nonhomogeneous PDE; by integrating that PDE over one period and imposing the condition that $\tilde{\Psi}^{(0)}$ be periodic the following secularity condition is obtained:

$$
F'' \int_0^{2\pi} [2(F-\cos\theta)]^{1/2}d\theta = H_0 \,. \tag{7.57}
$$

The constant H_0 can be shown to be zero by integrating along a streamline for which $\Psi^{(0)} \gg 1$, i.e., where F is known. Finally, (7.57) shows that $F'' = 0$ and it

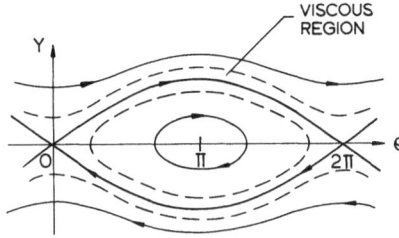

Y

O

θ

2π

Fig. 7.7. Streamline pattern in the nonlinear critical layer with viscous structure along the edges of the Kelvin cat's-eyes

follows that throughout the critical layer

$$\Psi^{(0)} = Y^2/2 + \cos\theta. \tag{7.58}$$

It is interesting to note that higher-order *inviscid* terms in the expansion do not yield a secularity condition. One must, in the steady case, follow the earlier ideas of Prandtl and bring in a slight viscosity in order to ensure a unique solution. The streamline pattern, according to (7.58), is the well-known Kelvin cat's-eye configuration illustrated in Fig. 7.7.

The next term of the inner expansion is determined in the same manner as was $\Psi^{(0)}$ and, again, turns out to be identical to its outer expansion. At $O(\varepsilon^{1/2})$, however, the principal results are obtained and due to the complexity of the analysis only the main conclusions will be summarized here. Outside of the closed streamline region, the vorticity is given by

$$\Psi_{YY}^{(2)} = M_s \int_1^{\Psi^{(0)}} \left\{ \int_0^{2\pi} [2(\eta - \cos\theta)]^{1/2} d\theta \right\}^{-1} d\eta + N_s, \tag{7.59}$$

where $s = \pm$ and the viscous secularity condition has been employed. Note that (7.59) involves all of the higher harmonics, i.e., $\cos 2\theta$, $\cos 3\theta$, etc.; a hint of that possibility is provided by (7.53), where it can be seen that for $(y - y_c) \sim O(\varepsilon^{1/2})$ the $\cos 2\theta$ term of ψ_{yy} is the same order of magnitude as the preceding $\cos\theta$ term.

Inside the cat's-eyes, the vorticity must be a constant according to the Prandtl-Batchelor theorem. However, the constants N_+ and N_- in (7.59) are unequal in the analyses of [7.61, 62], so the vorticity cannot be matched everywhere on the critical streamline $\Psi^{(0)} = 1$. These discontinuities can be smoothed out by adding viscous layers having a thickness of $O(\lambda^{1/2})$, as indicated in Fig. 7.7. The vorticity within these layers satisfies the heat-conduction equation and so a solution is readily obtained. At the cat's-eye corners, on the other hand, the analysis becomes quite involved and leads to a Wiener-Hopf problem which was formulated but not completely solved in [7.61]. That problem, as it turns out, was shown by *Brown* and *Stewartson* [7.63] to have no solution unless certain mean flow distortions are added, as discussed below.

To return to the larger-scale structure, a further integration with respect to Y yields the horizontal velocity[1] and matching $\Psi_Y^{(2)}$ for $|Y| \to \infty$ fixes the constants in the eigenvalue problem. It develops that the only outer solution compatible with a nonlinear critical layer has $A_- = A_+$. Thus, there is *zero* phase change and a new class of modes can be obtained from Rayleigh's equation.

Although these solutions, and indeed the entire analysis, are of considerable interest, there are some difficulties in applying the results to the flows discussed up to this point. Some of these were revealed in an extension of the theory reported by *Haberman* [7.64] to the case $\lambda \sim O(1)$. By solving the $O(\varepsilon^{1/2})$ problem numerically, it was shown in [7.64] that if we write $\log(y - y_c)$ as $\log|y - y_c| + i\theta$ below y_c, then the phase change θ_R is a continuous function of λ, varying from $-\pi$ to 0, as λ decreases from ∞ to 0. It was also found necessary to include certain jumps in the mean flow across the critical layer that are $O(\varepsilon^{1/2})$ in ψ, i.e., larger than the perturbation itself. With these mean flow distortions included, the vorticity is now continuous across the cat's-eye boundaries in the case $\lambda \ll 1$, although its derivative is still discontinuous in the absence of viscous layers. It is these jumps in the mean flow that are required in order for the Wiener-Hopf problem discussed above to have a solution.

The need for such large distortions in the mean flow seems a bit worrisome. This may be part of the price that one pays for insisting upon a steady flow and it is possible that the solutions with vorticity boundary layers would be more meaningful in a time-dependent problem. A further difficulty revealed by the study in [7.64] is that, except for long waves, the critical and wall layers are not usually distinct. Hence, a study of the $\lambda \sim O(1)$ problem with a nonlinear and diffusive wall layer would be more pertinent to real boundary layer and channel flows.

The most promising application of the nonlinear critical layer theory seems to be to geophysical flows. There, because of the presence of a stabilizing force, such as that due to gravity or rotation, large amplitude wave motions can occur at very high Reynolds numbers. The large length scales involved lead to the consideration of flows where $\text{Re} \sim O(10^6)$ or larger, so the situation $\lambda \ll 1$ is, more or less, the rule. These considerations were recognized early by *R. E. Kelly* who suggested to the author an application of the theory to stratified shear flows. The latter are relevant to clear air turbulence and to wave motion in the oceanic thermocline. A summary of the results, as well as a review of some earlier publications, is contained in [7.65]. A second application to geophysical fluid dynamics that is deserving of mention is the work of *Redekopp* [7.66] on Rossby wave solitons. He has suggested that the results are relevant to the Great Red Spot and other features of Jupiter's atmosphere. Finally, we note that ideas similar to those presented here have been applied to various problems in plasma dynamics (see, e.g., the analysis of the tearing-mode instability in [7.67]).

---

1 It was found in [7.61] that a continuous solution for the velocity $\Psi_Y^{(2)}$ exists even if the vorticity is allowed to be discontinuous.

7.4.3 Time Dependence and the Nonlinear Critical Layer

The critical point singularity is associated with neutral modes and does not occur if the perturbation is evolving rapidly in some sense. One might conclude therefore [see, e.g., (7.38)] that time dependence offers a third alternative to remedying the singular behavior of (7.7). The structure of the singular perturbation problem is somewhat different, however, because we do not have a small parameter multiplying higher-order derivative terms as was the case with the nonlinear and viscous terms. In the linear, inviscid IV problem, the effect of the singularity appears after some time as manifested by the neglected nonlinear terms becoming as important as the linear terms. The custom, until recently, has been to substitute conjecture for analysis at that point in order to describe the further development and eventual outcome of the phenomenon. A more satisfying approach has been pursued in work discussed below with interesting results.

As an illustrative example, we discuss the problem of a forced Rossby wave propagating toward the equator from some northern latitude in the presence of a zonal shear flow. The linear problem was formulated by *Dickinson* [7.68] for the case of a spatially periodic forcing switched on at $t=0$ and it was concluded that, due to the $-\pi$ phase change across the critical layer, the wave would be absorbed there. However, *Warn* and *Warn* [7.35] later showed that when $t \sim O(\varepsilon^{-1/2})$ the nonlinear terms have become the same order of magnitude as the linear terms within the critical layer, whose thickness at that point has decreased to $O(\varepsilon^{1/2})$. In order to follow the development for longer times, the results of the IV problem as $t \to \infty$ were used in [7.35] as initial conditions for a numerical study of the nonlinear evolution of the critical layer on the slow time scale $T=\varepsilon^{1/2}t$. The results for the phase shift, denoted θ_R, as a function of T are shown in Fig. 7.8. The departure from the linear result is seen to be substantial and the phase change passes through zero shortly before the calculation is terminated by numerical instabilities.

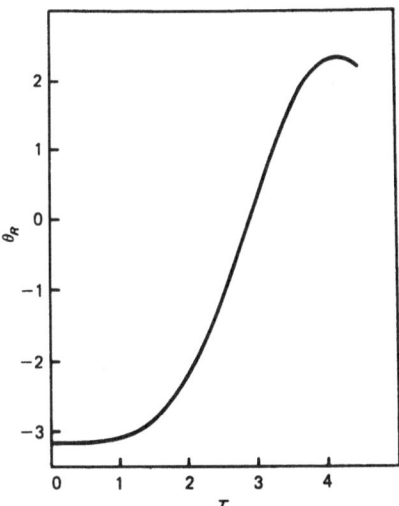

Fig. 7.8. Variation of the logarithmic phase shift as a function of the slow time scale $T=\varepsilon^{1/2}t$ for a forced Rossby wave [7.35]

Stewartson [7.69] has independently considered the same problem by analytical means. Employing a model equation (that becomes exact if the forcing is just the right distance from the critical point) he found results that are in substantial agreement with those of [7.35]. Moreover, the picture sketched above can be completed because the analysis is valid for larger values of T. Thus, it is found that the phase change oscillates for some time, finally tending to zero. That result can be interpreted as meaning that the wave is totally *reflected*, so the prediction of absorption in the linear theory, while valid for $t \sim O(1)$, is seen to be highly misleading for long times.

A direct comparison of the solution at large times with the steady modes of the nonlinear critical layer theory is a bit inappropriate because a steady state is never reached in the above analyses. The vorticity, in particular, continues to oscillate. However, the numerical results in [7.35] do show strong gradients of vorticity at the edges of the cat's-eyes, in agreement with the analytical work. Although there are a number of discrepancies in the detailed flow structure, it seems clear that including a slight viscosity would lead to closer agreement. That seems to be substantiated by the recent numerical work in [7.70].

Despite these discrepancies in detail, it can be said that the following principles have been established by the nonlinear critical layer studies: 1) The phase change across the singularity is not a universal constant as the linear theory suggests, but instead varies according to the relative importance of nonlinear, diffusive, and temporal effects; 2) Finite-amplitude waves in high Reynolds number flows have fine-scale structure in their critical layers that cannot be modelled by linear or quasi-linear analyses. A great deal of the published numerical work on geophysical flows with critical layers is inaccurate because the authors were not cognizant of the latter point. It is only when the computations are repeated using higher-resolution schemes, reflecting the relevant scales in the critical layer, that these errors become apparent.

7.5 Transition Experiments and some Theoretical Offspring

Experiments have played an important role in the subject, sometimes verifying (or disproving) theories, and often suggesting new theoretical developments. This experimental work has led to a quite comprehensive description of the transition process in many flows. However, significant portions of that process continue to defy all attempts to formulate a satisfactory theoretical description. Most of this section is devoted to a discussion of transition experiments on free shear layers and boundary layers. Theoretical work directed toward modelling certain of the observed phenomena will also be outlined.

7.5.1 Free Shear Layer Transition

The free shear layer (Fig. 7.2b) is a highly unstable flow having a maximum value of αc_i that is about 40 times the corresponding figure for a flat-plate boundary layer. Yet, the most unstable wave, in spite of its rapid amplification in the linear regime, is observed to equilibrate and nonlinear effects dominate the rest of the transition process, which encompasses several more stages. Before describing some of the nonlinear phenomena, we point out some interesting characteristics exhibited by the linear region of instability. It develops that these features can best be described by the theory of *spatially growing* waves.

To motivate that topic we note that in stability experiments it is often helpful to artificially excite disturbances by means of a loudspeaker or vibrating ribbon. In that way, there is a dominant frequency associated with the instability and it is easier to follow ensuing developments. During natural transition, on the other hand, a number of frequencies may be excited by free-stream turbulence and other noise in the system. A comparison of Figs. 6 and 8 of the paper by *Miksad* [7.71] clearly makes the point. It is particularly true for artificially introduced disturbances that a better description is provided by the spatial theory, in which one considers perturbations proportional to $\exp[i(\alpha x - \omega t)]$, where α is complex, ω is the real frequency, and $-\alpha_i$ the amplification factor (see, e.g., [7.72]).

The spatial stability calculations for a $\tanh y$ shear layer reported in [7.16] illustrate some significant differences with the temporal theory. Perhaps the most important is that the waves in the spatial theory are dispersive. For example, the profile $\bar{u} = 0.5(1 + \tanh y)$ has $c_r = 0.5$ at all wave numbers in the temporal theory. However, there are large variations in the quantity $c_r = \omega/\alpha_r$ in the spatial theory. That result is well supported by observation and, in fact, $c_r \to 1$ as $\alpha_r \to 0$. One consequence is that there is no longer symmetry about $y = 0$ in the modal structure of a spatially amplified wave and $|u'| = 0$ for one nonzero value of y, u' being the fluctuating component of the horizontal velocity. This "phase reversal" had previously been observed in hot-wire measurements but is not exhibited by the temporal theory. Another difference is that for the shear layer $\bar{u} = u_0 + \tanh y$, $-\alpha_i$ depends upon u_0 (and, in fact, vanishes in the case $u_0 = 0$), whereas αc_i is independent of u_0. The reader is referred to the survey article by *Michalke* [7.73] for further details of matters discussed up to this point.

It was mentioned above that the exponential growth of the linearly most unstable wave decreases as nonlinear effects come into play. As this mode tends toward equilibration, a *subharmonic* wave having one-half the frequency of the fundamental makes its appearance, as well as the expected higher harmonics. This most interesting phenomenon is observed in smoke-visualization experiments (see Fig. 7.9) and is also apparent from hot-wire signals. It is often described as "vortex pairing", i.e., adjacent pairs of vortices created by the initial periodic disturbance begin to roll around each other, finally coalescing into a single vortex. This process is now believed to occur even in turbulent shear layers where it may be an important factor contributing to the growth of the shear layer thickness [7.74, 75].

A quite successful theory has been proposed by *Kelly* [7.76] in connection with the initiation of vortex pairing. In that analysis, the stream function is

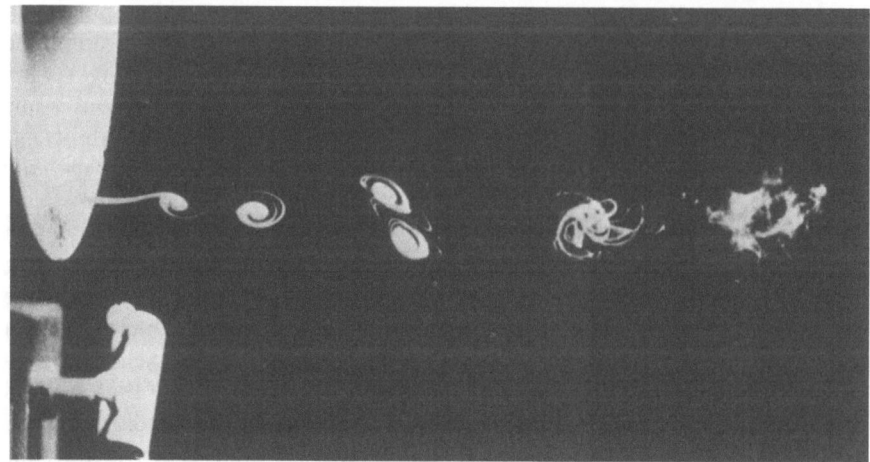

Fig. 7.9. Vortex pairing and coalescence in a transitional free shear layer as observed in smoke-flow experiments [7.91]

written as

$$\psi = \int \bar{u}(y)dy + \delta[\phi_{01}(y)e^{i(\beta x - \gamma t)} + *]$$
$$+ \varepsilon[B(X,\tau)\phi_{10}(y)e^{i(\alpha x - \omega t)} + *] + O(\varepsilon\delta) , \qquad (7.60)$$

where $X = \delta x$ and $\tau = \delta t$ (this is case A of [7.76]). The amplitude parameters are ordered such that $\varepsilon \ll \delta \ll 1$ and the $O(\delta)$ term, which represents the equilibrated fundamental mode, is taken to be periodic. Various weakly nonlinear interactions can occur depending upon the particular choice of the parameters in (7.60). For the tanh y shear layer, the case of greatest interest is when $\beta = 0.44$, corresponding to the linearly most unstable wave, and $\alpha = 0.22$. The complex conjugate of the $O(\varepsilon)$ term then interacts with the fundamental mode to produce a term that forces the subharmonic, i.e., $\beta - \alpha = \alpha$ and $\gamma - \omega = \omega$. As a consequence, the amplitude B can experience a growth rate that is enhanced in comparison with its linear value due to the subharmonic resonance. The incremental quantity depends upon δ, the amplitude of the fundamental. When $\delta \gtrsim 0.15$, the amplification rate of the subharmonic is greater than that of the linearly most unstable wave and it is suggested that only when δ exceeds 0.15 will the subharmonic emerge. The predicted value of δ corresponds to u' being about 12% of the mean velocity difference across the mixing layer. This is slightly larger than the experimental values reported at the time and is slightly smaller than the figure reported subsequently in [7.71].

There are other features of the analysis in [7.76] perhaps even more convincing. For example, destabilization of the 3/2 harmonic was predicted and later observed. It is also interesting that vortex pairing is *not* observed in symmetric flows such as jets and wakes. This, again, is in accord with the theory; it was found, for example, in [7.77] that even though two neutral modes of the jet profile $\bar{u} = \text{sech}^2 y$ satisfy the resonance conditions, no interaction takes place.

When this sort of interaction can take place in a shear flow it is potentially a very powerful mechanism because energy can be fed in from the mean flow without any decay of the interacting waves until, finally, $\bar{u}$ distorts appreciably. The foregoing analysis, however, is weakly nonlinear and only describes the initial "secondary bifurcation". A procedure that does allow one to follow the evolution of this process is to consider the instability of an array of point vortices. That idea was first employed in [7.75], although the analysis there contains some inconsistencies that can be remedied.

Returning now to the transition process, following the appearance of subharmonics and the fundamental mode equilibration, three-dimensional structure begins to form. This is in the nature of longitudinal vortices arranged periodically in the spanwise direction. The formation of these "Benney-Lin vortices" coincides roughly with a decay of the energy in the fundamental mode. Many features of this longitudinal vortex system can be modelled by considering the weak nonlinear interaction between a two-dimensional wave and an oblique wave having the same x component wave number; the latter behaves as a standing wave in the spanwise coordinate, but propagates in the x direction. Such a theory was applied by *Benney* [7.78] to the case of a tanh y shear layer. (Its application to boundary layers is somewhat more contentious as discussed in the following section.)

Once the streamwise vortex structure is established, a great deal of three-dimensional activity develops in the form of weak secondary instabilities. The frequency of these secondary instabilities is comparable to those of the primary modes, but they are of a smaller scale. Their appearance marks the final stage in the breakdown of the flow into turbulence. The entire process, from linear instability to turbulence, occurs in a distance of about five wavelengths of the fundamental according to the detailed study reported in [7.71].

7.5.2 Boundary Layer Transition

Research into the stability of boundary layers has been greatly influenced by experiments that were initiated at the U.S. National Bureau of Standards in the 1940's. These included the first laboratory observations of Tollmien-Schlichting waves, which were reported by *Schubauer* and *Skramstad* [7.79]. Before those experiments there was no direct link with stability theory. This breakthrough was achieved by reducing free-stream turbulence levels in the wind tunnel employed to values considerably lower than was the case in earlier experiments. Moreover, a vibrating ribbon traversing the boundary layer in the spanwise direction was employed to introduce two-dimensional perturbations at a controllable frequency. The location of the ribbon was sufficiently close to the leading edge so that the waves were initially stable. However, as they travelled downstream, the Reynolds number increased due to the growth of the boundary layer until, finally, the waves began to amplify at rates in reasonable agreement with the earlier calculations of Schlichting.

The initial instability of a boundary layer is two-dimensional in accordance with Squire's theorem. However, the two-dimensional waves are followed by a region of three-dimensional activity in which the flow is nearly periodic in the spanwise direction. The importance to transition of three-dimensional effects was not appreciated for some time. That recognition was preceded by *Emmons'* [7.80] discovery of turbulent spots; i.e., the final breakdown of the laminar motion occurs at isolated locations where spots (bursts) of turbulence are produced. The growth of these spots is rapid and they mix with other spots to produce a fully turbulent flow a short distance downstream of the initial breakdown. Following the discovery of the spots, more than five years elapsed before the complex three-dimensional events preceding the occurrence of breakdown were investigated. A particularly detailed description was provided by the experiments of *Klebanoff* et al. [7.81]. The principal results and related analytical work are summarized below. For more detailed reviews the reader is referred to the survey articles by *Stuart* [7.82], and *Lin* and *Benney* [7.60].

In the flat-plate boundary layer studies of [7.81], spanwise periodicity was imposed in a controlled manner by placing strips of tape at equidistant locations on the surface of the plate. A vibrating ribbon was again used to generate the waves and it was observed that a streamwise vortex system formed with adjacent pairs of vortices rotating in opposite directions. It was found that fluctuation amplitudes were largest at those spanwise locations between vortices where the flow was upward ("peaks") and smallest midway between peaks, where the flow was toward the wall ("valleys").

Further downstream, at the peaks, the instantaneous velocity profile was noted to develop regions of large shear. These appear in the oscillatory flow consisting of the mean part plus the primary wave. A somewhat weaker distortion occurs in the mean flow. However, it is the oscillatory profile that has received most attention, because during part of its cycle an inflection point occurs in the outer portion of the boundary layer at roughly $y = 0.6\delta$, where δ is the boundary layer thickness. A free shear layer instability then develops and it in turn leads to high-frequency fluctuations termed "hairpin eddies" in [7.81]. This secondary instability has a wavelength characteristic of the scale of the oscillatory shear layer, while its frequency is approximately ten times that of the primary wave. It seems that the bursting phenomenon is directly related to these secondary instabilities which develop once each cycle at the peaks.

Before reviewing more recent experimental work, we digress briefly to comment about theoretical work related to the three-dimensional aspects of transition just discussed. The Benney-Lin theory describing streamwise vortices was introduced in Sect. 7.5.1 and it was indicated that some controversy exists about its application to boundary layers [7.47, 60]. This theory reproduces so many features of the experiments that is seems highly likely that the basic idea is correct. For example, in contrast with some earlier theories, it permits the amplitude of the streamwise vortex component to be largest where the perturbed streamlines in the plane of the mean flow are convex (in terms of the cat's-eye pattern in Fig. 7.7 that would be near $\theta = \pi$). Although the latter is in accord with observation, the theory can only model this feature over distances on the order of a few wavelengths as the waves do not maintain the proper phase relationship indefinitely. Also, there are compatability conditions among the various wave number components and frequency that must be satisfied in order for both waves to have the same value of c_r as required by theory. In the case of Blasius flow these conditions are not satisfied.

A possible way out of this difficulty is through the theory of resonant interactions. *Craik* [7.83] has considered the weakly nonlinear interaction of three waves, two of them being oblique and inclined symmetrically about the x direction, and all having the same value of c_r. In the range of interest, say $2 \, \mathrm{Re_c} < \mathrm{Re} < 4 \, \mathrm{Re_c}$, the oblique waves would be slightly damped on a linear basis, whereas the two-dimensional wave can be chosen to be the most amplified of linear theory. Preliminary computations indicate that a strong instability exists (that is subcritical insofar as the oblique waves are concerned). The mean flow provides the energy for the growth as in the work of Kelly discussed earlier. Another possibility would be to consider resonant triads of neutral modes [7.47], but these seem to lack the symmetry necessary to model the longitudinal vortices.

A particularly important feature of transition, as discussed above, is the temporary formation of inflectional velocity profiles. A quasi-steady calculation was reported in [7.84] of the amplification due to the resultant secondary instability. During one half-period of the primary wave, the amplitude of the velocity perturbation was found to increase by a factor of ten, in good agreement with experiment. A model illustrating the processes leading to the formation of these small-scale shear layers has been formulated by *Stuart* [7.85] as a linear initial-value problem. The model employs a shear flow that is uniform in x; a streamwise vortex system periodic in z, the spanwise coordinate, is imposed on the basic flow $u = U(y)$ at $t = 0$. The analysis is inviscid so the transport of vorticity is due to convection and vortex stretching. The latter effect would be absent in a two-dimensional flow, but is found here to be an important factor in creating shear layers resembling those observed in experiments.

Turning now to current work, there seems to be a feeling at the present time that the great interest generated by the controlled experiments has caused the more basic problem of natural transition to be neglected and one should return to it. An important set of experiments oriented in that direction have been reported recently by *Gaster* and *Grant* [7.86]. These concerned the formation and evolution of a wave packet generated by a short-duration acoustic pulse. The latter was produced by a microphone placed beneath a plate with the perturbation introduced into the boundary layer through a vertical hole in the center of the plate. Despite the fact that the initial disturbance generates all

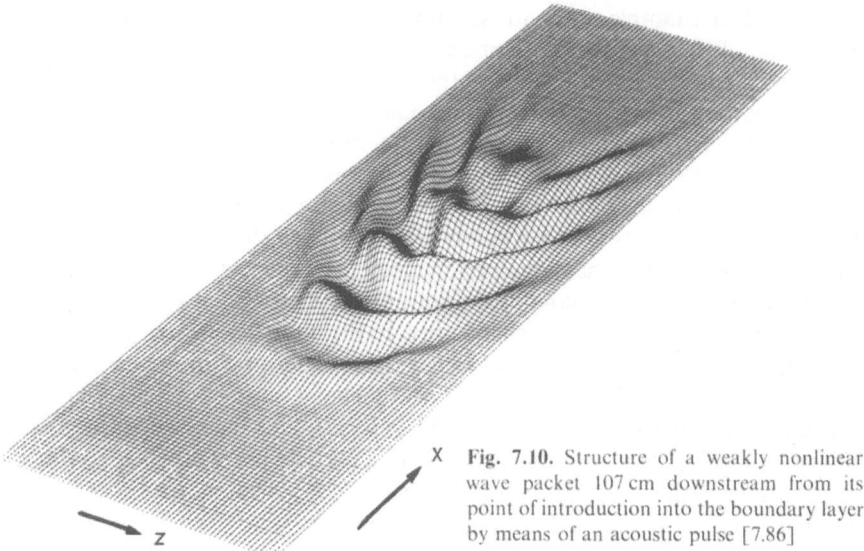

X   **Fig. 7.10.** Structure of a weakly nonlinear wave packet 107 cm downstream from its point of introduction into the boundary layer by means of an acoustic pulse [7.86]

possibles modes (as in natural transition), a wave packet of slowly varying amplitude emerges downstream as a result of the linear selection process. By means of a hot wire placed just outside the boundary layer, measurements were made of $(u'^2)^{1/2}$ at various downstream and spanwise locations. After digital filtering of the data from 256 samples, plots were constructed at a number of x stations showing the passage of the wave packet. A typical result is presented in Fig. 7.10. Further upstream the contours are smooth, but here the packet is showing the first signs of distortion due to nonlinearity.

More recent experiments by *Gaster* [7.87] deal with the effects of variations in initial amplitude and the nature of the forcing, i.e., sinuisoidal vs pulsed. Experiments on natural transition have also been conducted by *Arnal* et al. [7.88] and considerable data on the latter stages were obtained. Conditional sampling techniques were used to study turbulent spots in some detail and an interesting discussion is given of the difficulties encountered in the region of intermittency.

7.5.3 Poiseuille Flow

It was pointed out in Sect. 7.4 that the basic flow $\bar{u} = 1 - y^2$ is attractive as an example for application of nonlinear theories, but a practical difficulty exists in that transition occurs at values of Re far below the linear critical value of 5772. In a recent set of experiments, however, it was found possible to maintain a laminar flow for Re as high as 9000 by greatly reducing free-stream turbulence levels [7.51]. Thus, for the first time, it was possible to observe linearly unstable waves. More importantly, the notion of subcritical instability has been put on firm ground as a definite threshold amplitude was observed in a set of data taken at $Re = 5000$. Instability occurred for $u' \geq 1.2\%$ of the centerline value of $\bar{u}$.

Other features of interest included the observation that breakdown resulted from processes similar to those characteristic of Blasius flow, i.e., inflectional velocity profiles developed that were of a transient nature and these led to bursts. On the other hand, a significant difference *vis-à-vis* the boundary layer case was that higher harmonics were prominent near breakdown, whereas a rather surprising result noted in [7.81] concerned the absence of higher harmonics. A final puzzling note is that the mean flow distortion was much less than what one would have expected on the basis of current nonlinear theories.

7.6 Concluding Remarks

The preceding sections have reviewed both fundamental aspects of hydrodynamic stability theory and newer developments, particularly on the nonlinear side. Some important experimental studies were also reviewed, but on a less comprehensive basis, the object being to select a few representative investigations closely related to the theory presented earlier. We have focused upon boundary layers and free shear layers because these flows meet the following criteria: 1) They have been studied extensively both in the laboratory and analytically; and 2) They are important in practical applications. Although much of the research on these flows has been motivated by their importance in aeronautics (e.g., free shear layers are relevant to the jet noise problem), there are many other applications too numerous to cite here.

A significant contrast emphasized above is that the linear instability mechanism in a boundary layer is much weaker than that in a mixing layer. Consequently, transition of bounded flows does not seem to take place until localized free shear layers appear in the velocity profile. This illustrates a useful guiding principle, namely, that when αc_i is small linear instability does not lead directly to major changes in the flow. There are usually stabilizing factors present, such as nonlinearity and nonparallel effects, which counteract the instability. Thus, one should take into account the magnitude of αc_i before attributing dynamic phenomena to linear instability solely on the basis of a calculation yielding $c_i > 0$. Returning to the role of local inflectional profiles in boundary layer transition, there is an important practical application of the foregoing considerations. In situations where a turbulent boundary layer is desired, it is commonplace to "trip" the boundary layer by, for example, attaching a cylindrical wire to the wall. If the wire is large enough, the boundary layer separates and a free shear layer is produced behind the trip wire. A strong instability results that is clearly visible in the beautiful photographs taken by *Hall* [Ref. 7.89, Figs. 12 and 13] in a smoke tunnel.

A few remarks about promising areas for further research would perhaps be appropriate at this point. It seems to the author that nonlinear studies hold the key to understanding transition, but that for many flows we must get away from studying the immediate vicinity of the neutral curve. Series renormalization techniques offer a way to extend the validity of the rational theories developed until now, although one should be alert to the potential hazards in their use. An alternative is to combine empiricism with theory as was done in studies cited here of the secondary instabilities in both boundary layers and mixing layers. Other areas deserving of further study, especially in the context of boundary layers, include nonparallel and three-dimensional effects. The first area has received considerable attention in recent years (see, e.g., [7.90]), but the developments to date have not been dramatic, partly because most of the effort has been directed toward refining the linear theory. Concerning the three-dimensional effects, it is interesting to note that the critical point singularity occurs for an oblique wave even when $\bar{u}_c'' = 0$ as was pointed out in a nonlinear critical layer context in [7.61]. Finally, the application of finite-amplitude theories to compressible flows ought to be rewarding, but has received little attention.

Numerical simulations, i.e., solutions of the full Navier-Stokes equations, have been mentioned only briefly in this chapter. Their potential is obviously enormous, but, aside from a few notable investigations, results to date have not really justified the resources expended in such endeavors. Success in that field requires a rare combination of talents, namely, skill in the use of numerical methods plus a good understanding of the perturbation theories. The situation is bound to improve as more powerful computational tools become available, but it would be a mistake to assume that theoretical understanding will no longer be required.

Turning again to experiments, more studies are needed of the natural transition situation, as mentioned above, while the final stages of free shear layer transition seem to also require further scrutiny. An opportunity exists for a valuable contribution in the area of nonlinear critical layers. It should be possible with forced disturbances to alter the phase change across the critical layer by progressively increasing the wave amplitude and/or Reynolds number of the basic flow. While the theory has been substantiated in numerical simulations, it remains to confirm this important phenomenon in the laboratory.

Acknowledgement. The author is grateful to the John Simon Guggenheim Foundation for the award of a fellowship permitting him to write this article in London. During that time he has benefited from many stimulating discussions with Professor *K. Stewartson* of University College and Professor *J. T. Stuart* of Imperial College concerning the material presented here.

References

7.1   M. Gaster: "Series representation of the eigenvalues of the Orr-Sommerfeld equation", AGARD Conference Proceedings No. 224, Laminar-Turbulent Transition (1977) p. 2-1

7.2   A. H. Nayfeh: *Perturbation Methods* (Wiley-Interscience, New York 1973)

7.3   J. Kevorkian, J. D. Cole: *Perturbation Methods in Applied Mathematics* (Springer, Berlin 1981)

7.4   C. C. Lin: *The Theory of Hydrodynamic Stability* (Cambridge University Press 1955)

7.5   P. G. Drazin, L. N. Howard: Hydrodynamic stability of parallel flow of inviscid fluid. Adv. Appl. Mech. **9**, 1 (1966)

7.6   A. E. Gill: A mechanism for instability of plane Couette flow and of Poiseuille flow in a pipe. J. Fluid Mech. **21**, 503 (1965)

7.7   L. N. Howard: Note on a paper of John W. Miles. J. Fluid Mech. **10**, 509 (1961)

7.8   T. Tatsumi, K. Gotoh, K. Ayukawa: The stability of a free boundary layer at large Reynolds numbers. J. Phys. Soc. Jpn. **19**, 1966 (1964)

7.9   G. Carrier, M. Krook, C. Pearson: *Functions of a Complex Variable: Theory and Technique* (Mc Graw-Hill, New York 1966)

7.10 J. T. Stuart: On the role of Reynolds stresses in stability theory. J. Aero. Sci. **23**, 86 (1956)

7.11 L. Prandtl: "The Mechanics of Viscous Fluids", in *Aerodynamic Theory*, Vol. 3, ed. by W. F. Durand (Springer, Berlin 1935) p. 34

7.12 J. J. Stoker: *Water Waves* (Wiley-Interscience, New York 1957)

7.13 P. G. Drazin, L. N. Howard: Stability in a continuously stratified fluid. Trans. Am. Soc. Civil Eng. **128**, 849 (1963)

7.14 S. Leibovich: "Hydrodynamic Stability of Inviscid Rotating Coaxial Jets"; NASA CR-1363 (1969)

7.15 W. Blumen, P. G. Drazin, D. F. Billings: Shear layer instability of an inviscid compressible fluid, Part 2. J. Fluid Mech. **71**, 305 (1975)

7.16 A. Michalke: On spatially growing disturbances in an inviscid shear layer. J. Fluid Mech. **23**, 521 (1965)

7.17 J. W. Miles, L. N. Howard: Note on a heterogeneous shear flow. J. Fluid Mech. **20**, 331 (1964)

7.18 W. H. Reid: "The Stability of Parallel Flows", in *Basic Developments in Fluid Dynamics 1*, ed. by M. Holt (Academic Press, New York 1965) p. 249

7.19 W. P. Graebel: On determination of the characteristic equations for the stability of parallel flows. J. Fluid Mech. **24**, 497 (1966)

7.20 C. C. Lin, A. L. Rabenstein: On the asymptotic solutions of a class of ordinary differential equations of the fourth order. Trans. Am. Math. Soc. **94**, 24 (1960)

7.21 W. D. Lakin, B. S. Ng, W. H. Reid: Approximations to the eigenvalue relation for the Orr-Sommerfeld problem. Phil. Trans. R. Soc. London A **289**, 347 (1978)

7.22 R. Betchov, W. O. Criminale: *Stability of Parallel Flows* (Academic Press, New York 1967)

7.23 L. H. Thomas: The stability of plane Poiseuille flow. Phys. Rev. **91**, 780 (1953)

7.24 A. Davey: A simple numerical method for solving Orr-Sommerfeld problems. Q. J. Mech. Appl. Math. **26**, 401 (1973)

7.25 S. A. Orszag: Accurate solution of the Orr-Sommerfeld stability equation. J. Fluid. Mech. **50**, 689 (1971)

7.26 S. A. Maslowe: Weakly nonlinear stability of a viscous free shear layer. J. Fluid Mech. **79**, 689 (1977)

7.27 H. B. Keller: *Numerical Methods for Two-Point Boundary-Value Problems* (Ginn-Blaisdell, Waltham, Mass. 1968)

7.28 M. R. Osborne: Numerical methods for hydrodynamic stability problems. SIAM J. Appl. Math. **15**, 539 (1967)

7.29 J. Gary, R. Helgason: A matrix method for ordinary differential eigenvalue problems. J. Comp. Phys. **5**, 169 (1970)

7.30 L. N. Howard, S. A. Maslowe: Stability of stratified shear flows. Boundary-Layer Meteorol. **4**, 511 (1973)

7.31 C.S. Yih: *Fluid Mechanics* (McGraw-Hill, New York 1969) Chap. 9, pp. 481–483

7.32 A. Eliassen, E. Høiland, E. Riis: "Two-Dimensional Perturbation of a Flow with Constant Shear of a Stratified Fluid"; Inst. for Weather and Climate Res., Pub. No. 1, Oslo (1953)

7.33 K.M. Case: Stability of inviscid plane Couette flow. Phys. Fluids **3**, 143 (1960)

7.34 L. Engevik: "On the Stability of Plane Inviscid Couette Flow"; Rept. No. 12, Dept. Appl. Math., Univ. Bergen (1966)

7.35 T. Warn, H. Warn: The evolution of a nonlinear critical level. Stud. Appl. Math. **59**, 37 (1978)

7.36 G. Chimonas: Algebraic disturbances in stratified shear flows. J. Fluid Mech. **90**, 1 (1979)

7.37 C. Grosch, H. Salwen: The continuous spectrum of the Orr-Sommerfeld equation. Part 1. The spectrum and the eigenfunctions. J. Fluid Mech. **87**, 33 (1978)

7.38 J.W. Murdock, K. Stewartson: Spectra of the Orr-Sommerfeld equation. Phys. Fluids **20**, 1404 (1977)

7.39 C.C. Lin: Some mathematical problems in the theory of the stability of parallel flows. J. Fluid Mech. **10**, 430 (1961)

7.40 G.B. Whitham: *Linear and Nonlinear Waves* (Wiley-Interscience, Nw York 1974) pp. 371–374

7.41 D.J. Benney, A.C. Newell: The propagation of nonlinear wave envelopes. J. Math. Phys. **46**, 133 (1967)

7.42 L.A. Segel: Distant side-walls cause slow amplitude modulation of cellular convection. J. Fluid Mech. **38**, 203 (1969)

7.43 A.C. Newell, J.A. Whitehead: "Review of the Finite Bandwidth Concept", Proc. IUTAM Symp. Instab. Cont. Syste., 1969 (Springer, Berlin 1971) p. 284

7.44 K. Stewartson, J.T. Stuart: A nonlinear instability theory for a wave system in plane Poiseuille flow. J. Fluid Mech. **48**, 529 (1971)

7.45 D.J. Benney, S.A. Maslowe: The evolution in space and time of nonlinear waves in parallel shear flows. Stud. Appl. Math. **54**, 181 (1975)

7.46 M.A. Weissman: Nonlinear wave packets in the Kelvin-Helmholtz instability. Phil. Trans. R. Soc. London A **290**, 639 (1979)

7.47 J.T. Stuart: "Nonlinear Effects in Hydrodynamic Stability", Proc. Xth Intl. Congr. Appl. Mech. (Elsevier, Amsterdam 1962) p. 63

7.48 J.T. Stuart: On the nonlinear mechanics of wave disturbances in stable and unstable parallel flows. I. The basic behaviour in plane Poiseuille flow. J. Fluid Mech. **9**, 353 (1960)

7.49 P.M. Morse, H. Feshbach: *Methods of Theoretical Physics*, Vol. 1 (McGraw-Hill, New York 1953) pp. 870–877

7.50 W.C. Reynolds, M.C. Potter: Finite-amplitude instability of parallel shear flows. J. Fluid Mech. **27**, 465 (1967)

7.51 M. Nishioka, S. Iida, Y. Ichikawa: An experimental investigation of the stability of plane Poiseuille flow. J. Fluid Mech. **72**, 731 (1975)

7.52 H. Schade: Contribution to the nonlinear stability theory of inviscid shear layers. Phys. Fluids **7**, 623 (1964)

7.53 J.T. Stuart: On finite-amplitude oscillations in laminar mixing layers. J. Fluid Mech. **29**, 417 (1967)

7.54 P. Huerre: The nonlinear stability of a free shear layer in the viscous critical layer regime. Phil. Trans. R. Soc. London **293**, 643 (1980)

7.55 N. Itoh: Spatial growth of finite wave disturbances in parallel and nearly parallel flows. Part 2. The numerical results for the flat plate boundary layer. Trans. Jpn. Soc. Aero. Space Sci. **17**, 175 (1974)

7.56a A. Davey, H.P.F. Nguyen: Finite-amplitude stability of pipe flow. J. Fluid Mech. **45**, 701 (1971)

7.56b A. Davey: On Itoh's finite amplitude stability theory for pipe flow. J. Fluid Mech. **86**, 695–704 (1978)

7.57 T. Ellingsen, B. Gjevik, E. Palm: On the nonlinear stability of plane Couette flow. J. Fluid Mech. **40**, 97 (1970)

7.58 J.T. Stuart, R.C. DiPrima: The Eckhaus and Benjamin-Feir resonance mechanisms. Proc. R. Soc. London A **362**, 27 (1978)

7.59 L. M. Hocking, K. Stewartson: On the nonlinear response of a marginally unstable plane parallel flow to a two-dimensional disturbance. Proc. R. Soc. London A **326**, 289 (1972)

7.60 C. C. Lin, D. J. Benney: "On the Instability of Shear Flows and Their Transition to Turbulence", Proc. XIth Intl. Congr. Appl. Mech., ed. by H. Görtler (Springer, Berlin 1964) p. 797

7.61 D. J. Benney, R. F. Bergeron: A new class of nonlinear waves in parallel flows. Stud. Appl. Math. **48**, 181 (1969)

7.62 R. E. Davis: On the high Reynolds number flow over a wavy boundary. J. Fluid Mech. **36**, 337 (1969)

7.63 S. N. Brown, K. Stewartson: The evolution of the critical layer of a Rossby wave. Part II. Geophys. Astrophys. Fluid Dyn. **10**, 1 (1978)

7.64 R. Haberman: Critical layers in parallel flows. Stud. Appl. Math. **51**, 139 (1972)

7.65 S. A. Maslowe: Finite-amplitude Kelvin-Helmholtz billows. Boundary-Layer Meteorol. **5**, 43 (1973)

7.66 L. G. Redekopp: On the theory of solitary Rossby waves. J. Fluid Mech. **82**, 725 (1977)

7.67 P. H. Rutherford: Nonlinear growth of the tearing mode. Phys. Fluids **16**, 1903 (1973)

7.68 R. E. Dickinson: Development of a Rossby wave critical level. J. Atmos. Sci. **27**, 627 (1970)

7.69 K. Stewartson: The evolution of the critical layer of a Rossby wave. Geophys. Astrophys. Fluid Dyn. **9**, 185 (1978)

7.70 M. Beland: The evolution of a nonlinear Rossby wave critical level: effects of viscosity. J. Atmos. Sci. **35**, 1802 (1978)

7.71 R. W. Miksad: Experiments on the nonlinear stages of free shear layer transition. J. Fluid Mech. **56**, 695 (1972)

7.72 M. Gaster: The role of spatially growing waves in the theory of hydrodynamic stability. Prog. Aeronaut. Sci. **6**, 2510 (1965)

7.73 A. Michalke: The instability of free shear layers. Prog. Aeronaut. Sci. **12**, 213 (1972)

7.74 A. Roshko: Structure of turbulent shear flows: a new look. AIAA J. **14**, 1349 (1976)

7.75 C. D. Winant, F. K. Browand: Vortex pairing: the mechanism of turbulent mixing-layer growth at moderate Reynolds number. J. Fluid Mech. **63**, 237 (1974)

7.76 R. E. Kelly: On the stability of an inviscid shear layer which is periodic in space and time. J. Fluid Mech. **27**, 657 (1967)

7.77 R. E. Kelly: On the resonant interaction of neutral disturbances in two inviscid shear flows. J. Fluid Mech. **31**, 789 (1968)

7.78 D. J. Benney: A nonlinear theory for oscillations in a parallel flow. J. Fluid Mech. **10**, 209 (1961)

7.79 G. B. Schubauer, H. H. Skramstad: "Laminar Boundary Layer Oscillations and Transition on a Flat Plate"; NACA Tech. Rept. No. 909 (1947)

7.80 H. W. Emmons: The laminar-turbulent transition in a boundary layer. J. Aeronaut. Sci. **18**, 490 (1951)

7.81 P. S. Klebanoff, D. K. Tidstrom, L. M. Sargent: The three-dimensional nature of boundary layer instability. J. Fluid Mech. **12**, 1 (1962)

7.82 J. T. Stuart: Hydrodynamic stability. Appl. Mech. Rev. **18**, 523 (1965)

7.83 A. D. D. Craik: Nonlinear resonant instability in boundary layers. J. Fluid Mech. **50**, 393 (1971)

7.84 H. P. Greenspan, D. J. Benney: On shear layer instability, breakdown and transition. J. Fluid Mech. **15**, 133 (1963)

7.85 J. T. Stuart: "The Production of Intense Shear Layers by Vortex-Stretching and Convection"; AGARD Rept. 514 (1965)

7.86 M. Gaster, I. Grant: An experimental investigation of the formation and development of a wave packet in a laminar boundary layer. Proc. R. Soc. London A **347**, 253 (1975)

7.87 M. Gaster: The physical processes causing breakdown to turbulence. Twelfth Symp. Naval. Hydro., 1978 (U.S. Office of Naval Res., Washington, D.C.)

7.88 D. Arnal, J.-C. Juillen, R. Michel: "Analyse Experimentale et Calcul de l'apparition et du Developpement de la Transition de la Couche Limite"; AGARD Conference Proceedings NO. 224, Laminar-Turbulent Transition (1977) p. 13-1

7.89 G. R. Hall: Interaction of the wake from bluff bodies with an initially laminar boundary layer. AIAA J. **5**, 1386 (1967)

7.90 W. S. Saric, A. H. Nayfeh: Nonparallel stability of boundary-layer flows. Phys. Fluids **18**, 945 (1975)

7.91 P. Freymuth: On transition in a separated laminar boundary layer. J. Fluid Mech. **25**, 683 (1966)

8. Instabilities in Geophysical Fluid Dynamics

D. J. Tritton and P. A. Davies

With 23 Figures

Variability, familiar in the earth's atmosphere through the occurrence of weather systems, is an almost universal characteristic of natural flows: the oceans also contain weather systems, and indeed sea-surface temperature fluctuations may have effects on atmospheric weather comparable with the atmosphere's intrinsic variability; changes in the earth's magnetic field reflect changes in the earth's fluid core; the onset of the present epoch of continental drift with the opening up of the Atlantic Ocean basin around 10^8 years ago probably relates to a change in the creep pattern of the earth's mantle; examples of variability of the sun include solar flares, sunspots and changes in the pattern of granulation. The recent dramatic growth in our information about the atmospheres of other planets tells the same story: the appearance of Jupiter has changed far more than anticipated between the visits of the Pioneer probes in 1973–1974 and the Voyager probes in 1979; Mars has intermittent dust storms, fluctuating winds, occasional cloud patterns indicative of large cyclones; ultraviolet images of Venus give every indication of a complex meteorology. The major reason why these systems, extremely varied in their dynamical processes, all show this characteristic is the pervasiveness of instability in fluid motion.

As usual in fluid dynamics, understanding of these phenomena comes from an interplay of observations of the natural situations, mathematical theory (analytical or computational), and laboratory experiments. Usually the theory and experiments – particularly the latter – are aimed more at understanding basic fluid dynamical processes than at modelling all the complexities of a natural situation. Although many of the ideas, methods, and results developed in other contexts are applicable, geophysical[1] applications have brought a range of topics, with some coherence, to the fore. Broadly speaking, these concern flows that are affected by one or more of density stratification in a gravitational field, Coriolis effects, and electromagnetic processes. There are similarities in the action of all three (although the analogies are exact only in certain very restricted circumstances): for example, each allows the fluid to support wave motions – respectively, internal gravity waves, inertial waves, and Alfvén waves – that are very similar in their properties but that have no counterpart in the absence of all three.

---

1 We are using the word "geophysical" in a very broad sense. It is unfortunate that there is no single word meaning "geophysical, planetary physical, and astrophysical".

8.1 Overview

In the space available, we cannot attempt any complete discussion of instabilities occurring in such situations; we can only give an impression of the diversity and subtlety of the processes involved. (A long list of types of instability, with examples in nature, has been given in [8.1].) After some general remarks in Sect. 8.2 about the role of stability considerations in geophysics, we have selected four topics for discussion. The first two – shear flow instability in, respectively, a stratified fluid (Sect. 8.3) and a rotating fluid (Sect. 8.4) – illustrate how a "classical" situation is modified by "geophysical" effects. The others – baroclinic instability in a rotating fluid (Sect. 8.5) and multidiffusive instabilities (Sect. 8.6) – concern processes which arise specifically in geophysical fluid dynamics. Even within the four topics we have had to be very selective in our material, our selection being influenced in some cases by the existence of excellent reviews elsewhere [8.2–9]. We have chosen to emphasize laboratory experiments. In each case we precede the description of the results of these by an indication of the dynamical mechanism responsible for the instability – usually in terms of a "displaced particle" argument. We quote theoretical results that provide good illustrations of the mechanisms or that relate directly to the experiments, but we do not attempt to cover, except with references, the main body of available theoretical work. We have felt no need to explain the status of the different theoretical methods to which we refer, as the principles are featured in other chapters in this volume.

Stability studies should, of course, be seen in the context of other aspects of geophysical fluid dynamics, and the dimensionless parameters and other results we quote derive from appropriate formulations of the dynamical equations. Suggested references for readers wishing to fill in their background knowledge of these matters are [8.10–20; Ref. 8.12, Chaps. 8, 9].

The closeness of the correspondence between laboratory observations and geophysical applications is variable. We do not have space for systematic discussion of these applications, although we mention examples in passing. Suggested references for those wishing to read more about naturally occurring flows are [8.21–25].

Notation

The following notation is used throughout the chapter; other symbols are defined where they are used.

Cartesian coordinates (x, y, z) have z vertically upwards when gravity is relevant, and parallel to the rotation axis in rotating systems.

The basic flow of which the stability is under consideration is specified by velocity field U. In Cartesian coordinates $U \equiv (U, V, W)$. Components in other coordinate systems are specified by an appropriate suffix.

g acceleration due to gravity

$k \equiv (k_x, k_y, k_z)$ nondimensional wave number

L	length scale
N	Brunt-Väisälä frequency $\left(N^2 = -\dfrac{g}{\varrho_0}\dfrac{d\varrho_0}{dz}\right)$
Pr	Prandtl number (v/κ)
T	temperature
α	thermal density coefficient
γ	solute concentration density coefficient
κ	thermal diffusivity
κ_C	solute concentration diffusivity
v	kinematic viscosity
ϱ	density
ϱ_0	background density field
ΔC	solute concentration difference scale
ΔT	temperature difference scale
$\Delta\varrho$	density difference scale
Ω	angular velocity of system rotation

8.2 Consequences of Instabilities in Nature

Most geophysical flows occur at very high Reynolds numbers. One might therefore expect them to be fully turbulent, with stability theory having little to tell us beyond this general fact. This is indeed sometimes the case. For example, the atmospheric boundary layer is observed to be turbulent, but we do not find in nature the sequence of events by which boundary layers can be observed in the laboratory to become turbulent [8.11]. In everyday observation this is partly because of the turbulence generated in the flow past buildings, trees, hills – roughness of a wide range of scales. But the general comment remains true over very flat ground and over the sea. The essential point is just that, since mechanisms exist for the maintenance of turbulence in a shear flow [8.11, 12], one does not need to invoke stability theory to understand why such a flow should be turbulent. The details of the processes by which a corresponding laminar flow becomes turbulent are not very relevant.

However, there are many cases in nature where remarkably regular flow structures are observed instead of or in addition to turbulent motion. For example, clouds frequently form regular patterns – parallel rows or cellular structures. As an example of a rather different type, some of the boundaries between the light and dark bands of Jupiter are contorted by a form of turbulence but others are remarkably sharp (Fig. 8.1). There are two principal reasons for these and similar, at first sight surprising, regularities.

Firstly, instabilities occur over a wide range of length scales. Small-scale turbulence may produce an "eddy viscosity" [8.21] which greatly reduces the effective Reynolds number of a larger scale flow. (The concept of eddy viscosity must be invoked with caution. As a quantitative tool for the calculation of

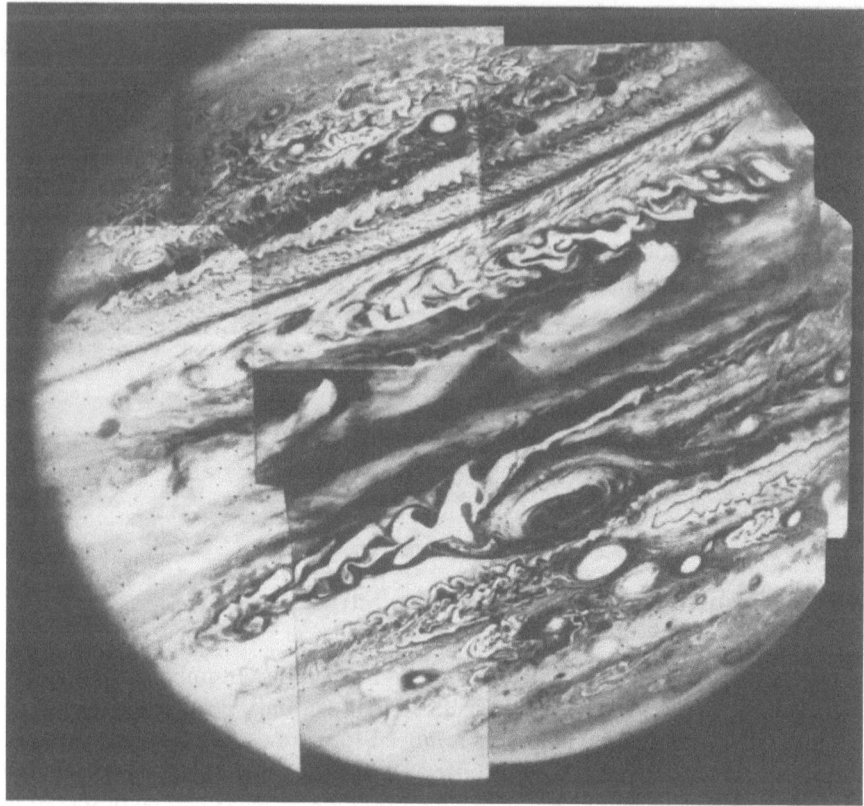

Fig. 8.1. Mosaic of Jupiter in violet light from Voyager 1 spacecraft (JPL/NASA photo)

turbulent flows it involves serious oversimplification, and it has been suggested [8.26] that in some particular geophysical situations, turbulence may produce countergradient transport of momentum. However, often the *qualitative* mixing action of small-scale turbulence may be to stabilize large-scale motion in a way similar to the effect of a greatly increased viscosity.)

Secondly, stratification, rotation, and a magnetic field can each be stabilizing. They exert constraints which prevent or inhibit certain types of motion. As a result, flows in the presence of these influences may remain laminar even at very high values of the Reynolds number. The constraining processes give rise to the wave motions mentioned in Sect. 8.1. If additionally some destabilizing influence is present (shear, convection, differential rotation), then the result may be the spontaneous amplification of wavelike motions.

There are, however, some complications arising from this. One should not necessarily interpret all observed wave motions as consequences of instability. There may be some disturbance present generating the waves directly; i.e., there is no solution of the equations for the appropriate conditions *without the wave*

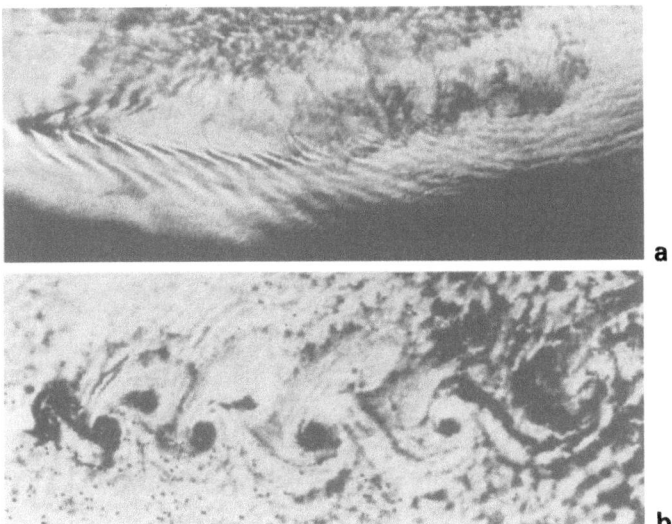

Fig. 8.2a, b. Cloud patterns in the lee of Jan Mayen Island (71°N, 8°W) photographed by NOAA5 and NOAA4 satellites. (Photos provided by Department of Electrical Engineering, Dundee University.) **(a)** "Ship-wave" pattern. **(b)** "Vortex street"

[8.27]. Stability considerations enter only if there is a waveless solution of the equations of motion, which may or may not occur depending on whether superimposed waves (or other perturbations) decay or amplify. Figure 8.2 shows two photographs of cloud patterns revealing flow features downwind of Jan Mayen Island in conditions of stable stratification; in broad terms they may be summarized as a "ship-wave" pattern and a "vortex-street" pattern. Both evidently result from the disturbance of the wind by the island, but it is likely that only the second is a consequence of instability. One would expect (by analogy with true ship waves) that the former is a wave train generated at the island. In contrast, the vortex street presumably arises through instability of the wake behind the island [8.28, 29]. In this example it is not difficult to deduce whether each pattern involves stability considerations, but in some cases one may observe waves without this being immediately apparent. A meteorologically important example concerns waves in the general circulation of the earth's atmosphere, such as those shown in Fig. 8.3 [8.30]. We shall be discussing, in Sect. 8.5, the instability mechanism that frequently gives rise to baroclinic waves in the atmosphere. However, the isobar pattern will also contain contributions from wave modes which occur because of restoring forces in the system and which may be forced by mountain ranges or localized heat sources [8.31, 32].

The general situation is yet further complicated by the fact that wave motions may themselves become unstable – whether or not the original process

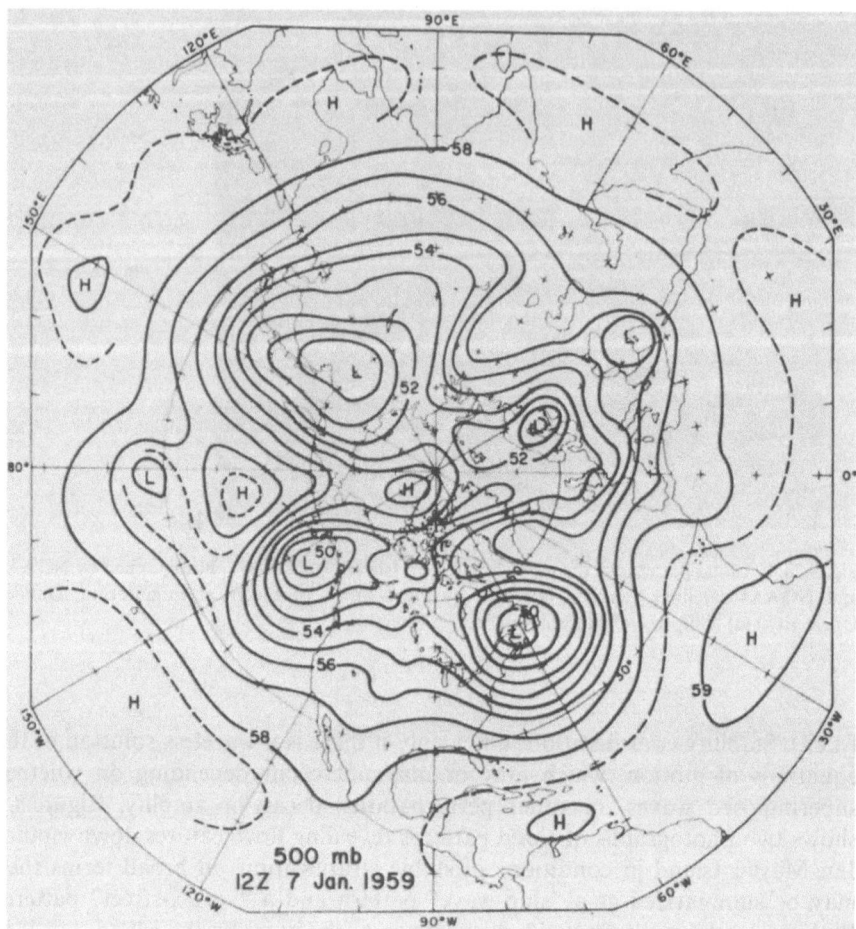

Fig. 8.3. Contours of the height (in units of 100 m) of the 500 mb surface in the northern hemisphere on 7 Jan 1959 [8.30]

generating the waves was one of instability. In particular, wave breaking, arising from strongly nonlinear steepening of the waves [8.27], usually produces regions of turbulence. We shall be seeing an example later in Fig. 8.8.

8.3 Stratified Shear Flow

A flow is stratified when the density of the fluid has a basic vertical variation $d\varrho_0/dz$ [Ref. 8.10, Chap. V, C; Ref. 8.11, Chap. 16; Ref. 8.12, Chap. 8; 8.13, 14]. Such variations often arise in nature owing to variations of temperature or of salt concentration, and they may similarly be produced in the laboratory. For the thermal case, it is strictly the variation of the specific entropy rather than that of the density which determines the stratification, owing to the thermo-dynamic effects of the hydrostatic pressure gradient. In natural situations the distinction is usually significant; in the laboratory it is almost always negligible. We shall formulate the discussion in the way appropriate to the latter, and we refer the reader to other sources [Ref. 8.11, Sect. 14.3 and appendix to Chap. 14; Ref. 8.33, Chap. 3 and Sect. 8.3] for consideration of just how one makes quantitative comparisons between laboratory and natural flows.

Qualitatively, the effect of stratification on shear flow instability is readily understood. Positive $d\varrho_0/dz$ (z being directed vertically upwards) implies that any vertical motion releases potential energy; thus, such motions occur more readily than in the corresponding unstratified flow and this will enhance any instability. Conversely, negative $d\varrho_0/dz$ tends to suppress vertical motions and so may inhibit the development of instabilities. Whether this changes the critical condition or only the subsequent evolution depends on the stage at which the unstratified instability generates vertical motion. For example, the marginally supercritical instability of a horizontal flow varying in the other horizontal direction (dU/dy) generates only horizontal perturbations and will thus be unaffected by stable stratification; its subsequent three-dimensional evolution into a turbulent flow must, however, be inhibited. In contrast, for a vertically varying horizontal flow (dU/dz) the marginally supercritical flow does generate vertical motions and so all stages of the growth of perturbations are modified by stratification. It is the latter geometry that arises most frequently in geophysical applications, and we shall thus focus our discussion on this, although we shall return to the other case briefly at the end of this section.

8.3.1 The Richardson Number

Quantitatively, the effect of stratification is indicated by some form of the Richardson number [8.13, 34; Ref. 8.10, Sect. V.12; Ref. 8.11, Sects. 16.1, 22.8; Ref. 8.12, Sects. 8.1, 8.4; Ref. 8.14, Sect. 4.7]. The most direct measure of the relative influences of the density and velocity gradients is provided by the gradient form

$$\text{Ri}_g = -\frac{g(d\varrho_0/dz)}{\varrho_0(dU/dz)^2}. \tag{8.1}$$

In general, this varies with z, and for any particular flow, it may be convenient to introduce a bulk form, such as

$$\text{Ri}_b = -gL\Delta\varrho/\varrho_0 U^2 \tag{8.2}$$

(the precise definition depending on the particular flow), where L is the length scale of the velocity and/or density variations. Usually Ri_b will provide a

a

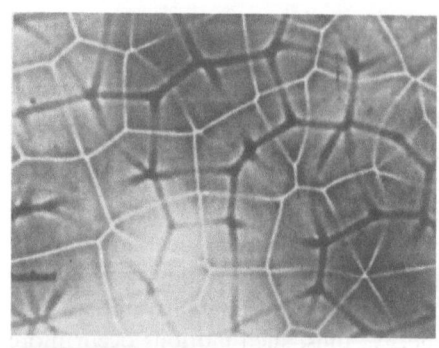

b

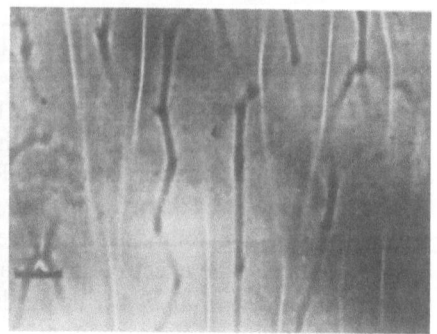

c

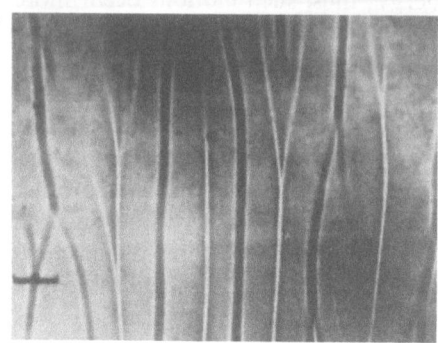

Fig. 8.4a–c. Convection in plane Couette flow; alignment of irregular cellular pattern. Sequence of photos shows the pattern (**a**) before, (**b**) shortly after, and (**c**) a longer time after one boundary of a Bénard layer starts to move in its own plane. Prandtl number $= 8600$; Rayleigh number $= 1.4 \times 10^5$; Péclet number, $UL/\kappa = 717$ ([8.37]; copyrighted by American Geophysical Union)

measure of a typical value of Ri_g. Much of the discussion of the significance of the Richardson number applies to either form (so we shall refer simply to Ri), although we shall see later that there are situations where it is important to maintain the distinction.

The Richardson number is defined so that negative values correspond to destabilizing stratification and positive to stabilizing. The limiting case of $Ri \to -\infty$ corresponds to free convectional instability, treated by Busse in Chap. 5.

When $-Ri$ is large but finite, one may think of the situation as a free convectional instability modified by a superimposed shear. The consequences depend on the convective regime. For example, if this is cellular, one may observe alignment of the cells in the flow direction [8.35–38], as illustrated by Fig. 8.4. (The most obvious geophysical application of this is as one mechanism for the formation of cloud streets [Ref. 8.13, Sect. 7.2.1], although the work from which Fig. 8.4 is taken was motivated by possible application to convection in the earth's interior.) Or, if the convection is turbulent, one may observe the development of this turbulence with distance downstream – a quick appearance of large-scale "thermals" with subsequent gradual intensification of smaller eddies [8.39]. There are indications that, in this transitional zone, the mean flow may be acclerated by the convection [8.40], providing a simple example of a "negative eddy viscosity" phenomenon [8.26]; as mentioned in

Sect. 8.2, phenomena of this general category are thought to account for a range of geophysical observations, but laboratory examples are sparse.

In the opposite extreme of high positive Ri, instability must always ultimately be suppressed; the shear is a minor perturbation on a stably stratified fluid.

8.3.2 Stably Stratified Free Shear Layers

Conversely, for relatively small |Ri| one may regard the shear flow instability as being modified by the stratification. We shall consider in particular the case of a free shear layer with positive Ri, where the velocity and density profiles have the general forms shown in Fig. 8.5. There are three reasons why this is an interesting case: inviscid stability theory provides a good demonstration of the effect of increasing Ri; sufficient experiments have been performed for a cogent, although probably incomplete, description of the flow development; and some of the observations have rather direct geophysical applications. However, since there are already several reviews of the topic, our treatment will be quite terse [8.2–4; Ref. 8.13, Chap. 4].

Inviscid linear stability theory is informative because the unstratified flow is unstable according to such theory as indicated by the Rayleigh inflection point criterion (see Sect. 7.2). Thus, one may ask, when does the flow become stable as Ri is increased? An important theorem, due to *Miles* [8.41], states that one always has stability if $Ri_g > 0.25$ at all z. This is a sufficient but not necessary condition for stability; there are cases of stable flows where Ri_g falls below 0.25. However, *Hazel* [8.42] points out that in all the particular cases that have been analyzed, one has instability if Ri_g at the inflection point is less than about 0.2. The significance of Miles' theorem depends on the detailed velocity and density

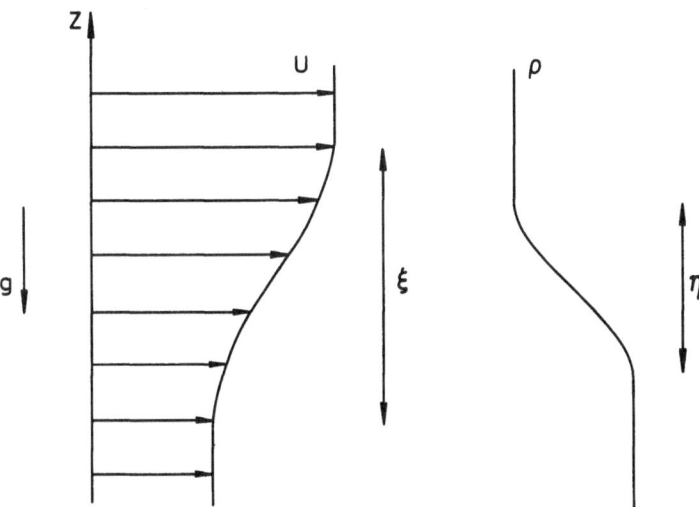

Fig. 8.5. Schematic of velocity and density profiles under discussion

profiles. In some cases it may enable one to say that the flow is stable (to small perturbations) when Ri_b exceeds a critical value of order 1. However, it must be remembered that the theorem fails if Ri_g falls below 0.25 at any point. An example, which relates to experimental observations described below, occurs when the density change takes place over a narrower region than the shear ($\eta \ll \xi$ in Fig. 8.5). Ri_g on the fringes may be small even when Ri_b is large and instability is then still possible.

Miles' theorem is, of course, only the starting point of a large body of theory. A detailed review of other ramifications of linear inviscid theory has been given by *Drazin* and *Howard* [8.34], with more recent developments being described by *Hazel* [8.42]. Inclusion of either diffusion (at low Prandtl number) [8.43] or nonlinearity [8.44, 45] can lead to instability for conditions of stability according to Miles' theorem, although they do not change the order of magnitude of the critical Richardson number. References [8.45–47] are recent papers that provide good points of entry into the extensive theoretical literature on this topic. In general, these theories provide substantial understanding of the earlier stages of the flow development, but its later stages are known primarily through experimental work.

There have been various experimental studies of the instability of profiles generally similar to those in Fig. 8.5, with both $\eta \sim \xi$ and $\eta \ll \xi$ [8.48–53]. In either case, when the Richardson number is sufficiently low[2], the first instability produces cross-flow vortices initially not significantly different from those in an unstratified shear layer resulting from the Kelvin-Helmholtz instability (see Chap. 7), the subsequent evolution being modified by the stratification. For higher Ri and $\eta \sim \xi$, the instability is suppressed, but for $\eta \ll \xi$ a new type of instability arises that has no counterpart in an unstratified flow.

Very regular arrays of the cross-flow vortices, often known as billows, have been observed. In part, this may be because one of the experimental techniques available in a stratified fluid [8.48] generates a shear that is very uniform in its flow direction, as well as because the stratification has a direct effect on the instability. Similar regular arrays are observed – in nature – in cloud patterns [Ref. 8.13, Sect. 4.1.4], in "clear air turbulence" as revealed by radar (Fig. 8.6 [8.54]), and in the ocean thermocline [Ref. 8.13, Sect. 4.3.3]. Even though such arrays may arise transiently without stratification, the common occurrence of regular structures is almost certainly a consequence of the natural situations being stably stratified.

In the laboratory, the observed effects of stratification on the evolution of the instability are related to the thickening of the shear layer by the fluctuations. This increases Ri_b, since L increases while $\Delta\varrho$ and U remain constant. Hence, the growth is ultimately limited, with detailed consequences that depend on how far the evolution towards turbulent flow has proceeded when this happens, i.e., that depend on the Reynolds and Richardson numbers. Figure 8.7

---

2 We refer the reader to the original papers or to the reviews mentioned above for the quantitative details, as these depend on the profiles and on the Reynolds number.

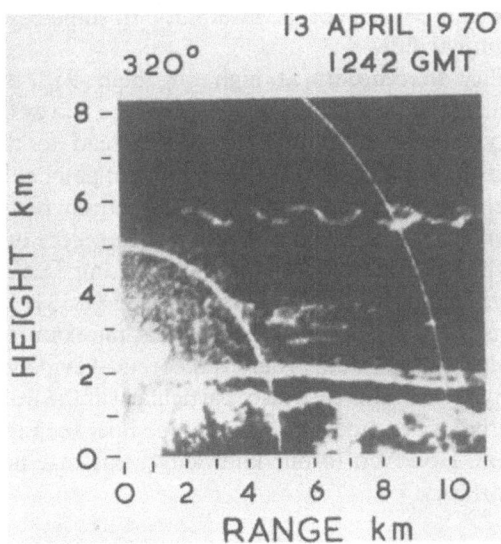

Fig. 8.6. Kelvin-Helmholtz billows in clear air shown by a radar echo. The echo is produced by scattering from inhomogeneities in the refractive index (associated with temperature and humidity variation) due to small-scale turbulence within the billows [8.54]

Fig. 8.7a–d. Dye patterns at different distances downstream in stably stratified free shear layer [8.52]

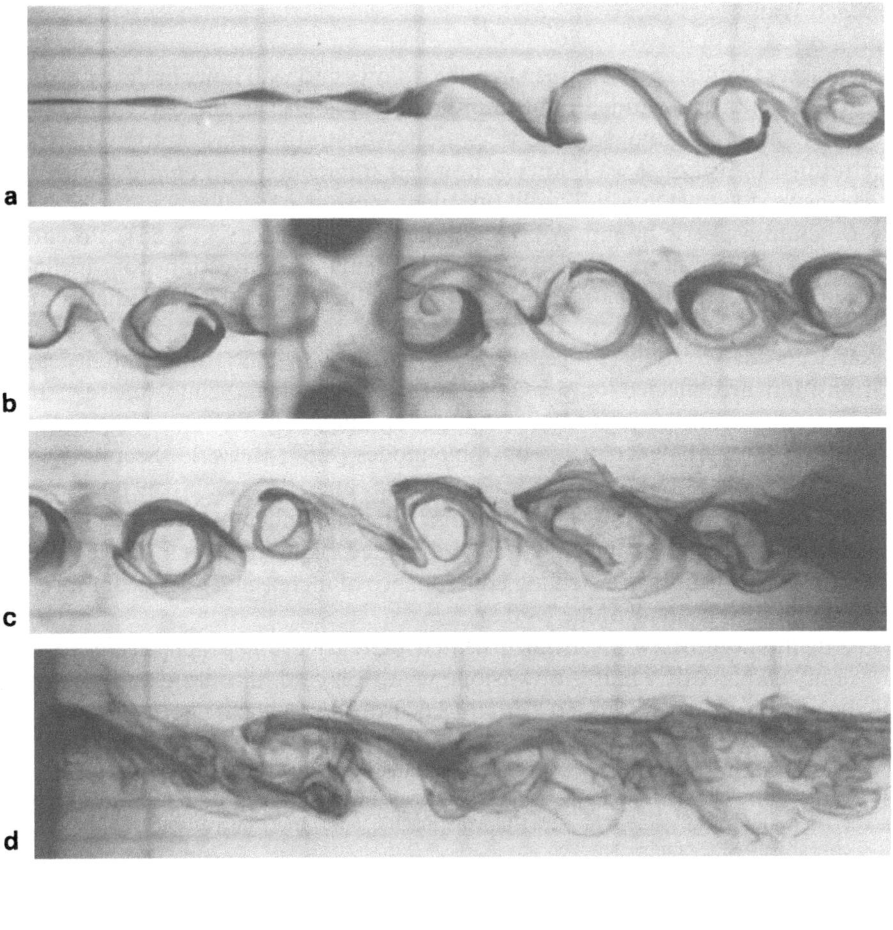

illustrates the growth and subsequent collapse of the instability. In some cases the layer ultimately reverts to laminar flow.

In the new type of instability that occurs at high ξ/η, high $\mathrm{Ri_b}$, the disturbances that grow on the fringes of the velocity profile generate waves on the comparatively sharp density interface. Nonlinear processes lead to the steepening of these waves and thus to breaking and the generation of patches of turbulence (Fig. 8.8). The steepening is particularly strong for certain phase relationships of the two waves that arise, with different phase speeds, on the two sides of the profile; there are thus particular sites for the breaking [8.52]. Ultimately, the turbulence is again suppressed by the stratification [8.53].

An obvious question concerns what happens when $\xi/\eta \ll 1$, as, for example, in a shear layer within a uniformly stratified fluid. There is theoretical evidence [8.55] that new types of instability arise in this case also, particularly if the fluid is bounded so that resonant internal wave modes may be trapped between the boundary and the shearing region. However, to our knowledge, this case has not been investigated in the laboratory.

8.3.3 Wall Flows

The marked differences between free shear flows and wall flows in the transition process of an unstratified fluid imply differences also in the effect of stratification, and we note some observations illustrating this. Information on stratified transitional wall flows is however fragmentary (in contrast with the extensive documentation of fully turbulent motion, particularly in the context of the atmospheric boundary layer on the earth's surface [8.12, 56]). Potential sources of such information are the experiments on convection from inclined heated surfaces and in inclined layers between heated and cooled surfaces. It is tempting to suppose that the component of the buoyancy parallel to the surface(s) serves only to generate the mean flow and that stability and transition are affected only by the normal component. Unfortunately, this supposition is not always valid; longitudinal buoyancy may enter significantly into the energy balance of the unstable modes [8.57; Ref. 8.11, Sect. 14.5].

Sparrow and *Husar* [8.58], and *Lloyd* and *Sparrow* [8.59] observed longitudinal vortices (Fig. 8.9) replacing Tollmien-Schlichting waves as the first stage of transition in the flow above a heated plate at a small angle to the vertical[3]. One thus has a striking consequence of increasing $-\mathrm{Ri_b}$ from zero to a comparatively small value (~ 0.1).

In the flow below an inclined heated plate (relevant geophysically to, for example, katabatic winds – downslope winds on cold hillsides – and turbidity currents on the ocean floor), *Tritton* [8.60] observed a range of conditions in

---

3 Interpretation is complicated by the fact that, although the Reynolds number at which the vortices were observed was lower than that at which Tollmien-Schlichting waves were observed on a vertical plate in the same series of experiments, it was much higher than the theoretical critical Reynolds number for such waves.

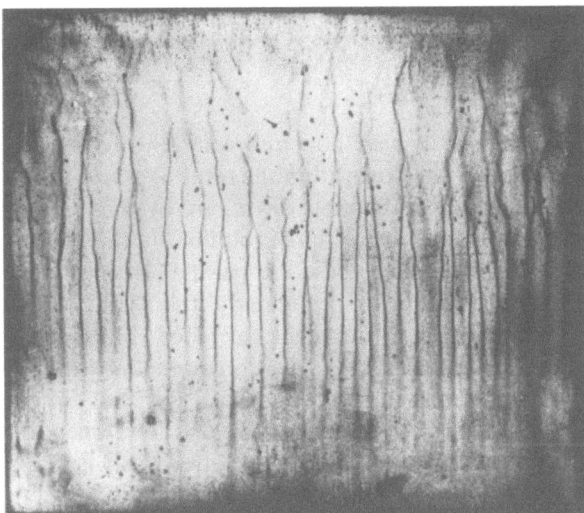

Fig. 8.8. Dye pattern in stably stratified free shear layer (at higher Richardson number than in Fig. 8.7), showing breaking of interfacial waves [8.52]

Fig. 8.9. Longitudinal vortices in convection of water below a heated surface inclined at 35° to the vertical. Note that dye is produced (electrochemically) *uniformly* over the surface of the plate [8.58]

which the effect of inclination on the onset of instability was comparatively slight but the effect on complete transition was very marked. Intermittent turbulence appeared as in unstratified flow on a vertical plate, but the increase in intermittency factor with distance downstream was inhibited. Either entrainment by turbulent spots was reduced or some spots were relaminarizing (with some evidence towards the latter interpretation). The result is a very long region of intermittent turbulence. Possibly, in some cases, the intermittency factor never reaches 1, and in that sense transition is never completed.

8.3.4 Horizontal Shear

We pointed out earlier in this section that the action of stratification on a horizontal shear (dU/dy) is inherently different from that on a vertical shear (dU/dz), as in principle it arises only after the first instability. It is interesting to inquire what sort of motion results from the instability that will exist even at very high positive values of the Richardson number. However, this case appears to have been little investigated, presumably because it has no obvious direct geophysical counterpart. It does, nevertheless, have indirect ones, the island wake of Fig. 8.2b being an example. Presumably a three-dimensional "obstacle" has generated a wake instability similar to that usually found behind two-dimensional obstacles [Ref. 8.11, Chap. 3], because stratification has suppressed the instability associated with the vertical shear but not that associated with the horizontal shear. Laboratory experiments have demonstrated the occurrence of a vortex street behind a sphere in a stratified fluid [8.29, 61]. The three-dimensional structure of this flow is, however, obscure, though some information is provided by *Brighton* [8.62].

8.4 Shear Flows in Rotating Fluids

The form of the interaction between a shear flow and rotation of the whole system depends on the orientation of the flow within the rotating frame. This is conveniently expressed in terms of the relative directions of the vorticity associated with the shear and that associated with the rotation. In this chapter, we shall restrict attention to the case when these are parallel or antiparallel (though there are other cases of interest, particularly the Ekman layer instabilities described in [8.63–65]). In a frame of reference rotating with angular velocity Ω about the z axis, there is a flow $U(y)$, predominantly in the x direction with its principal variation in the y direction. Since, in the laboratory, the rotation axis is normally vertical, this is a horizontal flow with perpendicular horizontal variation. For example, the geometries of a free shear flow and plane Poiseuille flow (topics to be discussed shortly) are indicated in planform in Fig. 8.10. In each case, the depth of the system normal to the plan view would be large compared with the width of the shearing region.

8.4.1 Stabilizing and Destabilizing Effects of Rotation

The general effect of rotation is less physically obvious than that of stratification. One can readily see that rotation will often be a stabilizing influence; the constraints which give rise to the Taylor-Proudman theorem [Ref. 8.11, Sect. 15.3; Ref. 8.15, Sect. 7.6] reduce the range of modes which can amplify – much as the rotation of a solid gyroscope constrains the ways in which it can move. It is less obvious (but nevertheless true) that rotation can sometimes be a destabilizing influence. Moreover, the criterion that determines whether it is stabilizing or destabilizing is not immediately apparent. It is useful therefore to formulate a simple "displaced particle" argument that demonstrates these points. Figure 8.11 compares the Coriolis force $2\varrho\Omega U_2$ acting on an undisplaced particle with the force $2\varrho\Omega U_1'$ acting on a particle which has been displaced a distance ζ from a position where the velocity is U_1. The former will be balanced by a pressure gradient in the y direction present in the unperturbed flow, and this will act also on the displaced particle[4]. Hence, this particle is

---

4 Strictly, to eliminate the complication of longitudinal pressure gradients generated by the displacement, the "particles" should be long thin strips in the x direction, and the comparison made between displaced and undisplaced strips at the same y but different z. However, this is more difficult to show schematically.

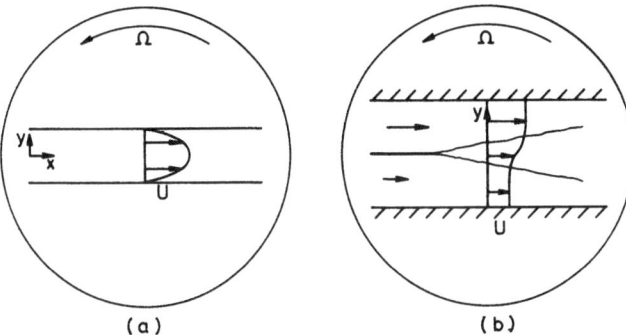

Fig. 8.10a, b. Schematic sections, perpendicular to the rotation axis, of (**a**) plane Poiseuille flow and (**b**) a free shear layer in a rotating system, such that the shear vorticity and system vorticity are parallel or antiparallel

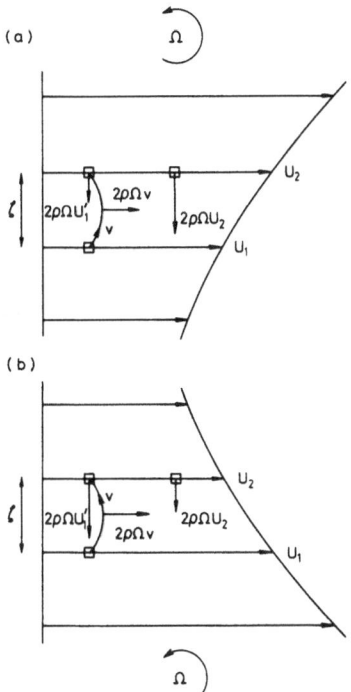

Fig. 8.11a, b. Sketch relating to displaced particle discussion – see text

displaced further from its original position, giving instability, if $U'_1 < U_2$ (for positive ζ) as in Fig. 8.11a, while it is restored towards its original position, giving stability, if $U'_1 > U_2$. However, $U'_1 \neq U_1$, because of the Coriolis acceleration $2\Omega v$ in the x direction that has acted on the particle while it was being displaced with velocity v in the y direction. This produces a total velocity change

$$U_1' - U_1 = \int 2\Omega v\, dt = 2\Omega\zeta \tag{8.3}$$

(independent of v). But

$$U_2 - U_1 = \zeta \partial U/\partial y. \tag{8.4}$$

Hence, $U_1' < U_2$ only if $\partial U/\partial y > 2\Omega$. In case a (shear vorticity and background vorticity of opposite signs), the Coriolis effect provides a destabilizing mechanism if and only if the shear is large enough; otherwise, the acceleration during displacement reverses the sign of $(U_2 - U_1')$. In case b (vorticities of the same sign), the mechanism is always stabilizing; the acceleration during displacement reinforces the initial difference between U_1 and U_2. An informative way of expressing this is to note that one has destabilization when the absolute vorticity $(2\Omega - \partial U/\partial y)$ is of opposite sign from the background vorticity (2Ω). If we now translate this result to a situation in which one has a given shear $\partial U/\partial y$ (taken positive) and then investigate the effect of rotation, one finds that the rotation is destabilizing if 2Ω lies in the range $0 < 2\Omega < \partial U/\partial y$; stabilizing otherwise. Note that large enough Ω of either sign is stabilizing, in line with our earlier comment about "gyroscopic" constraints.

The above conclusions, inferred from displaced particle arguments, can be confirmed in various ways. One is inviscid linear stability theory which will be considered below. Another is to note that the turbulence energy equations apply also to a transitional flow. Hence, the formulation of these by *Johnston* et al. [8.66] and discussed also by *Tritton* [8.67] is relevant.

It is also informative to note that circular Couette flow (see Chap. 6) becomes an example of this process in the limits of small gap and small differential rotation. The former makes it approximate to a plane shear. The latter makes it appropriate to use a rotating reference frame, and the destabilizing mechanism normally associated with centrifugal effects is then associated with Coriolis effects. It may be shown that the Rayleigh criterion for Couette flow then becomes just the above criterion.

The processes we have been considering depend essentially on the presence of both shear and rotation. In the destabilized case, the Coriolis effect provides a new mechanism for the instability of shear flows. But, of course, shear flows are prone to instability in the absence of rotation – through the Kelvin-Helmholtz or Tollmien-Schlichting mechanisms (see Chap. 7), which can still act in a rotating fluid. A measure of the relative importance of the Coriolis mechanism to them is provided by the parameter B introduced by *Bradshaw* [8.68] and *Johnston* et al. [8.66],

$$B = -\frac{2\Omega(\partial U/\partial y - 2\Omega)}{(\partial U/\partial y)^2} = S(S+1), \tag{8.5}$$

where $S = -2\Omega/(\partial U/\partial y)$ and is the reciprocal of a form of the Rossby number. B plays a role for these flows similar to that of the Richardson number for stratified flows: small B indicates that the stability is little changed from the nonrotating situation; negative B indicates that the rotation is destabilizing and positive stabilizing; and large B (which can occur only with $B > 0$) indicates dominance of rotational stabilization. However, the analogy must be used with caution; the impossibility of large negative B is not the only difference from stratified flow, as we shall see below.

8.4.2 Theoretical and Experimental Examples

Johnson [8.69] has carried out an extensive analysis of the linear inviscid stability problem with particular reference to the profile

$$U = U_0 \tanh(y/b). \tag{8.6}$$

This has an inflection point giving instability when the system is not rotating. Figure 8.12 shows stability boundaries in the (k_x, k_z) plane, where k is the wave number nondimensionalized by $1/b$, for the two senses of rotation. (*Johnson* also considered other axes of rotation in the $x-z$ plane but we shall not discuss these.) Different lines are for different values of the Rossby number, $Ro = U_0/b\Omega$, which is equal to $2/|S|$ at the position of maximum $\partial U/\partial y$. For positive S, the unstable regions are entirely contained within the corresponding region [the semicircle $(k_x^2 + k_z^2) < 1$, $k_x > 0$] for $\Omega = 0$; rotation is entirely stabilizing. For negative S, in contrast, there are unstable modes outside the nonrotating instability zone. However, at large negative S the loops become very similar to those for large positive S, corresponding to the fact that then $B \sim S^2$.

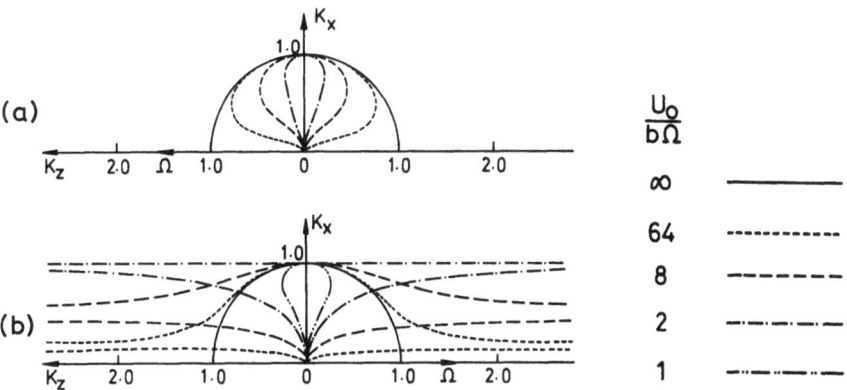

Fig. 8.12a, b. Neutral stability curves with respect to x and z components of the wave number for the situation described in the text. Shear vorticity and system vorticity (a) parallel; (b) antiparallel [8.69]

Throughout these developments, the instability range on the x axis remains $0 < k_x < 1$. There are thus always some unstable modes no matter how small the Rossby number, i.e., no matter how high B. There is no counterpart of Miles' theorem. The Rayleigh inflection point criterion remains applicable even in a rapidly rotating system.

Thus, there arises the question, what are the consequences of the instability when the Rossby number is low? It seems likely that, when the rotation suppresses many modes of instability, the remaining unstable ones must give rise to highly regular flow structures. But a full answer to the question requires information from both nonlinear viscous theory and from experiments. To our knowledge, there exists no relevant nonlinear theory. There are some experiments, which we shall consider below, but they provide a very incomplete picture.

Contributions to linear viscous stability theory are contained in [8.70–73]. We shall not discuss these systematically, but we shall quote examples of the results particularly where they relate to experimental results.

Plane Poiseuille flow in a rotating system has been studied both experimentally and theoretically [8.70, 71]. Its profile involves both signs of $\partial U/\partial y$ and thus both signs of S. For small rotation rates there are stable and unstable halves of the channel. As the rotation rate is increased the unstable region occupies less of the channel width, d, until, at $\mathrm{Ro} = U_{\mathrm{av}}/\Omega d = 1/3$, it vanishes. Linear stability theory and its experimental confirmation indicate that in the inviscid limit (Ekman number, $\mathrm{Ek} = \nu/\Omega d^2 \to 0$), the Coriolis mechanism generates instability whenever $\mathrm{Ro} > 1/3$ (i.e., whenever there is even the smallest unstable region) but that larger values of Ro are required for the instability when Ek is finite [8.70]. In consequence there is a minimum critical Reynolds number, $U_{\mathrm{av}}d/\nu$, of about 90 at $\mathrm{Ro} = 2$ [8.71][5]. This compares with a critical Reynolds number for the Tollmien-Schlichting instability for $\Omega = 0$ of 7700 [8.74]. (The general problem for $\Omega \neq 0$ in which the two instability mechanisms are both present has not been treated.)

Near this minimum one can observe experimentally the flow produced when the Coriolis mechanism is the primary cause of instability [8.70]. Longitudinal rolls are generated, analogous to those in Fig. 8.9.

Free shear layers are perhaps of rather more direct relevance than Poiseuille flow to geophysical applications, although it happens that they have been studied also for their engineering applications. *Rothe* and *Johnston* [8.75] carried out an experimental investigation of the free shear layer between a separation point at a sharp corner and reattachment. The flow is transitional and turbulent in the absence of rotation, and the modifications to these processes by rotation were investigated. Rossby numbers were comparatively high, so that rotation in one sense gave positive B and the other negative. The contrast was apparent. An interesting feature was that the stabilized case could

---

5 Readers using this reference should note that the symbol Ro used there refers to a rotation number which is the reciprocal of this Rossby number.

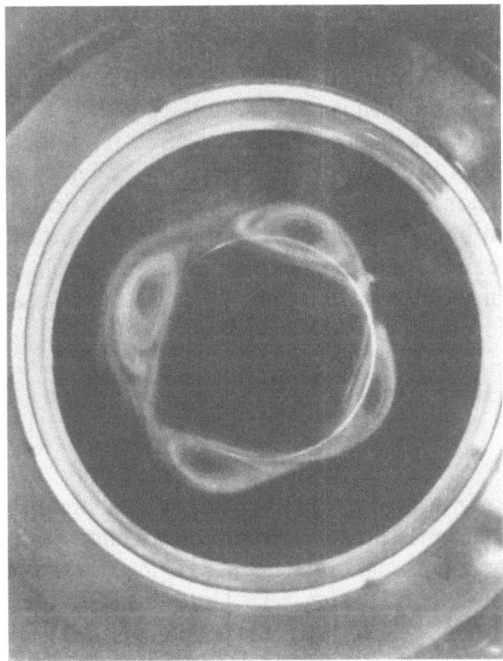

Fig. 8.13. Vortices arising from instability of annular Stewartson layer. View is parallel to axes of rotation and differential rotation [8.76]

at first sight appear as the one in which the more vigorous disturbance was developing. This is because the cross-stream vortices which arise from the instability are not directly affected by the rotation. Normally, these are subsequently disrupted by three-dimensional disturbances, but these latter are inhibited by the rotation. Hence, the vortices can grow into more vigorous and more organized structures.

The above illustrates one experimental approach; one takes a flow that has proved of interest in the absence of rotation and then investigates how this is modified by rotation. An alternative approach is to investigate the stability of a basic flow that is itself characteristic of rotating fluids. Stewartson layers are shear layers of this type and instabilities of these, as illustrated by Fig. 8.13, have been observed by *Hide* and *Titman* [8.76]. A cylindrical tank rotates about its axis (vertical) with angular velocity Ω. A horizontal disc at the center of the tank rotates relative to the laboratory with angular velocity $\Omega(1+\varepsilon)$, ε being small. This induces an angular velocity $\Omega(1+\varepsilon/2)$ in the body of fluid closer to the axis than the disc edge, while fluid further out retains the angular velocity Ω. The two regimes are separated by a cylindrical Stewartson layer, which may become unstable and, for $\varepsilon > 0$, break up into a number of vortices as shown in the figure. The critical condition appears to be independent of the sign of ε, but the experiments suggest that the consequence of the instability is different in the two cases.

Busse [8.73] has applied viscous linear stability theory to show that this instability is essentially a shear layer instability. It should be noted that, although the basic flow is a low Rossby number flow (Ro=ε), the Rossby number based on shear layer thickness, which is relevant to the stability, is not so low – in the range 10^{-1} to 1. In Busse's theory, the viscous effects limiting the instability arise not in the shear layer itself but in the Ekman layers generated on the end walls by the perturbation. This illustrates a general complicating property of rotating fluids: distant boundaries may be surprisingly important; one cannot automatically suppose that, so long as the size of the system is large compared with the length scale of the shearing region, one is dealing with an effectively infinite expanse of fluid.

8.4.3 The β Effect

There are, of course, various ways in which any geophysical flow differs from those considered above. However, one factor is common to almost any geophysical application and thus requires mention here. This is the so-called β effect [8.18, 19; Ref. 8.15, Sect. 7.7; Ref. 8.21, Sect. 7.3]. The effect of rotation often enters the dynamics through the vertical component, $\Omega \cos\theta$, (where θ is the co-latitude) of the angular velocity. The fact that this varies with θ can give rise to new effects. One defines the Coriolis parameter, $f = 2\Omega \cos\theta$. For theoretical purposes, this situation is often modelled by considering a plane horizontal layer rotating about a vertical axis but with a variable value of f used in the Coriolis term in the equations. The name β plane for such a system derives from the conventional notation $df/dy = \beta$, where the y coordinate corresponds to the north direction.

An important result concerning the stability of a shear flow $\partial U/\partial y$ is that the Rayleigh inflection point criterion is now modified. Instead of requiring a maximum in $|\partial U/\partial y|$ for instability, it requires a maximum in $|(\partial U/\partial y - f)|$, which corresponds, of course, to

$$\partial^2 U/\partial y^2 = \beta. \tag{8.7}$$

In principle, this modification may be either stabilizing or destabilizing with respect to a corresponding hypothetical flow with $\beta = 0$. In practice, in the situations to which one wishes to apply it, it is stabilizing. A zonal flow with velocity varying with latitude on a planet is likely to have points of inflection, $\partial^2 U/\partial y^2 = 0$; however, depending on its speed and length scale, $\partial^2 U/\partial y^2$ may always be $< \beta$.

The cases with interesting consequences are, of course, those in which the instability is not completely suppressed but is modified, i.e., in which $\partial^2 U/\partial y^2$ is $\sim \beta$ but in some places $> \beta$. The shift in the most unstable region away from the inflection point seems likely to change the structure of the motion that results.

This is a difficult configuration to simulate in the laboratory, and little is known about it.

The relevance of these ideas to the earth's atmosphere has been considered by *Kuo* [8.77]. *Ingersoll* and *Cuzzi* [8.78], and *Ingersoll* [8.79] have applied them to Jupiter's atmosphere. In particular, they relate the observation, noted in Sect. 8.2 (Fig. 8.1), that some boundaries are sharp and others turbulent to this modification of shear flow instability by the β effect. It may, however, be a serious oversimplification to leave buoyancy effects out of account in interpreting Jupiter's atmospheric turbulence.

8.5 Baroclinic Instability in a Rotating Fluid

We turn now to a situation involving both density variations and rotation in which, moreover, both are intrinsic to the establishment of the basic state as well as to the instability. We are concerned with instabilities[6] that arise essentially from an inclination θ of constant density surfaces ϱ (isopycnals) as shown schematically in Fig. 8.14. $\varrho_1 < \varrho_2 < \varrho_3 < \ldots$, so the fluid is stably stratified, but there is also a horizontal density gradient. In a nonrotating fluid, such a density distribution would necessarily (as a consequence of non-equilibrium rather than instability) give rise to circulation in the xz plane. In a fluid rotating sufficiently rapidly about the vertical (z) axis, however, it may be associated with a flow only in the y direction – the tendency for the horizontal density gradient to generate vorticity in the y direction may be balanced by

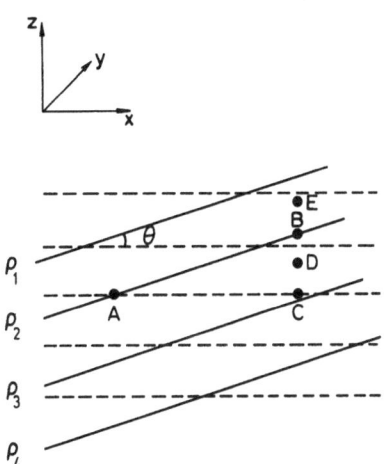

Fig. 8.14. Schematic diagram illustrating the baroclinic instability mechanism for a rotating fluid in which isopycnals (——) are inclined to horizontal surfaces (– – –)

---

6 In common with meteorologists and other geophysicists, we shall classify such instabilities as "baroclinic" instabilities, remembering that this is a somewhat specialized use of a term with a more general meaning.

twisting of the background vorticity, as indicated by the thermal wind equation

$$g\partial\varrho/\partial x = 2\varrho\Omega\partial V/\partial z \tag{8.8}$$

(the vorticity equation when buoyancy and Coriolis terms dominate over inertial and viscous forces). The flow velocity must thus vary vertically but not necessarily in its flow direction. It is thus consistent with maintenance of the density distribution ($U \cdot \nabla\varrho = 0$).

Density and velocity fields of this general form are generated in a variety of natural situations. In the atmosphere, for example, horizontal density gradients are maintained on a global scale by differential heating between equator and poles; in the oceans, not only the above thermal forcing but also mechanical forcing from surface winds contribute to the form of the basic state.

The stability of the basic state can be understood qualitatively by referring back to Fig. 8.14 and considering the consequences of exchanging a typical fluid particle A, in turn, with others labelled B, C, D, and E. For cases B and C (particles on the same isopycnal and same horizontal level, respectively) no work is required to conduct the exchange and no energy is released. Work must be done to exchange particles A and E, however, since this involves lifting a relatively heavy particle and lowering a relatively light one. Only in the case of particles such as A and D is energy actually released by the exchange. Infinitesimal disturbances can therefore grow and gain kinetic energy at the expense of the potential energy of the basic state only if the two particles lie on planes inclined to the horizontal at an angle less than θ. Consideration of the dynamical equations shows that the slope of fluid particle trajectories will be of the order of the Rossby number, so that increasing the rotation rate of the system will reduce this slope to a value less than θ (which itself increases with increasing rotation rate). Baroclinic instability will therefore occur whenever a critical rotation rate is exceeded. The particular critical value will be determined primarily by the density distribution in the system and the presence of damping agencies such as viscosity and thermal diffusion. However, because of the rapid background rotation it will also be determined by the orientation of the bounding surfaces; indeed, in general, instability is determined not by θ but by the ratio ψ/θ, where ψ is the slope of both bounding surfaces to the horizontal. For some values of ψ/θ the system is stable for all Ω [8.6].

8.5.1 The Eady Problem

The earliest linear stability analysis of a baroclinic flow, due to *Eady* [8.80], considered the stability of an inviscid stratified shear flow $V(z)$, of constant buoyancy (Brunt-Väisälä) frequency N [Ref. 8.11, Sect. 16.4; Ref. 8.21, Sect. 9.4], rotating rapidly about a vertical axis with angular velocity Ω. The flow, which is driven by an externally imposed temperature gradient $\partial T/\partial x$, is in thermal wind balance so that the shear in the zonal flow $\partial V/\partial z$ satisfies (8.8).

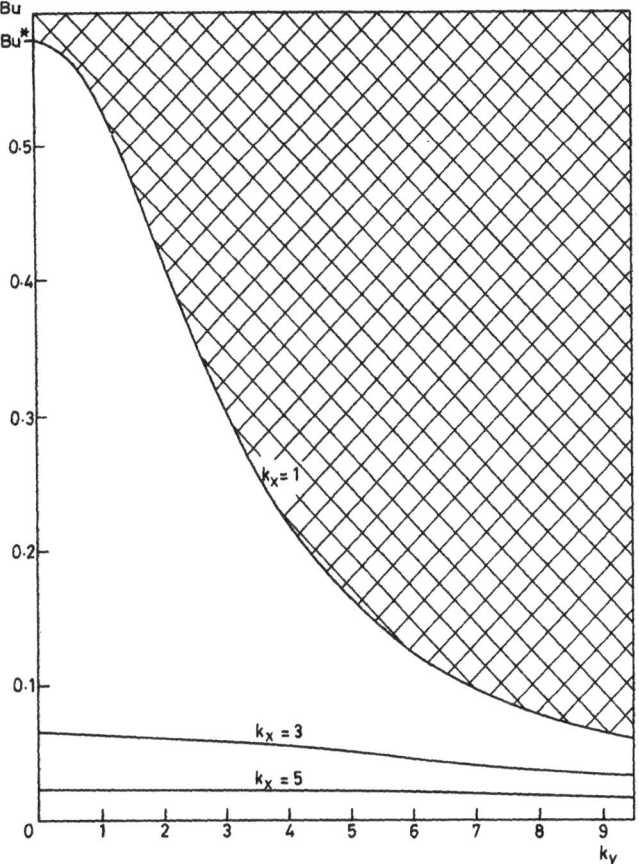

Fig. 8.15. Plots of Burger number Bu against wave number k_y for three values of wave number k_x. Unshaded area represents the regime for which the fluid is baroclinically unstable

Such a flow, bounded above and below by horizontal planes and bounded laterally (in the x direction only) by vertical planes, has a density distribution of the form shown in Fig. 8.14. Linear theory shows that the flow is stable or unstable depending upon the value of the Burger number Bu, where

$$Bu = (NH/2\Omega L)^2 \tag{8.9}$$

and H and L are vertical and horizontal length scales, respectively.

As shown on Fig. 8.15, the flow is stable [8.7] for all modes whenever Bu exceeds a critical value Bu* of 0.58, but below Bu* only modes having wave number less than a particular value are unstable. The $k_x = 1$ mode has the greatest growth rate and only modes having wavelength λ greater than λ_c where

$$\lambda_c = 2\pi/(k_y)_c = 2[(Bu^*/Bu) - 1]^{-1/2} \tag{8.10}$$

can grow. (Additionally, as we anticipate from our earlier discussion, the relative orientation of the bounding surfaces to the horizontal and to each other has a noticeable effect upon the above results [8.5, 6].) The effects of the boundaries become important, of course, whenever viscous effects are considered, particularly in a rapidly rotating system when Ekman boundary layers are formed on all nonvertical boundaries. If viscosity is considered solely through its effects in these regions, the linear stability characteristics of the Eady inviscid flow are modified [8.81, 82]. Neutral stability curves constructed on the basis of such modifications are in good qualitative agreement [8.95] with data obtained from laboratory experiments (see Sect. 8.5.3).

As we have seen, for $0 < \mathrm{Bu} < \mathrm{Bu}^*$, there still exist inviscid short wave modes which are stable according to linear theory. This property of the flow does not extend to cases in which the β effect is included [8.83–85]. *Green* [8.83] has shown that even for small values of β all modes are unstable. This destabilizing effect on short wave modes has been discussed thoroughly, particularly in relation to laboratory experiments [8.86] which have attempted to model the β effect for such flows by containing the fluid between boundaries inclined relative to each other.

8.5.2 Symmetric Baroclinic Instability

Modes independent of the y direction, in the form of rolls, can also become unstable under certain circumstances. *Stone* [8.87] has demonstrated that such cases of symmetric baroclinic instability will occur whenever $\mathrm{Ri} < 1$, where Ri is the Richardson number, defined in this context as

$$\mathrm{Ri} = N^2/(\partial V/\partial z)^2 , \tag{8.11}$$

but only for cases in which $0.25 < \mathrm{Ri} < 0.95$ will symmetric baroclinic instability modes dominate over Eady ($\mathrm{Ri} > 0.95$) and/or Kelvin-Helmholtz ($\mathrm{Ri} < 0.25$) modes. Several studies have attempted to illustrate the occurrence of symmetric baroclinic instability in laboratory flows and the mechanism has been proposed as the explanation for the occurrence of zones and belts in the Jovian atmosphere. Thus far, the evidence for either is inconclusive [8.88–90].

8.5.3 Annulus Experiments

The majority of laboratory studies of baroclinic instability have been concerned with baroclinic waves of the Eady type. Such studies still not only constitute the main tests for both linear and nonlinear theories of baroclinic instability but also offer a means of observing the transition from the basic flow through the various stages of instability to an irregular state.

Laboratory investigations of baroclinic waves have been of two main types; those in which a rotating annulus of fluid is subjected to differential heating

[8.5, 6,91–96] and those [8.97–99] in which a rotating two-layer fluid is driven by differential rotation of a solid boundary in contact with the surface of the upper fluid. Experiments of the first type have been much more numerous, and comparison of theoretical results with experiment have most often been made in the context of this configuration.

The fluid annulus is formed between two bounded concentric cylinders, across which is imposed a constant temperature difference ΔT. There is no differential rotation between the two cylinders but the whole system rotates about a vertical axis with angular velocity Ω. In the absence of rotation the impressed temperature difference drives a closed meridional convective circulation in the fluid, but, as the system is rotated, Coriolis forces convert the meridional motion into an axisymmetric zonal flow in thermal wind balance. Such a flow, commonly referred to as a baroclinic circular vortex, possesses vertical shear in the azimuthal velocity U_ϕ and satisfies the thermal wind equation

$$\frac{\partial U_\phi}{\partial z} = \frac{g\alpha}{2\Omega} \frac{\partial T}{\partial r}, \tag{8.12}$$

where (r, ϕ, z) denotes the cylindrical coordinate system used. The basic flow therefore closely resembles the Eady flow except for the presence in the experiments of curvature effects – which are known to exert a weak destabilizing influence [8.7]. The form of the instability which develops in the annulus was first investigated in a systematic manner by *Hide* [8.91]. In general, for an effectively inviscid fluid contained in a given annulus with a fixed temperature contrast applied across it, the instability appears as a steady system of waves as a critical value of the rotation rate is exceeded. Subsequent detailed investigations [8.6, 86, 92, 93] have confirmed the correlation between the onset of baroclinic instability and the appearance of waves (see Fig. 8.16a) and have been able to discriminate with some precision between the various changes in their subsequent development. Here we have space only to indicate the general features of the flows and their transitions. It is found experimentally that the transition of an annulus flow from a symmetric zonal motion in thermal wind balance to a fully developed baroclinic wave system is determined primarily by the values of the thermal Rossby number Θ and the Taylor number Ta, defined by

$$\Theta = \frac{g\alpha(\Delta T)H}{\Omega^2 L^2} \left[= \frac{4\mathrm{Bu}(\Delta T)}{H(\delta T/\delta z)} \right], \tag{8.13}$$

$$\mathrm{Ta} = 4\Omega^2 L^5/v^2 H [= L/H(\mathrm{Ek})^2]. \tag{8.14}$$

From Fig. 8.16b it can be seen that the stable regime is split into upper symmetric and lower symmetric portions, the former corresponding to the situation discussed earlier, where, for moderate temperature differences, ro-

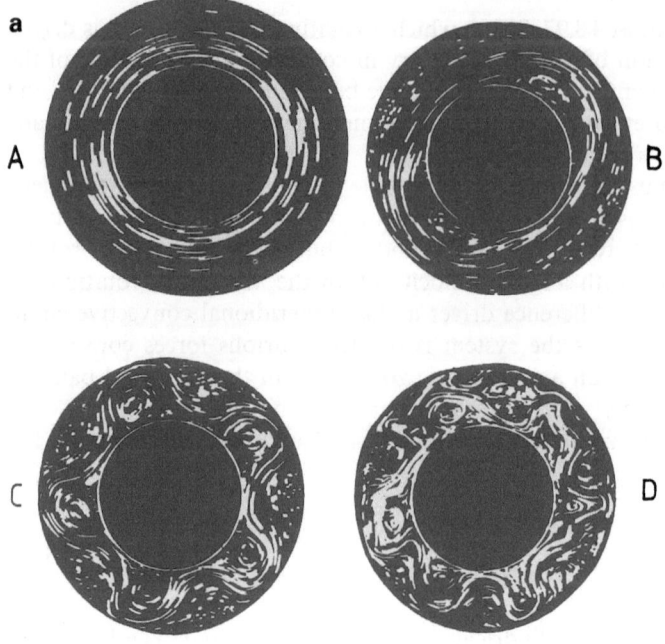

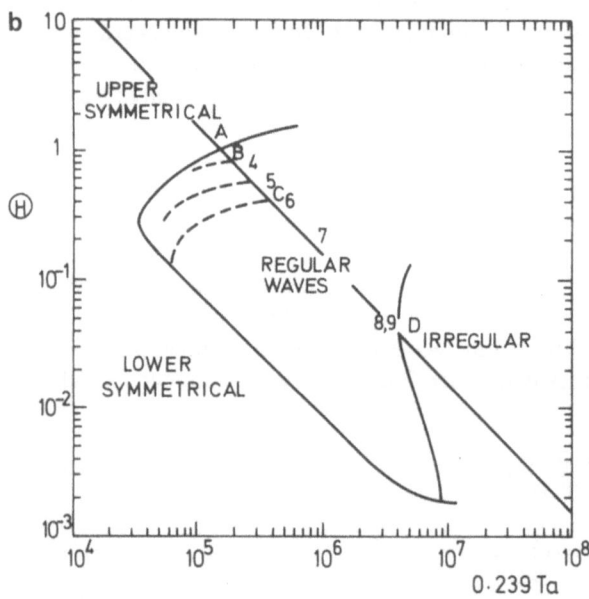

Fig. 8.16. (a) Plan photographs of characteristic flows observed in the annulus; (b) regime diagram showing their occurrence as a function of Θ and Ta [8.6]

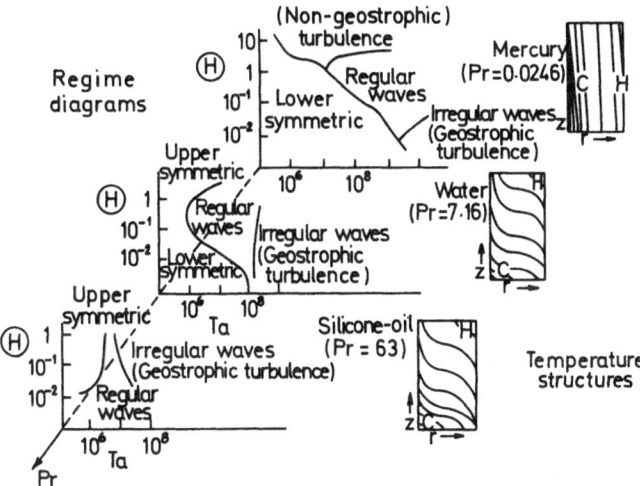

Fig. 8.17. (Θ, Ta, Pr) regime diagram and thermal structure for a rotating differentially heated fluid annulus [8.100]

tation effects are insufficient to cause particle trajectories to be inclined to the horizontal at an angle less than θ (see Fig. 8.14), and the latter corresponding to cases in which instabilities are unable to develop because of damping by diffusive processes. In view of the importance of diffusive effects in these circumstances it is clear that the character of the flow will be Prandtl number dependent also. The modifications to the stability diagram arising out of this dependence [8.100] are shown in Fig. 8.17.

Within the unstable regime the initial system of baroclinic waves is observed to change character as the external conditions are changed (see Fig. 8.16a). The full sequence of changes can be delineated most effectively by choosing a path such as ABCD on the stability diagram. Consideration of the significance of Θ and Ta will show that following this path from A to D is equivalent to choosing a preset temperature difference across a given annulus and observing the flow as the rotation rate is increased. Near the marginal stability boundary the steady wave system consists of a regular array of waves (see Fig. 8.16a) interspersed with closed vortices, and a jet stream meandering between the hot and cold sides of the annulus.

Theoretically, the number of waves present in the system is not uniquely determined by the values of Θ and Ta, though geometrical constraints impose an upper limit on the maximum wave number which can develop. On the stability diagram, therefore, wave numbers having *maximum* probability of formation are indicated. As the value of Θ decreases the value of the most probable wave number increases. In the wave regime, systematic scanning [8.101] of all regions of the fluid indicates that the wave system consists of a dominant wave number m as indicated, but, in addition, a significant portion of

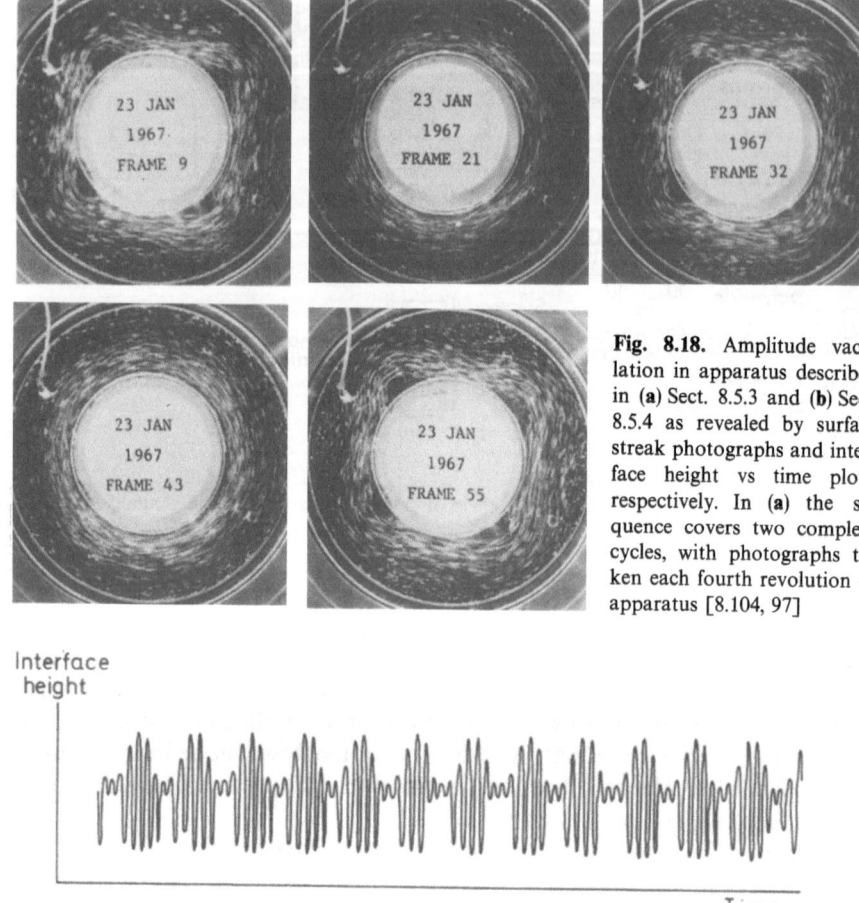

Fig. 8.18. Amplitude vacillation in apparatus described in (**a**) Sect. 8.5.3 and (**b**) Sect. 8.5.4 as revealed by surface streak photographs and interface height vs time plots, respectively. In (**a**) the sequence covers two complete cycles, with photographs taken each fourth revolution of apparatus [8.104, 97]

the wave energy is shared by harmonics of the main wave and sideband modes $m \pm 1$. The individual contribution of each mode to the overall wave pattern varies as Θ is varied.

The transition from the steady wave pattern, which is observed near marginal stability, to a fully irregular regime in which there is no recognizable wave structure to the flow ("geostrophic turbulence" [8.6]) is a complicated one. During the intermediate stages between the two regimes the observed pattern, though becoming more irregular, retains a wavelike form. The waves are said to "vacillate". Vacillation can take the form of shape, or tilted-trough, vacillation [8.91] in which the waves tilt periodically without change in amplitude, amplitude vacillation [8.102] in which the amplitude of the wave changes in a periodic manner (see Fig. 8.18a) [8.104], wave number vacillation [8.5] in which the dominant wave number fluctuates, or wave dispersion [8.103] in which two component waves drift at different speeds.

The transition from the vacillating wave regime to the irregular flow regime (in which there is no longer any wavelike form to either the flow itself or the interior temperature field – the two primary observables) is less well understood, though it has been suggested [8.20] that the baroclinic flow becomes barotropically unstable, the dominant wave interacting nonlinearly with other modes in such a way as to transfer kinetic energy to other scales.

8.5.4 Two-Layer Flows

Finally, we mention briefly the experiments using mechanical forcing of a two-layer system. In this configuration the basic flow is again a rotating stratified shear flow. The surface forcing causes the interface between the layers to distort, and this distortion may be sufficient to release the available potential energy in the density distribution. Such a flow can be shown to be baroclinically unstable according to linear theory (see Fig. 8.19a) for certain values of a Burger number Bu_b, defined in a two-layer fluid as

$$Bu_b = g(\Delta\varrho)H/4\varrho\Omega^2 L^2 , \tag{8.15}$$

and a Rossby number $Ro = \omega/2\Omega$, based upon the differential angular velocity ω. The procedure followed in such experiments is usually to keep Ω fixed and increase ω until the flow instability appears as an oscillation in the interface height. Near marginal stability the oscillations are regular, as with the temperature oscillations in the related annulus flow, and aspects of their occurrence accord well with linear theory (see Fig. 8.19b). Nonlinear interactions become important for supercritical conditions and vacillations appear in the flow (see Fig. 8.18b). Fourier decomposition of the interfacial

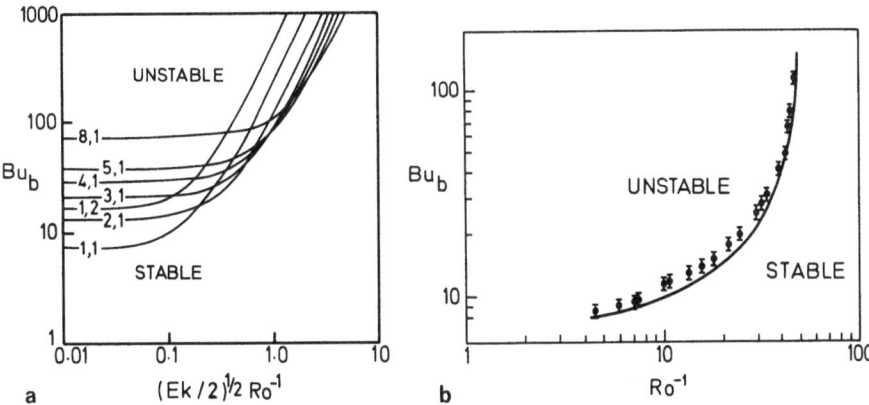

Fig. 8.19. (a) Linear stability diagram for various (k_ϕ, k_r) modes present in apparatus of Sect. 8.5.4 plotted in terms of Bu_b and $(Ek/2)^{1/2}Ro^{-1}$ and (b) comparison of experimental data (O) with linear stability curve (−) [8.97]

oscillations shows [8.99] that many features of the fully developed baroclinic waves (such as the presence of a dominant mode, harmonic modes, and sideband modes) exhibit qualitative similarities with those in the annulus. Important quantitative differences between the two systems exist, however, particularly in the distribution of energy between the various modes in the time-averaged amplitude spectra [8.99].

8.6 Multidiffusive Instabilities

Instabilities which arise solely as a result of differential diffusion of two or more components of a multicomponent fluid system are of considerable geophysical relevance. Perhaps the best known example of this type of instability occurs in a fluid stratified jointly by heat and salt. Instabilities are able to develop as a result of the disparity in diffusion rates of salt and heat even when the overall density distribution is stable. This is just one example of a double diffusive instability; others of the same general type can occur in fluid systems stratified by one agency alone; for example, as a result of differential diffusion of momentum and heat or momentum and salinity. Indeed, in general, it is likely that many diffusion coefficients characterize a natural fluid system. In these cases, as we indicate later, multidiffusive instabilities having hybrid properties are able to develop.

For our purpose, it is convenient to discuss multidiffusive instabilities primarily in terms of the thermohaline system described above. It is the case which has been investigated most extensively [8.8, 9, 13, 105–108], particularly with regard to laboratory experiments, and it is the case for which there are oceanographic data [8.109–114] to confirm its importance in natural situations. The nature and form of thermohaline instability is determined by which of the components is stabilizing and which is destabilizing. In either case the instability is driven by the release of the potential energy associated with the unstable gradient of the destabilizing component.

Consider the consequences of displacing a fluid particle vertically in such a system. When the temperature field is destabilizing (cold fresh water above warm salty water) a fluid particle displaced upwards will adjust its temperature to the local value relatively quickly. However, because of the disparity in diffusion coefficients between salt and heat (typically a factor of 10^2) the particle remains locally heavy by virtue of its excess salinity and it is forced to sink to its original level and overshoot. In its new position it experiences an upward force because of its local salinity deficiency so that it begins to oscillate about its original position, experiencing buoyancy restoring forces at the extreme points in its motion.

Because of the lag in temperature between the particle and its surrounding fluid (κ, though *relatively* large, is still finite), the particle always leaves either of the limiting positions faster than it arrives. This results in an *increase* in the amplitude of the oscillations with time with the instability being maintained by the potential energy in the unstable temperature field.

Such a motion is not possible in the case where the salinity profile is destabilizing. Here, by the same differential diffusivity argument as above, if a fluid particle is displaced vertically, buoyancy forces always act to keep the particle moving in the direction of the displacement. That is, any displacement will grow directly with time, the instability being driven on this occasion by the potential energy in the salinity gradient.

8.6.1 Linear Stability Theory

The instability modes described above are predicted by stability analyses of each of the double-diffusive systems, as shown in the linear stability regime diagram in Fig. 8.20. Following *Turner* [8.8] the principal boundaries between unstable and stable regimes are delineated by the straight lines XW and XZ. Excluding cases in which thermal and salinity gradients are both stabilizing or both destabilizing (i.e., negative Ra/positive Rs or positive Ra/negative Rs, respectively) we are interested primarily in the upper right and lower left quadrants of the figure. Note that PQ denotes the line of neutral buoyancy where equal and opposite contributions to the density profile are made by ΔT and ΔC, respectively, so that points to the right of this line represent statically stable Ra : Rs combinations. Purely thermohaline diffusive instabilities, therefore, are observed only in the regimes indicated.

In the upper right quadrant (the so-called diffusive regime) double-diffusive instability occurs above a critical value of Ra which itself depends linearly upon Rs. For values of Rs greater than that corresponding to the intersection of PQ and XW instability occurs when the fluid is statically *stable*. Linear stability analyses [8.115–118] confirm the earlier prediction that in marginal conditions the form of the instability is oscillatory, the frequency of the most unstable mode decreasing with increasing Ra for constant Rs. Such overstable oscillations are therefore to be expected in the region formed by PQ and XW, as indicated, and also (according to linear theory) in the regime defined by XV and XW. Along XV, the most unstable mode changes from oscillatory to direct [8.118] and beyond XV only directly growing modes are possible.

For all combinations of negative Ra and negative Rs to the left of ZX, only directly growing unstable modes can exist. As before, such instabilities (which are of the type discussed for this case earlier) are found in a regime where the overall density gradient is stable. Since the form of the most rapidly growing mode has a high length-to-width aspect ratio in the vertical direction (at least at high Ra), this regime is known as the finger regime.

The above stability boundaries are, of course, constructed on the basis of linear theory, so that caution must be employed when attempting to draw conclusions from regions of the diagram far from the principal stability boundary. Indeed, in laboratory experiments, as we see below, finite amplitude instabilities often develop preferentially, with the result that agreement between linear theory and experiment is only moderate. Despite these difficulties (and others associated with different boundary conditions for the two situations) laboratory experiments have been very successful in demonstrating the existence of these types of instabilities.

8.6.2 Diffusive Layering

The oscillatory instability characterizing the diffusive regime near marginal stability has been investigated by *Shirtcliffe* [8.119]. In his experiments a fluid layer stably stratified by sugar was heated uniformly from below. The temperature of the bottom boundary increased with time and the thermal state of the fluid interior was monitored continuously. As the marginal state was reached, an oscillatory instability was detected near the heated surface, the amplitude of the oscillation growing with time. As the imposed temperature difference increased further (equivalent to moving towards XV in Fig. 8.20) the oscillations took on an increasingly irregular form and a steady cellular structure became visible in the lower parts of the fluid region. The occurrence of the cellular steady mode was interpreted by *Shirtcliffe* as evidence of the preferential appearance of the subcritical finite amplitude instability predicted for this case by nonlinear theory [8.116]. From these and other studies it is observed that as a result of the appearance of this cellular convective mode near the heated boundary, a mixed layer develops within which there is vigorous overturning, with thermal boundary layers above and below. The mixed layer thickens with time as more fluid is entrained from above, until the upper thermal boundary layer (itself growing with time) becomes unstable and forms a second mixed layer in which there is also strong overturning. By this process a stack of sharply bounded mixed layers is built up (see Fig. 8.21a), with the lower

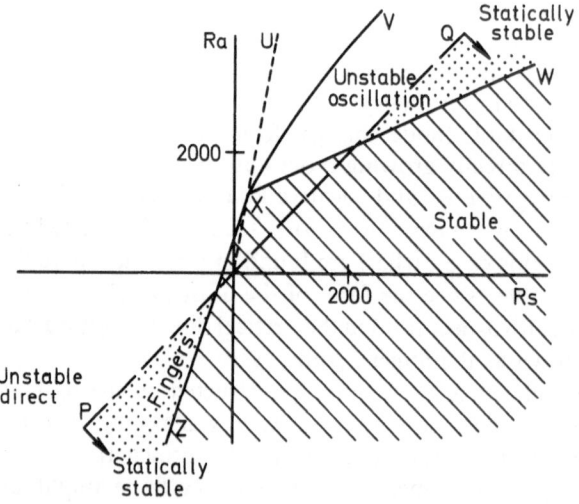

Fig. 8.20. Regime diagram showing theoretically determined linear stability boundaries for a thermohaline fluid system in terms of Ra $(=g\alpha H^3 \Delta T/\nu\kappa)$ and Rc $(=g\delta H^3 \Delta C/\nu\kappa)$ (reproduced, with permission, from [8.8])

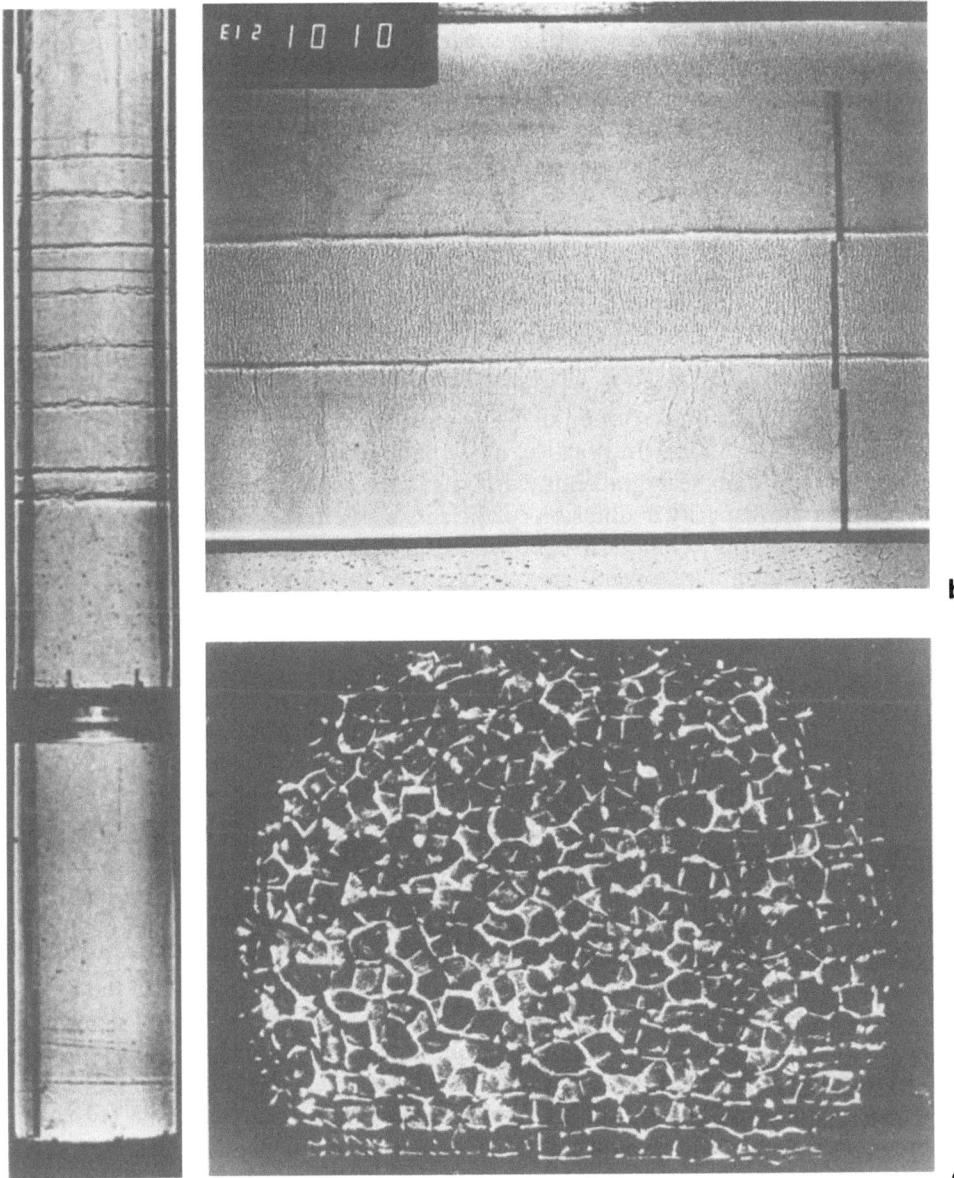

Fig. 8.21a–c. Photographs showing (a) layering in a stable salt gradient heated from below, (b) vertical section, and (c) planform views of fingers in a sugar (Su) salt (Sa) system. (a) Photograph supplied by Dr. P. F. Linden; (b) from [8.108], copyrighted by American Geophysical Union; (c) from [8.13]

layers sometimes merging together as new upper layers are formed. As a result of the vigorous convection within separate layers, both heat and solute are redistributed throughout the fluid. Temperature and solute profiles are steplike, illustrating the essentially isothermal, isohaline character of the layers and the existence of large thermal and haline gradients at the interfaces. Through these regions the convective motions are unable to penetrate.

8.6.3 Salt Fingers

Layering – a characteristic consequence of double-diffusive instability in the diffusive regime – is observed also in the finger regime, though the form of the causal instability is quite different. As predicted by linear theory the instability takes the form of thin "fingers" having a planform which is approximately square, in which fluid rises up one side of each finger and descends down the other (see Fig. 8.21c). (In practice, in the laboratory it is often convenient to replace salt and heat gradients with sugar and salt gradients, respectively, utilizing the disparity in diffusion coefficients between the two solutes.) Fingers produced by this technique are shown in Fig. 8.21b. Again by a mechanism of successive instability, several sharply bounded mixed layers, each containing overturning finger-type motions, are formed in the fluid. Important quantitative data on the process of layer formation, and measurements of the fluxes of salt and heat across both types of convecting system, have been obtained; here we have space only to refer the reader to the studies by *Turner* [8.8, 9] and *Linden* [8.107, 108] for full details.

8.6.4 Sideways Diffusive Instability

Interesting experiments related to those described above have investigated the effects of horizontal differences in salt and heat in a fluid system stratified by either or both components ("sideways diffusive instability"). Examples of such experiments are those of *Thorpe* et al. [8.120] and *Huppert* and *Turner* [8.121], while related theoretical work has been performed by *Hart* [8.122]. In the first group of experiments a stable salinity gradient was subjected to sidewall heating. In this case double-diffusive instability takes the form of a stack of two-dimensional convective rolls having their long axes perpendicular to both gravity and the direction of the impressed temperature gradient. The layered system which develops is different in many respects from the layered systems described earlier. In particular, the layers appear simultaneously near the heated wall and grow inwards from it. The instability, while still a double-diffusive instability, arises essentially from the instability of the sidewall thermal boundary layer. A fluid particle near the wall is forced to move upwards by thermally induced buoyancy, but the distance it can travel in this direction is limited by its salinity which, because κ_C/κ is of order 10^{-2}, remains essentially unaltered. It is therefore forced to move horizontally into the interior of the fluid, cooling as it does so, until it returns in a cellular manner to the hot wall.

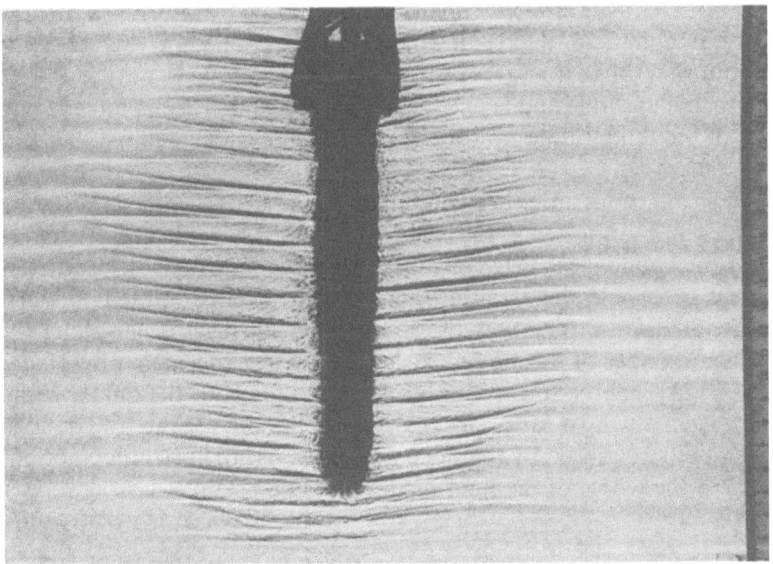

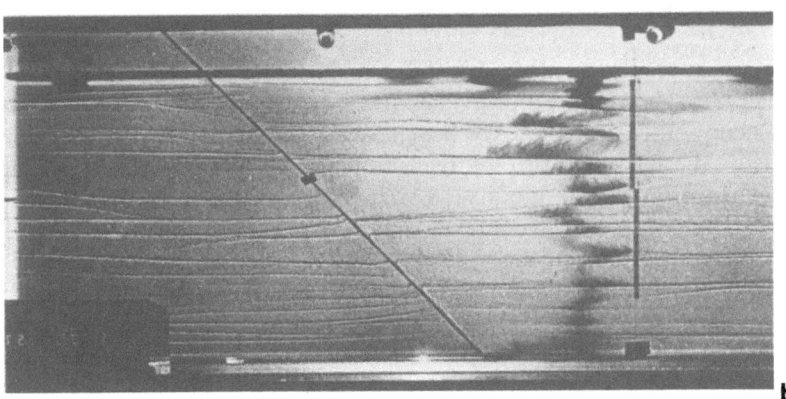

Fig. 8.22a, b. Shadowgraphs showing layers produced (**a**) by the insertion of a block of ice into a salt-stratified fluid reservoir, and (**b**) by having one solid boundary of a salt-sugar fluid system inclined at 45° to the horizontal. In (**b**) the concentrations of salt and sugar increase and decrease with height, respectively [8.121, 123]

The process is identical for all fluid particles in the vicinity of the wall so that a stack of two-dimensional longitudinal rolls is formed simultaneously, each having a depth which is typically small compared with the overall fluid depth. As the heating continues the depth of the layers remains the same, but they grow in horizontal extent (tilting downwards as they cool), and so intrude further into the interior of the fluid.

Huppert and *Turner* [8.121] have recently investigated a problem similar in several respects to that described above. In their experiments horizontal gradients of both temperature *and* salinity were imposed by inserting a melting block of ice into a fluid region stratified by salt alone. In this case the differential diffusion of heat and salt causes the boundary layer next to the ice surface to become unstable and layers form. As above, these layers intrude into the interior of the fluid, tilting upwards this time as heat is acquired (see Fig. 8.22a).

Horizontal variations in either component of a thermohaline system can also be imposed by enclosing the fluid in a container having a sloping boundary. If the boundary is thermally insulating and impermeable, isotherms and isohalines must both distort in order to become normal at the boundary, and a steady, diffusion-driven flow of each component must occur along the boundary. The stability of this boundary layer flow was considered by *Linden* and *Weber* [8.123] who performed an investigation for both the diffusive and finger regimes, employing a range of slope angles. In both regimes the double-diffusive instability produced by the slope flow gives rise to horizontal motions and layering in the whole of the fluid region (see Fig. 8.22b).

8.6.5 Nonthermohaline Double Diffusion

Thus far we have discussed double-diffusive instabilities solely in terms of thermohaline systems. As we indicated earlier, however, the class of instabilities contains many other examples. We have space to mention only two of these, albeit briefly: the axisymmetric Goldreich-Schubert instability [8.124, 125] and the viscous overturning instability [8.126], both of which arise in several geophysical and astrophysical contexts as a result of the disparity in the rate of diffusion of momentum and heat. In the first example (which occurs when $\mathrm{Pr} \ll 1$), we consider a differentially rotating self-gravitating spherical fluid body thermally stably stratified in the radial direction. In such a system the radial stratification would normally be expected to exert a stabilizing influence on the centrifugal instability which would develop in the homogeneous case if the inviscid Rayleigh criterion (see Sect. 6.1) were satisfied. If $\mathrm{Pr} \ll 1$, however, (as in the sun, for example) differential diffusion of momentum and heat can nullify the stabilizing action of the stratification. In this case a ring element displaced radially will adjust quickly to the local temperature and will therefore not experience the presence of the stratification. It will not acquire the local angular momentum however, and in the absence of magnetic fields and/or concentration gradients, its stability will be determined by the Rayleigh criterion for the homogeneous case. A discussion of the types of double and multidiffusive instabilities which can develop in low Prandtl number situations in which several diffusion scales are present has recently been presented by *Acheson* [8.127].

In the second example ($\mathrm{Pr} \gg 1$) the roles of heat and momentum will be reversed; a displaced particle will adjust quickly to the velocity field but slowly to the local temperature field. For the case of a baroclinic vortex (see Sect. 8.5)

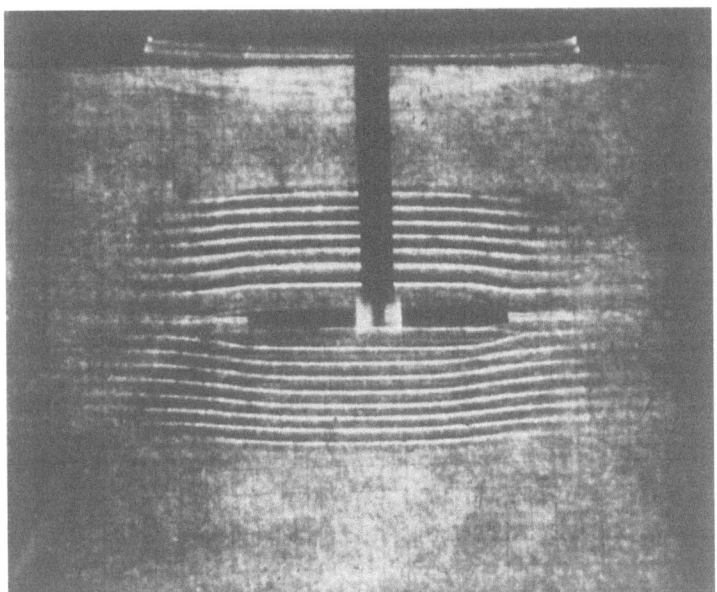

Fig. 8.23. Photograph illustrating the occurrence of layers caused by the relative motion of a counterrotating disc in a rotating, stably stratified salt solution. ([8.128]; copyrighted 1971 by the American Association for the Advancement of Science)

the vorticity generated by the net excess of thermal buoyancy can drive "viscous overturning" [8.126]. Laboratory experiments [8.128, 129] have shown that this type of instability does indeed occur for such a flow, with the characteristic results of layering in the body of the fluid (see Fig. 8.23). The data obtained present evidence for yet another distinctive mechanism of importance in layer formation in natural systems.

References

8.1   J.G.Charney: "Planetary Fluid Dynamics", in *Dynamical Meteorology*, ed. by P.Morel (Reidel, Dordrecht 1973) pp. 97–352

8.2   S.A.Thorpe: Turbulence in stably stratified fluids; a review of laboratory experiments. Boundary-Layer Meteorol. **5**, 95–119 (1973)

8.3   T.Maxworthy, F.K.Browand: Experiments on rotating and stratified flows; oceanographic application. Annu. Rev. Fluid Mech. **7**, 273–305 (1975)

8.4   F.S.Sherman, J.Imberger, G.M.Corcos: Turbulence and mixing in stably stratified waters. Annu. Rev. Fluid Mech. **10**, 267–288 (1978)

8.5   R.Hide: "Some laboratory experiments on free thermal convection in a rotating fluid subject to a horizontal temperature gradient and their relation to the theory of the global atmospheric circulation", in *The Global Circulation of the Atmosphere*, ed. by G.A.Corby (R. Meteorol. Soc., London 1969) pp. 196–221

8.6   R.Hide, P.J.Mason: Sloping convection in a rotating fluid. Adv. Phys. **24**, 47–100 (1975)
8.7   P.G.Drazin: "Variations on a theme of Eady". In Ref. 8.17, pp. 139–169
8.8   J.S.Turner: Double diffusive phenomena. Annu. Rev. Fluid Mech. **6**, 37–56 (1974)
8.9   J.S.Turner: Laboratory experiments on double-diffusive instabilities. Adv. Chem. Phys. **32**, 135–149 (1975)
8.10   L.Prandtl: *Essentials of Fluid Dynamics* (Blackie, London 1952)
8.11   D.J.Tritton: *Physical Fluid Dynamics* (Van Nostrand Reinhold, Wokingham 1977)
8.12   A.A.Townsend: *The Structure of Turbulent Shear Flow*, 2nd ed. (Cambridge University Press, Cambridge 1976)
8.13   J.S.Turner: *Buoyancy Effects in Fluids* (Cambridge University Press, Cambridge 1973)
8.14   C-S.Yih: *Dynamics of Nonhomogeneous Fluids* (Macmillan, New York 1965)
8.15   G.K.Batchelor: *An Introduction to Fluid Dynamics* (Cambridge University Press, Cambridge 1967)
8.16   H.P.Greenspan: *The Theory of Rotating Fluids* (Cambridge University Press, Cambridge 1968)
8.17   P.H.Roberts, A.M.Soward (eds.): *Rotating Fluids in Geophysics* (Academic Press, London 1978)
8.18   L.N.Howard: Fundamentals of the theory of rotating fluids. J. Appl. Mech. **30**, 481–485 (1963)
8.19   M.J.Lighthill: Dynamics of rotating fluids; a survey. J. Fluid Mech. **26**, 411–431 (1966)
8.20   R.Hide: Dynamics of rotating fluids. Q. J. R. Meteorol. Soc. **103**, 1–28 (1977)
8.21   J.R.Holton: *An Introduction to Dynamic Meteorology* (Academic Press, New York 1972)
8.22   O.M.Phillips: *The Dynamics of the Upper Ocean*, 2nd ed. (Cambridge University Press, Cambridge 1978)
8.23   M.E.Stern: *Ocean Circulation Physics* (Academic Press, New York 1975)
8.24   R.Hide: Motions in planetary atmospheres. Q. J. R. Meteorol. Soc. **102**, 1–23 (1976)
8.25   P.H.Stone: "The meteorology of the Jovian atmosphere", in *Jupiter, Studies of the Interior, Atmosphere, Magnetosphere, and Satellites*, ed. by T.Gehrels (University of Arizona Press, Tucson 1976)
8.26   V.P.Starr: *Physics of Negative Viscosity Phenomena* (McGraw-Hill, New York 1968)
8.27   J.Lighthill: *Waves in Fluids* (Cambridge University Press, Cambridge 1978)
8.28   H.G.Moll: Die atmosphärische Umströmung Madeiras. Beitr. Phys. Atmos. **44**, 227–244 (1971)
8.29   E.Berger, R.Wille: Periodic flow phenomena. Annu. Rev. Fluid Mech. **4**, 313–340 (1972)
8.30   E.Palmén, C.W.Newton: *Atmospheric Circulation Systems* (Academic Press, New York 1969)
8.31   G.W.Platzmann: The Rossby wave. Q. J. R. Meteorol. Soc. **94**, 225–248 (1968)
8.32   J.R.Bates: Dynamics of stationary ultra-long waves in middle latitudes. Q. J. R. Meteorol. Soc. **103**, 397–430 (1977)
8.33   J.T.Houghton: *The Physics of Atmospheres* (Cambridge University Press, Cambridge 1977)
8.34   P.G.Drazin, L.N.Howard: Hydrodynamic stability of parallel flow of inviscid fluid. Adv. Appl. Mech. **9**, 1–89 (1966)
8.35   Y.Mori, Y.Uchida: Forced convective heat transfer between horizontal flat plates. Int. J. Heat Mass Transfer **9**, 803–817 (1966)
8.36   M.Akiyama, G.J.Hwang, K.C.Cheng: Experiments on the onset of longitudinal vortices in laminar forced convection between horizontal plates. J. Heat Transfer **93**, 335–341 (1971)
8.37   F.M.Richter, B.Parsons: On the interaction of two scales of convection in the mantle. J. Geophys. Res. **80**, 2529–2541 (1975)
8.38   R.M.Clever, F.H.Busse, R.E.Kelly: Instabilities of longitudinal convection rolls in Couette flow. Z. angew. Math. Phys. **28**, 771–783 (1977)
8.39   D.J.Tritton: Turbulent free convection above a heated plate inclined at a small angle to the horizontal. J. Fluid Mech. **16**, 282–312 (1963)
8.40   A.A.Townsend: Mixed convection over a heated horizontal plane. J. Fluid Mech. **55**, 209–228 (1972)

8.41   J.W.Miles: On the stability of heterogeneous flows. J. Fluid Mech. **10**, 496–508 (1961)

8.42   P.Hazel: Numerical studies of the stability of inviscid stratified shear flows. J. Fluid Mech. **51**, 39–62 (1972)

8.43   K.S.Gage: Linear viscous stability theory for stably stratified shear flow; a review. Boundary-Layer Meteorol. **5**, 3–17 (1973)

8.44   S. A. Maslowe: Finite-amplitude Kelvin-Helmholtz billows. Boundary-Layer Meteorol. **5**, 43–52 (1973)

8.45   S. A. Maslowe: Weakly non-linear stability theory of stratified shear flows. Q. J. R. Meteorol. Soc. **103**, 769–783 (1977)

8.46   P.C.Patnaik, F.S.Sherman, G.M.Corcos: A numerical simulation of Kelvin-Helmholtz waves of finite amplitude. J. Fluid Mech. **73**, 215–240 (1976)

8.47   S.N.Brown, K.Stewartson: The evolution of a small inviscid disturbance to a marginally unstable stratified shear flow; stage two. Proc. R. Soc. London A **363**, 175–194 (1978)

8.48   S. A. Thorpe: Experiments on the stability of stratified shear flows; miscible fluids. J. Fluid Mech. **46**, 299–320 (1971)

8.49   R.S.Scotti, G.M.Corcos: An experiment on the stability of small disturbances in a stratified free shear layer. J. Fluid Mech. **52**, 499–528 (1972)

8.50   F.K.Browand, C.D.Winant: Laboratory observations of shear-layer instability in a stratified fluid. Boundary-Layer Meteorol. **5**, 67–77 (1973)

8.51   D.P.Delisi, G.Corcos: A study of internal waves in a wind tunnel. Boundary-Layer Meteorol. **5**, 121–137 (1973)

8.52   C.G.Koop: "Instability and Turbulence in a Stratified Shear Layer"; Univ. Southern Calif., Dept. Aerospace Eng. Rep. USCAE 134 (1976)

8.53   C.G.Koop, F.K.Browand: Instability and turbulence in a stratified fluid with shear. J. Fluid Mech. **93**, 135–160 (1979)

8.54   K. A. Browning: Structure of the atmosphere in the vicinity of large-amplitude Kelvin-Helmholtz billows. Q. J. R. Meteorol. Soc. **97**, 283–299 (1971)

8.55   P. A. Davis, W. R. Peltier: Resonant parallel shear instability in the stratified planetary boundary layer. J. Atmos. Sci. **33**, 1287–1300 (1976)

8.56   J.L.Lumley, H. A. Panofsky: *The Structure of Atmospheric Turbulence* (Interscience, New York 1964)

8.57   J.E.Hart: Stability of the flow in a differentially heated inclined box. J. Fluid Mech. **47**, 547–576 (1971)

8.58   E.M.Sparrow, R.B.Husar: Longitudinal vortices in natural convection flow on inclined plates. J. Fluid Mech. **37**, 251–256 (1969)

8.59   J.R.Lloyd, E.M.Sparrow: On the instability of natural convection flow on inclined plates. J. Fluid Mech. **42**, 465–470 (1970)

8.60   D.J.Tritton: Transition to turbulence in the free convection boundary layers on an inclined heated plate. J. Fluid Mech. **16**, 417–435 (1963)

8.61   W.Debler: "The Towing of Bodies in a Stratified Fluid"; Univ. Michigan, Dept. Engng. Mech., Tech. Rep. EM-71-1 (1971)

8.62   P.W.M.Brighton: Strongly stratified flow past three-dimensional obstacles. Q. J. R. Meteorol. Soc. **104**, 289–307 (1978)

8.63   A.J.Faller, R.Kaylor: Instability of the Ekman spiral with application to the planetary boundary layers. Phys. Fluids **10**, Suppl. 212–219 (1967)

8.64   P.R.Tatro, E.L.Mollo-Christensen: Experiments on Ekman layer instability. J. Fluid Mech. **28**, 531–543 (1967)

8.65   D.R.Caldwell, C.W.Van Atta: Characteristics of Ekman boundary layer instabilities. J. Fluid Mech. **44**, 145–160 (1970)

8.66   J.P.Johnston, R.M.Halleen, D.K.Lezius: Effects of spanwise rotation on the structure of two-dimensional fully developed turbulent channel flow. J. Fluid Mech. **56**, 533–558 (1972)

8.67   D.J.Tritton: "Turbulence in rotating fluids". In Ref. 8.17, pp. 105–138

8.68   P.Bradshaw: The analogy between streamline curvature and buoyancy in turbulent shear flow. J. Fluid Mech. **36**, 177–192 (1969)

8.69   J. A. Johnson: The stability of shearing motion in a rotating fluid. J. Fluid Mech. **17**, 337–352 (1963)

8.70   J. E. Hart: Instability and secondary motion in a rotating channel flow. J. Fluid Mech. **45**, 341–352 (1971)

8.71   D. K. Lezius, J. P. Johnston: Roll-cell instabilities in rotating laminar and turbulent channel flows. J. Fluid Mech. **77**, 153–175 (1976)

8.72   M. C. Potter, M. D. Chawla: Stability of boundary layer flow subject to rotation. Phys. Fluids **14**, 2278–2281 (1971)

8.73   F. H. Busse: Shear flow instabilities in rotating systems. J. Fluid Mech. **33**, 577–589 (1968)

8.74   C. E. Grosch, H. Salwen: The stability of steady and time-dependent plane Poiseuille flow. J. Fluid Mech. **34**, 177–205 (1968)

8.75   P. H. Rothe, J. P. Johnston: "The Effects of System Rotation on Separation, Reattachment, and Performance in Two-Dimensional Diffusers"; Stanford Univ., Dept. Mech. Engng., Rep. PD-17 (1975)

8.76   R. Hide, C. W. Titman: Detached shear layers in a rotating fluid. J. Fluid Mech. **29**, 39–60 (1967)

8.77   H. Kuo: Dynamic instability of two-dimensional non-divergent flow in a barotropic atmosphere. J. Meteorol. **6**, 105–122 (1949)

8.78   A. P. Ingersoll, J. N. Cuzzi: Dynamics of Jupiter's cloud bands. J. Atmos. Sci. **26**, 981–985 (1969)

8.79   A. P. Ingersoll: Pioneer 10 and 11 observations and the dynamics of Jupiter's atmosphere. Icarus **29**, 245–253 (1976)

8.80   E. A. Eady: Long waves and cyclone waves. Tellus **1**, 33–52 (1949)

8.81   V. Barcilon: Role of the Ekman layers in the stability of the symmetric regime obtained in a rotating annulus. J. Atmos. Sci. **21**, 291–299 (1964)

8.82   J. Brindley: Stability of flow in a rotating viscous incompressible fluid subject to differential heating. Phil. Trans. R. Soc. London A **253**, 1–25 (1960)

8.83   J. S. A. Green: A problem in baroclinic instability. Q. J. R. Meteorol. Soc. **86**, 237–251 (1960)

8.84   F. P. Bretherton: Critical layer instability in baroclinic flows. Q. J. R. Meteorol. Soc. **92**, 325–334 (1966)

8.85   F. P. Bretherton: Baroclinic instability and the short wavelength cut-off in terms of potential vorticity. Q. J. R. Meteorol. Soc. **92**, 335–346 (1966)

8.86   R. Hide, P. J. Mason: On the transition between axisymmetric and non-axisymmetric flow in a rotating liquid annulus subject to a horizontal temperature gradient. Geophys. Astrophys. Fluid Dyn. **10**, 121–156 (1978)

8.87   P. H. Stone: On non-geostrophic baroclinic instability. J. Atmos. Sci. **23**, 390–400 (1966)

8.88   P. H. Stone, S. Hess, R. Hadlock, P. Ray: Preliminary results of experiments with symmetric baroclinic instabilities. J. Atmos. Sci. **26**, 997–1001 (1969)

8.89   T. Maxworthy: A review of Jovian atmospheric dynamics. Planet. Space Sci. **21**, 623–641 (1973)

8.90   I. C. Walton: The viscous non-linear symmetric baroclinic instability of a zonal shear flow. J. Fluid Mech. **87**, 65–84 (1975)

8.91   R. Hide: An experimental study of thermal convection in a rotating liquid. Phil. Trans. R. Soc. London A **250**, 441–478 (1958)

8.92   W. W. Fowlis, R. Hide: Thermal convection in a rotating annulus of liquid: effect of viscosity on the transition between axisymmetric and non-axisymmetric flow regimes. J. Atmos. Sci. **22**, 541–558 (1965)

8.93   R. L. Pfeffer, W. W. Fowlis, J. S. Fein, J. Buckley: Experimental determinations of the transitions between the symmetrical and wave regimes in a rotating differentially heated annulus of fluid. Pure Appl. Geophys. (PAGEOPH) **81**, 263–271 (1970)

8.94   J. A. C. Kaiser: Rotating deep annulus convection. I. Thermal properties of the upper symmetric regime. Tellus **21**, 789–804 (1969)

8.95   J. A. C. Kaiser: Rotating deep annulus convection. II. Wave instabilities, vertical stratification and associated theories. Tellus **22**, 275–287 (1970)

8.96   C. B. Ketchum: An experimental study of baroclinic annulus waves at large Taylor number. J. Atmos. Sci. **29**, 665–679 (1972)

8.97   J. E. Hart: A laboratory study of baroclinic instability. Geophys. Fluid Dyn. **3**, 181–210 (1972)

8.98   R. Krishnamurti: "Experiments in Ocean Circulation Modelling"; Tech. Rep. 14, Dept. Oceanography, Florida State University (1977)

8.99   J. C. King: An experimental study of baroclinic wave interactions in a two-layer system. Geophys. Astrophys. Fluid Dyn. **13**, 153–168 (1979)

8.100 J. S. Fein, R. L. Pfeffer: An experimental study of the effects of Prandtl number on thermal convection in a rotating differentially heated cylindrical annulus of fluid. J. Fluid Mech. **75**, 81–112 (1976)

8.101 R. Hide, P. J. Mason, R. A. Plumb: Thermal convection in a rotating fluid subject to a horizontal temperature gradient: spatial and temporal characteristics of fully developed baroclinic waves. J. Atmos. Sci. **34**, 930–950 (1977)

8.102 R. L. Pfeffer, Y. Chiang: Two kinds of vacillation in rotating laboratory experiments. Mon. Weather Rev. **95**, 75–82 (1967)

8.103 R. L. Pfeffer, W. W. Fowlis: Wave dispersion in a rotating differentially heated cylindrical annulus of fluid. J. Atmos. Sci. **25**, 361–371 (1968)

8.104 W. W. Fowlis, R. L. Pfeffer: Characteristics of amplitude vacillation in a rotating differentially heated fluid determined by a multi-probe technique. J. Atmos. Sci. **26**, 100–108 (1969)

8.105 H. Stommel, A. B. Arons, D. C. Blanchard: An oceanographical curiosity: the perpetual salt fountain. Deep Sea Res. **3**, 152–153 (1956)

8.106 M. E. Stern, J. S. Turner: Salt fingers and convecting layers. Deep Sea Res. **16**, 497–511 (1969)

8.107 P. F. Linden: The formation and destruction of fine structure by double-diffusive processes. Deep Sea Res. **23**, 895–908 (1976)

8.108 P. F. Linden: The formation of banded salt finger structure. J. Geophys. Res. **83**, 2902–2912 (1978)

8.109 R. I. Tait, M. R. Howe: Some observations of thermohaline stratification in the deep ocean. Deep Sea Res. **15**, 275–280 (1968)

8.110 B. Magnell: Salt fingers observed in the Mediterranean outflow using a towed sensor. J. Phys. Oceanogr. **6**, 511–523 (1976)

8.111 R. W. Schmitt, D. L. Evans: An estimate of the vertical mixing due to salt fingers based on observations in the North Atlantic Central Water. J. Geophys. Res. **83**, 2913–2919 (1978)

8.112 A. E. Gargett: An investigation of the occurrence of oceanic turbulence with respect to fine structure. J. Phys. Oceanogr. **6**, 139–156 (1976)

8.113 A. J. Williams: Salt fingers observed in the Mediterranean outflow. Science **185**, 941–943 (1974)

8.114 H. E. Huppert, J. S. Turner: Double diffusive convection and its implications for the temperature and salinity structure of the ocean and Lake Vanda. J. Phys. Oceanogr. **2**, 456–461 (1972)

8.115 M. E. Stern: The salt fountain and thermohaline convection. Tellus **12**, 172–175 (1960)

8.116 G. Veronis: A finite amplitude instability in thermohaline convection. J. Mar. Res. **23**, 1–17 (1965)

8.117 G. Veronis: Effect of a stabilizing gradient of solute on thermal convection. J. Fluid Mech. **34**, 315–336 (1968)

8.118 P. G. Baines, A. E. Gill: On thermohaline convection with linear gradients. J. Fluid Mech. **37**, 289–306 (1969)

8.119 T. G. L. Shirtcliffe: Thermosolutal convection: observation of an overstable mode. Nature **213**, 489–490 (1967)

8.120 S. A. Thorpe, P. K. Hutt, R. Soulsby: The effect of horizontal gradients on thermohaline convection. J. Fluid Mech. **28**, 375–400 (1969)

8.121 H. E. Huppert, J. S. Turner: On melting icebergs. Nature **271**, 46–48 (1978)

8.122 J. E. Hart: On sideways diffusive instability. J. Fluid Mech. **49**, 279–288 (1971)

8.123 P. F. Linden, J. E. Weber: The formation of layers in a double-diffusive system with a sloping boundary. J. Fluid Mech. **81**, 757–773 (1977)

8.124 P.Goldreich, G.Schubert: Differential rotation in stars. Astrophys. J. **150**, 571 (1967)

8.125 C.A.Jones: The onset of shear instability in stars. Geophys. Astrophys. Fluid Dyn. **8**, 165–184 (1977)

8.126 M.E.McIntyre: Diffusive destabilization of the baroclinic circular vortex. Geophys. Fluid Dyn. **1**, 19–57 (1970)

8.127 D.J.Acheson: On the instability of toroidal magnetic fields and differential rotation in stars. Phil. Trans. R. Soc. London A **292**, 459–500 (1978)

8.128 D.J.Baker: Density gradients in a rotating stratified fluid: experimental evidence for a new instability. Science **172**, 1029–1031 (1971)

8.129 J.Calman: Experiments on high Richardson number instability of a rotating stratified shear flow. Dyn. Atmos. Ocean. **1**, 277–297 (1977)

9. Instabilities and Chaos in Nonhydrodynamic Systems

J. M. Guckenheimer

With 7 Figures

In this chapter, we shall examine several examples of unstable systems with chaotic dynamics which occur outside the context of hydrodynamics. To illustrate the wide range of areas in which such examples exist, we shall draw upon the fields of geophysics, physics, chemistry, ecology, and electrical circuit theory. In each of the examples, we shall describe a mathematical model phrased in terms of differential or difference equations which possess chaotic solutions. The analyses of these models and of the relationships of the models and the phenomena they mimic are incomplete and in some instances just beginning. In most instances, we still depend heavily upon numerical computations together with a healthy dose of geometric intuition concerning the models.

In Chap. 4 *Yorke* and *Yorke* describe the Lorenz system of ordinary differential equations which provide a model for convective fluid motion in a thin loop. The solutions of the Lorenz system display an instability which makes prediction of some details of their long-term behavior impossible. At the same time, one can describe other features of the long-term behavior of the solutions and certain statistical features of their asymptotic behavior. All of these characteristics seem seem to be present in a wide variety of natural phenomena. We shall discuss them further after examining a few specific models and their properties.

9.1 The Rikitake Dynamo Model for the Earth's Magnetic Field

The first example that we consider involves models of the gross features of the earth's magnetic field. The polarity and magnitude of the earth's magnetic field varies in an irregular manner on a time scale of millions of years. At irregular intervals, the polarity of the field undergoes relatively rapid reversals. It is possible that these reversals are part of the magnetohydrodynamics of a rotating, conducting fluid. As *Busse* [9.1] discussed, evidence in support of this possibility has come from models of disc dynamos which lead to systems of ordinary differential equations very much like those studied by Lorenz in relation to problems of hydrodynamics. An approximation procedure is used to replace a system with infinitely many degrees of freedom, described by a partial

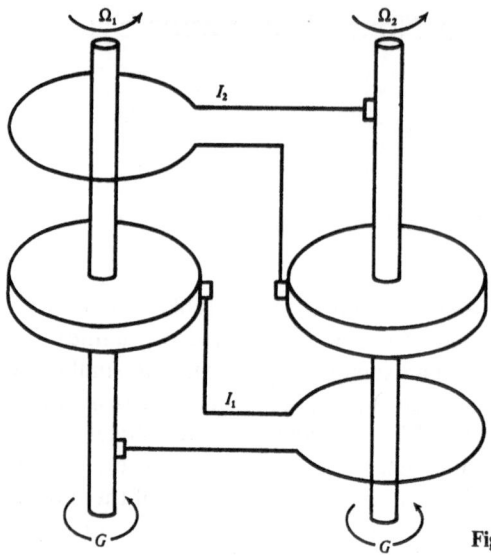

Fig. 9.1. The Rikitake two-disc dynamo [9.2]

differential equation, by a system with a finite number of degrees of freedom.
Cook and Roberts [9.2], and Robbins [9.3] have studied two different models of
this sort.

Cook and Roberts studied the Rikitake two-disc dynamo illustrated in
Fig. 9.1. Each dynamo is driven with the same torque G, and one assumes that
the state of the system is determined by the currents I_1 and I_2 in the coils of
each dynamo and the angular velocities Ω_1 and Ω_2 in each dynamo. The
differential equations which result are

$$L\dot{I}_1 + RI_1 = M\Omega_1 I_2 \,,$$
$$L\dot{I}_2 + RI_2 = M\Omega_2 I_1 \,,$$
$$C\dot{\Omega}_1 = G - MI_1 I_2 \,,$$
$$C\dot{\Omega}_2 = G - MI_1 I_2 \,.$$

(9.1)

The coefficients in the equation are the self-inductance L of each dynamo, the
mutual inductance M between the two dynamos, the resistance R within each
dynamo circuit, and the moment of inertia C of each dynamo. The equations
are not independent of one another since $\dot{\Omega}_1 - \dot{\Omega}_2 = 0$. Thus $\Omega_1 - \Omega_2$ is a
constant of the motion. If we fix this constant as A and rescale the variables,
(9.1) can be written as

$$\dot{X}_1 + \mu X_1 = YX_2 \,,$$
$$\dot{X}_2 + \mu X_2 = (Y - A)X_1 \,,$$
$$\dot{Y} = 1 - X_1 X_2 \,.$$

(9.2)

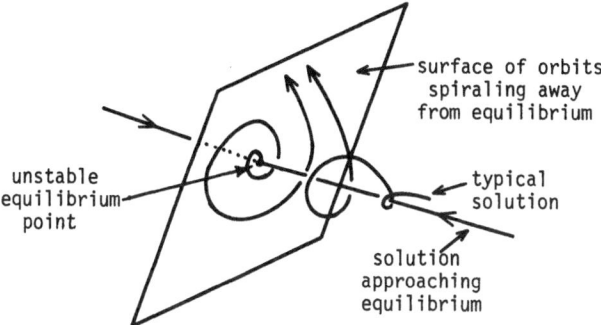

surface of orbits
spiraling away
from equilibrium

unstable
equilibrium
point

typical
solution

solution
approaching
equilibrium

Fig. 9.2. Phase space trajectories of Rikitake dynamo near an unstable equilibrium point

The *equilibrium* (or *rest* or *singular*) points are determined by solving the equations $\dot{X}_1 = \dot{X}_2 = \dot{Y} = 0$. There are two equilibrium points for this system of equations, for each of which $Y > 0$ and $X_1 X_2 = 1$. The local behavior of solutions near equilibrium can be studied by looking at the linear system of equations obtained from the first-order terms in the Taylor expansion there [9.4]. At each of the singular points, the linearization of (9.2) is given by a matrix with one negative eigenvalue and a pair of pure imaginary eigenvalues. A more refined analysis of the equations in the eigenspace of the imaginary eigenvalues shows that the equilibrium points are unstable unless μ or A is 0. Thus solutions of (9.2) which start near one of the equilibrium points are drawn toward a two-dimensional surface, along which they slowly spiral away from the equilibrium point. The behavior near an equilibrium point is illustrated in Fig. 9.2. When a trajectory has travelled far enough from one equilibrium along the surface passing through that equilibrium, it is drawn to the surface passing through the other equilibrium. It then spirals away from the second equilibrium until it is far enough away, at which time it returns to the surface through the first equilibrium. The timing of the transitions from one surface to the other is irregular and eventually sensitive to changes in initial conditions. The solutions of the system (9.2) seem to behave in much the same way as those in the Lorenz system, but the third critical point of the Lorenz system lies at "infinity" along the Y axis in this case.

The solutions of the Rikitake two-disc dynamo are suggestive of the irregular pattern of reversals of the earth's magnetic field. Other dynamo models of the same order of complexity also allow chaotic solutions. *Robbins* [9.3] has investigated numerically a model proposed by Malkus of a single-disc dynamo which includes a shunt or external load. The equations governing this model include those of the Lorenz system in the limiting case that the shunt inductance tends to zero. The solutions display the same kinds of irregularity. As with the Lorenz equations viewed as a model for two-dimensional convective fluid rolls, these examples are useful indicators that the pattern of irregularity may be inherent in the complicated dynamics of the solution of the underlying system of deterministic equations rather than being caused by random influences or a breakdown in the solutions to the equations.

Let us describe briefly the nature of the irregularity displayed by the solutions of these systems of differential equations. Two familiar types of limiting behavior for solutions are asymptotic approaches to a stable equilibrium and a stable periodic oscillation (called a *limit cycle;* see Chap. 2). In conservative systems, one also frequently encounters quasi-periodic motion which is represented by solutions which are the superposition of several periodic functions of different periods. The irregularity encountered here is more extreme than quasi-periodic motion. There is a complicated set, called an *attractor*, to which nearby solutions are drawn (see Chap. 2). Inside the attractor, the solutions behave in a complicated fashion which makes it impossible to predict the location of the solution in the attractor at a large time in the future based upon an approximate knowledge of its current value. This feature of these systems we call *chaotic*.

Geophysics and fluid dynamics are not the only fields in which approximations lead to systems of differential equations in three-dimensional space which have chaotic solutions. *Haken* [9.5] has derived such equations in a model for the undamped spiking behavior of lasers. Again the suggestion is that the irregular patterns observed in laser emissions are an intrinsic feature of deterministic equations describing laser phenomena.

There are other areas in which the relationship between the mathematics of unstable dynamical systems and irregular phenomena has not followed the pattern of these physical examples. With physical problems a principal obstacle to further understanding lies in the mathematics. We can assume the validity of a system of partial differential equations as a model but need to know more about the faithfulness of ordinary differential equation approximations to the system of partial differential equations.

Here are two ways of carrying out such approximations. The first consists of discretizing the physical system itself: a continuous mass distribution is replaced by a finite number of point masses, an electrical field is assumed to be generated by point charges, etc. Another approach consists of representing the solutions of the partial differential equations in terms of a well-chosen basis for the appropriate function space (think of Fourier series) and then assuming that the interesting dynamics of the system can be expressed in terms of a finite number of the basis functions. The Lorenz system was obtained in this way. Little can be said about the effect of these approximations on the asymptotic properties of solutions. In a number of other applications of the mathematics of deterministic chaos, the emphasis is different.

9.2 The Belousov-Zhabotinskii Chemical Reaction

Homogeneous chemical reactions are described mathematically by systems of ordinary differential equations. A natural question asks whether these systems of differential equations include ones which have chaotic solutions. Another question asks whether one can find real reactions which display the kind of irregularity seen in the equations. The answer to both of these questions seems to be yes. Work on these questions has begun only recently, and the first results have appeared in recent years. The initial attention by *Olsen* and *Degn* [9.6], *Schmitz* et al. [9.7], and *Rössler* and *Wegmann* [9.8], has focussed on the horseradish peroxidase catalyzed oxidation of NADH and the Belousov-Zhabotinskii reaction. Here we shall discuss the Belousov-Zhabotinskii re-action since there have been substantial theoretical studies of this reaction (see [9.9, 10] in addition to the experimental evidence of irregular oscillations.

The Belousov-Zhabotinskii (BZ) reaction has been the object of intense interest for a decade. The reaction itself involves potassium, bromine, and malonic acid in the presence of a cerium catalyst. In a well-stirred solution, the reaction undergoes sustained oscillations for a substantial period of time which can be seen easily through the color changes in an indicator. When the solution is poured into a thin layer, the oscillations give rise to striking patterns of ring or spiral-shaped waves propagating outward from various points [9.11]. These patterns in the BZ reaction have been intensively investigated in terms of their relationship to diffusion-reaction equations, particularly by *Kopell* and *Howard* [9.12]. This work has been of particular interest to biologists curious about the role of diffusion in pattern formation during morphogenesis of developing organisms. Our interest in the BZ reaction here comes from recent observations that the oscillations of the well-stirred reaction can be irregular in appropriate circumstances. The first observations were made by *Schmitz* et al. [9.7], and these were confirmed by *Rössler* and *Wegmann* [9.8].

The statistical character of these irregular chemical oscillations has not been studied. No analysis of the sort employed by *Gollub* and *Swinney* (see Sect. 6.5.2) in determining the onset of turbulence in fluid mechanical experi-ments has been made which would distinguish almost periodic oscillations from something chaotic. Nonetheless, the oscillations are very complicated, and most of the accounts of chaos give intuitive arguments for suspecting that the kinetics of the reaction are indeed chaotic.

Tyson [9.9, 10] has carried this argument further for the BZ reaction. Spurred by the interest in pattern formation, a great deal of effort has gone into the elucidation of the kinetics of the BZ reactions. Various reaction schemes have been proposed, including a system of three ordinary differential equations proposed by *Field* and *Noyes* [9.13] and popularly called the Oregonator. The Oregonator is based upon a much larger system of equations describing the kinetics of a reaction scheme involving some 15 chemical species. *Tyson* [9.9, 10] has modified the Oregonator to take better account of some of the experimental data concerning particular reactions within the whole system. In dimensionless form, the equations of this modified Oregonator are

$$\varepsilon(dx/dt) = \mu y - xy + x(1-z) - rx^2 ,$$
$$\varepsilon'(dy/dt) = -\mu y - xy + fz , \qquad\qquad (9.3)$$
$$dz/dt = x(1-z) - z ,$$

with the coefficients specified in the ranges $\varepsilon' \sim 5 \cdot 10^{-6}$, $\varepsilon \sim 10^{-3}$, $\mu \sim 2 \cdot 10^{-5}$, $r \sim 1$, and $f \sim 1$. This is once again a system of three differential equations given by quadratic polynomials. Numerical results suggest that it has chaotic solutions.

The dynamics of this system of equations has been studied less thoroughly than that of the other examples. The large differences in the coefficients of the equations result in a "stiff" system which is difficult to integrate numerically. The stiffness problem is that an integration procedure which uses fixed time steps must use very small ones to obtain reliable approximate solutions. There has been some theoretical work devoted to analyzing the chaotic behavior of stiff systems. The example which has received the most attention is the forced van der Pol equation which is discussed below. Clearly, the study of chaotic chemical reactions has only begun, and the near future is likely to see a substantial increase in our understanding of such systems.

9.3 A Model for Population Dynamics

Let us turn now to an example in which the mathematical analysis is better than the available data. The population dynamics of an ecological system are often modeled by systems of differential or difference equations. Let us describe one of the simplest such models (see the related discussion in Sect. 4.4). Consider an insect population which has one generation per year. A typical life history is that pupae overwinter and emerge in the spring or early summer as adults. The adults lay their eggs, the eggs hatch into larvae and develop into pupae by the start of the next winter.

A simple model of the population dynamics of such a species assumes that the population size each year depends only on the population size the preceding year (see [9.14]). We write N_t for the population size in year t, and $N_{t+1} = f(N_t)$ for some function f. The function f should satisfy certain properties if it is to be biologically reasonable. First, $f(0)=0$ since we rule out spontaneous generation. Second, $f'(0)$ will be determined by the maximum reproductive ability of single individuals. There is a biological limit to the number of eggs one female can lay, but one still expects $f'(0)$ to be quite large for insects. Third, f will be bounded because resources available to the species are limited. Finally, we assume that once the population size exceeds that required to produce the maximum population size, in the next generation effects of overcrowding result in a function f which decreases with population size.

Models of this sort have severe limitations in describing biological reality accurately. Nonetheless, they are suggestive of the way density dependence of the population dynamics can lead to complicated behavior. The simplest functions which satisfy the assumptions of the previous paragraph are functions like $f(N_t)=rN_t \exp(-N_t)$ or $f(N_t)=rN_t(1-N_t)$ which have a single nondegenerate maximum (see Sect. 4.4). We shall not be overly concerned with the specific function f which describes the population dynamics. Difference equations on the real line have been a subject of mathematical investigation. Indeed, the first use of the word "chaos" in the sense used in this volume was in the paper "Period three implies chaos" by *Li* and *Yorke* [9.15].

For the theory of difference equations on the line, let us describe the nature of the chaos one finds. The orbit of a point consists of a sequence $\{x_i\}$ with $x_{i+1} = f(x_i)$. A periodic point for the function f of period k is a population size N_0 such that $N_k = N_0$ but $N_i \neq N_0$ for $0 < i < k$. The functions we have written above have many periodic points if r is not too small. For example, the function $f(N_t) = rN_t(1 - N_t)$ has 2^i periodic points whose period divides k, provided $r \geq 4$. There may also be orbits for the function f which do not tend asymptotically to a periodic orbit. These orbits are "chaotic" or "aperiodic". The aperiodic orbits usually display the kind of sensitivity of asymptotic behavior to initial conditions that we have seen in the systems of differential equations considered earlier. The set of these orbits is often unstable in another way, however. Orbits which start near the set of aperiodic orbits need not remain there; they may tend asymptotically to a stable periodic orbit. The mathematical theory has been unable to sort out thus far whether the aperiodic behavior is "almost always" unstable in this sense. Indeed, different technical interpretations of "almost always" may yield different answers. Nevertheless, there is a precise criterion as to whether there are aperiodic orbits: aperiodic orbits exist for a function f if and only if there is a periodic point whose period is 2^k for every $k \geq 0$. In addition, there are aperiodic orbits for f if f has a periodic point whose period is not a power of 2.

It is not difficult to build deterministic population models which have chaotic dynamics (see [9.16]). It is difficult to gather the data to test such models. There are few systems in which accurate, automatic population counts are possible. The time scale for the dynamics of most ecosystems is measured in years and there are few sets of data which extend long enough to make sensible statistical tests of whether the population is described well by a chaotic model. Furthermore, models typically contain many parameters which have not been measured. Fitting model and data requires extensive computation which is seldom justified and can be relied upon to overtax the patience of all but the most persistent investigators. All of these practical problems can be expected to place fundamental limitations on the detailed simulation of ecosystems by models with chaotic asymptotic behavior.

9.4 The van der Pol Equation

Let us turn to a final example drawn from electrical circuit theory in which the relationship between model and physical system is very good. This example occupies an important historical position as the first system (apart from conservative mechanical systems) in which chaotic behavior was observed and

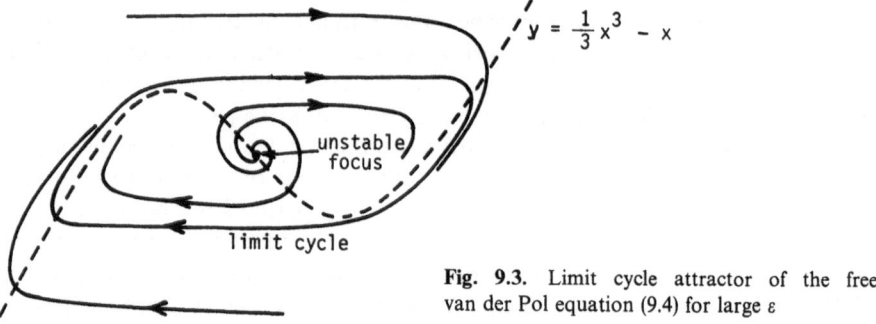

$$y = \frac{1}{3}x^3 - x$$

unstable focus

limit cycle

Fig. 9.3. Limit cycle attractor of the free van der Pol equation (9.4) for large ε

studied mathematically. The work of *Cartwright* and *Littlewood* [9.17] on the forced van der Pol equations was a motivating force in the development of a modern mathematical theory of dynamical systems during the 1960s [9.18]. It also provides another example of a system in which experimental observation provided the basis for new mathematical theory.

To describe the example mathematically, let us introduce the (free) van der Pol equation. This can be written as a second-order equation,

$$\ddot{x} + \varepsilon(x^2 - 1)\dot{x} + x = 0 , \tag{9.4}$$

or as an equivalent pair of first-order equations,

$$\dot{x} = \varepsilon\left[y - \left(\frac{x^3}{3} - x\right)\right] ,$$
$$\dot{y} = -\varepsilon^{-1}x . \tag{9.5}$$

Here $\varepsilon > 0$ is a parameter, not necessarily small. Equations (9.5) have a single singular point at the origin which is an unstable focus. There is a single limit cycle. When ε is small, this limit cycle lies close to a circle centered at the origin. We shall be mainly interested in the case where ε is large. Then the limit cycle lies close to a piecewise smooth curve built from two horizontal line segments and two segments of the curve defined by $y = (x^3/3) - x$ (see Fig. 9.3). The motion along the limit cycle is very far from uniform in the case ε is large. The speed of the trajectory on the portions which are nearly horizontal is very large, so that a point moving along the limit cycle seems to jump between the segments of the curve $y = (x^3/3) - x$. This behavior was called a relaxation oscillation by van der Pol. It is easy to build an electrical circuit (say on an analog computer) which is described very well by the van der Pol equation.

We now want to consider the behavior of the van der Pol equation when it is modified to include a sinusoidal forcing term;

$$\ddot{x} + \varepsilon(x^2 - 1)\dot{x} + x = b\varepsilon\omega\cos\omega t , \tag{9.6}$$

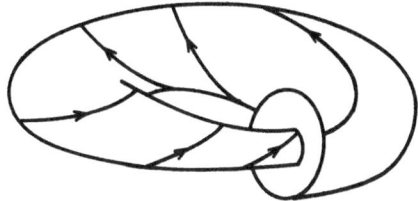

Fig. 9.4. The image of a disc for the return map of the flow of the forced van der Pol equation (9.8) on a plane of constant θ

or equivalently

$$\dot{x} = \varepsilon \left[y - \left(\frac{x^3}{3} - x \right) + b \sin \omega t \right],$$

$$\dot{y} = -\varepsilon^{-1} x .$$

(9.7)

Here b and ω are parameters. There are two ways of interpreting this system geometrically.

The theory of dynamical systems has been developed in terms of *autonomous* differential equations, that is, for systems which do not explicitly involve time. To obtain an autonomous system of equations from (9.7), introduce another variable θ which is regarded as the periodic angular variable on the unit circle. Then the forcing term $b \sin \omega t$ is $b \sin \theta$ if $\theta = \omega t$. In other words, system (9.7) is replaced by the following system

$$\dot{x} = \varepsilon \left[y - \left(\frac{x^3}{3} - x \right) + b \sin \theta \right],$$

$$\dot{y} = -\varepsilon^{-1} x ,$$

$$\dot{\theta} = \omega ,$$

(9.8)

defined on $\mathbb{R}^2 \times S^1$ since θ is a periodic variable. Looking at the intersection of trajectories of the flow of (9.8) with a plane of constant θ leads to the second interpretation of (9.7). The map $T: \mathbb{R}^2 \to \mathbb{R}^2$ which follows trajectories of (9.7) for a time $2\pi/\omega$ is independent of t. It is also given by the return map of the flow of (9.8) on a plane of constant θ. The image of a disk for this return map is illustrated in Fig. 9.4.

The asymptotic behavior of this map T is considerably more complicated for some choices of parameters b, ε, ω than in the unforced case. This discovery was made by *Cartwright* and *Littlewood* during experimental investigations of electrical circuits during World War II. Recently, numerical and singular perturbation studies have delineated quantitatively the regions in parameter space which lead to the behavior we shall describe.

The asymptotic behavior of flow trajectories for an autonomous system of equations in the plane $\mathbb{R}^2$ satisfies topological constraints embodied in the

Poincaré-Bendixon theorem [9.4]. This states that the only possibilities for the limit set of a trajectory are single points and simple closed curves. The intuitive reason is that each trajectory has a left side and a right side. If a trajectory comes back close to itself on the left side, there is no way for it later to return on the right side. Flows in higher dimensional spaces or maps of the plane need not satisfy the Poincaré-Bendixon theorem. Chaotic behavior is possible. The map of the forced van der Pol equations displays such chaotic behavior, but often in sets which are not attracting. Most points tend to stable periodic orbits, but there are sheets of orbits on which the motion is chaotic. In this regard, the map is more like the ecological difference equation studied above than the Lorenz equations. The picture that we describe below for the asymptotic behavior of the forced van der Pol equation is based upon observation rather than proof. Establishing the estimates necessary to show that the picture we describe is accurate is a formidable task which deserves attention. *Levi* [9.19] has carried through these estimates for a piecewise linear modification of the van der Pol equation similar to that studied by *Levinson* [9.20].

When a nonlinear oscillator such as that described by the van der Pol equation is driven with a periodic forcing term, a phenomenon known as *phase locking* often occurs. Phase locking means that the forced oscillator has a stable limit cycle whose period is an integer multiple n of the period τ of the forcing term. These solutions are called *subharmonic* because their frequency $2\pi/n\tau$ divides the frequency of the forcing term $2\pi/\tau$. The time τ flow map T has a stable periodic orbit of period n formed from n points along the limit cycle.

The key observation initially made by *Cartwright* and *Littlewood* was that there are parameter values for which two different subharmonic solutions occur simultaneously with distinct periods. Along with these stable limit cycles come additional unstable chaotic trajectories. The full extent of this additional chaotic behavior has not been described before, but its presence has been proved by *Levinson* [9.20] for an equation which is very similar to the forced van der Pol equation.

9.5 A Dynamical Systems Analysis of the van der Pol Model

In much of the remainder of this chapter we shall show how techniques from the modern theory of dynamical systems can lead geometrically to a qualitative understanding of the chaotic behavior generated by systems of ordinary differential equations. The following is adapted from *Guckenheimer* [9.21].

Consider an annulus $D \subset \mathbb{R}^2$ of moderate size which contains the limit cycle of the van der Pol equation (9.5). The time $2\pi/\omega$ map T of the forced van der Pol equations (9.7) maps this annulus into itself in such a way that the image is very thin. Figure 9.5 gives a rough sketch of D and $T(D)$. Two important features of $T(D)$ are that it is very thin and that there are "kinks". We

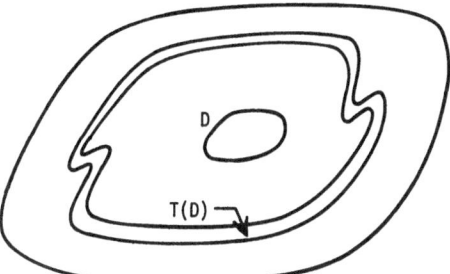

curve of constant θ

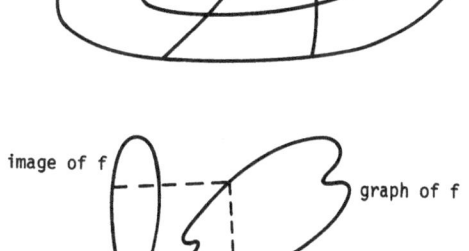

Fig. 9.5. The annulus D contains the limit cycle of the van der Pol equation (9.5). $T(D)$ is the image of the annulus given by the map T which follows trajectories for a time $2\pi/\omega$

Fig. 9.6. Schematic representation of the map F (which maps the disc into itself) and the map f (which maps the circle into itself)

want to exploit these properties by carrying the first to the extreme. In doing so, we shall be able to study the asymptotic behavior of a map of the circle rather than T. In particular, let $F:D\to D$ be a map which is close to T and has a one-dimensional image. More precisely, assume that D is partitioned into smooth curves which join the inside and outside boundaries of D and have the property that F maps each one of these curves to a single point. These curves establish an "angular" coordinate θ on D, and F can be viewed as a map which depends only on this angular coordinate. Regard θ as lying in the circle S^1 and write $f:S^1\to S^1$ for the map which takes the variable whose value at θ is given by the angular coordinate of $F(\theta)$. See Fig. 9.6 for a representation of F, θ, and f.

The asymptotic behavior of F is clearly determined by f, so we focus our attention on f now[1]. A full enumeration of the different possibilities for the asymptotic behavior of f has not yet been made. Rather than try to describe general results, we shall examine one particular example which illustrates the kinds of things which occur. The type of analysis we shall give is often called *symbolic dynamics*. It employs a strategy which allows us to characterize a lot of

---

1 For a continuous map h of the circle which has no folds, there is an old theory of rotation numbers due to Poincaré. This theory characterizes many of the asymptotic properties of a map in terms of a single real number. It is defined as follows. Pick a point x and count the number of times $h^i(x)$ lies in the arc from x to $h(x)$ for $1 < i \le n$. Call this number $\alpha(n)$. The rotation number of h is the limit of $\alpha(n)/n$. It exists and is independent of x. One of the main results of this theory is that a homeomorphism of the circle has a periodic orbit if and only if the rotation number is rational. The map f we are interested in studying does have folds, so this classical theory of rotation numbers does not apply.

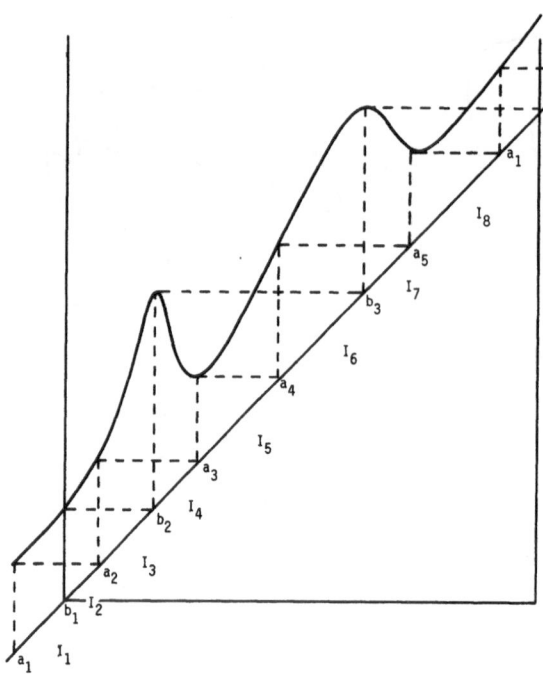

Fig. 9.7. Graph of a map $f:S^1 \to S^1$. Two of the four critical points lie in periodic orbits of period 5 (labeled a_1, a_2, a_3, a_4, a_5) and period 3 (labeled b_1, b_2, b_3). The I_k $(k=1,...,8)$ denote partitions of the circle into 8 arcs

qualitative information about the chaotic behavior of an example in a small amount of data. Figure 9.7 shows the graph of a map $f:S^1 \to S^1$ which has four critical points. We assume that two of these critical points lie in a periodic orbit of period 5 and two lie in a periodic orbit of period 3. The points of these periodic orbits are labeled a_1, a_2, a_3, a_4, a_5 and b_1, b_2 b_3, respectively. We assume that the ordering of these points on the circle is $a_1 b_1 a_2 b_2 a_3 a_4 b_3 a_5 a_1$ and that f cyclically permutes the points of each orbit in the order they are numbered.

These assumptions allow us to construct a symbolic scheme for identifying all of the asymptotic behavior expected of f. Think of each interval joining an adjacent pair of periodic points as "state". There are eight states. The trajectory of each point will be associated to a sequence of states. We call such a sequence a symbolic description of a trajectory. The set of sequences which occur in this representation will be identified. The representation will preserve many of the aspects of the asymptotic behavior of f. For instance, periodic orbits will be associated to periodic sequences.

Let us proceed with the symbolic analysis of the example depicted in Fig. 9.7. Partition the circle into the eight arcs $I_1=[a_1,b_1]$, $I_2=[b_1,a_2]$, $I_3=[a_2,b_2]$, $I_4=[b_2,a_3]$, $I_5=[a_3,a_4]$, $I_6=[a_4,b_3]$, $I_7=[b_3,a_5]$, and $I_8=[a_5,a_1]$. Since the a's and b's are in periodic orbits, the image $f(I_j)$ is a union of these I_k. Specifically, $f(I_1)=I_3$, $f(I_2)=I_4$, $f(I_3)=I_5\cup I_6$, $f(I_4)=I_6$, $f(I_5)=I_6\cup I_7$, $f(I_6)=I_8\cup I_1$, $f(I_7)=I_1$, and $f(I_8)=I_1\cup I_2$. The fact that f maps the set of partition points to itself is called the *Markov* property. This information about images can be embodied in the *transition matrix*,

$$A = \begin{pmatrix} 0 & 0 & 1 & 0 & 0 & 0 & 0 & 0 \\ 0 & 0 & 0 & 1 & 0 & 0 & 0 & 0 \\ 0 & 0 & 0 & 0 & 1 & 1 & 0 & 0 \\ 0 & 0 & 0 & 0 & 0 & 1 & 0 & 0 \\ 0 & 0 & 0 & 0 & 0 & 1 & 1 & 0 \\ 1 & 0 & 0 & 0 & 0 & 0 & 0 & 1 \\ 1 & 0 & 0 & 0 & 0 & 0 & 0 & 0 \\ 1 & 1 & 0 & 0 & 0 & 0 & 0 & 0 \end{pmatrix} \tag{9.9}$$

which is defined by

$$a_{ij} = \begin{cases} 0 & \text{if } I_j \not\subset f(I_i) \\ 1 & \text{if } I_j \subset f(I_i) . \end{cases} \tag{9.10}$$

The transition matrix A allows us to calculate the image $f^k(I_i)$ for the kth iterate of f.

The map f^k sends I_i back and forth across certain of the I_j. There are no turning points in the middle of an I_j because the extrema of f are partition points, and the partition is Markov. The number of times $f^k(I_i)$ crosses I_j can be calculated from A^k. To see this, consider the formula for A^k,

$$(A^i)_{ij} = \sum_{i_1,\dots,i_{k-1}=1}^{\infty} a_{i_0 i_1} a_{i_1 i_2} \dots a_{i_{k-1} i_k} \quad \text{with } i_0 = i \text{ and } i_k = j . \tag{9.11}$$

Each term is zero unless all factors are 1, and then $I_{l+1} \subset f(I_l)$ for $0 \leq l < k$. This implies $I_j \subset f^k(I_i)$. There is one such factor with all 1's each time $f^k(I_i)$ crosses I_j. Therefore, $(A^k)_{ij}$ is the number of times $f^k(I_i)$ covers the arc I_j.

Each time $f^k(I_i)$ covers itself, there will be at least one unstable fixed point of f^k in I_i because there are points x, y in I_i with $f(x)$ the left end point of I_i and $f(y)$ the right end point of I_i. Then $f(x) < x$ and $f(y) > y$, so there is a z between x and y with $f(z) = z$. Thus the number of unstable fixed points of f^k is at least $\sum_{i=1}^{\infty} (A^k)_{ii}$, the trace of A^k. If the only stable periodic orbits of f are the two described above, then the trace of A^k is the exact number of unstable fixed points of f^k. Table 9.1 gives the number of unstable fixed points of f^k for $k \leq 15$.

The transition matrix A allows us to determine more than just the periodic orbits of f. All of the asymptotic behavior of f can be characterized in terms of

Table 9.1. The number of unstable fixed points of the map f^* for $k \leq 15$

k	1	2	3	4	5	6	7	8	9	10	11	12	13	14	15
No. unstable fixed points f^*	0	0	3	16	5	3	28	64	39	45	176	307	260	392	1028
No. unstable periodic orbits, period k	0	0	1	4	1	0	4	6	4	4	16	24	20	26	68

sequences of the "state" symbols $1, \ldots, 8$. To do so, we introduce the *subshift of finite type* Σ with transition matrix A. The subshift consists of a set of sequences, which we again call Σ, and a map $\sigma : \Sigma \to \Sigma$. The set Σ consists of some sequences $\underline{i} = \{i_l\}$ with $1 \leq i_l \leq 8$ and the index l running from 0 to ∞. The sequences which occur in Σ are precisely those which have the property that $a_{i_l i_{l+1}} = 1$ for all $l \geq 0$. The map σ is defined by shifting the indices on a sequence, omitting the first term: $\sigma(\underline{i}) = \underline{j}$ with $j_l = i_{l+1}$.

If we once again assume that f has just two stable periodic orbits, and further assume f is "generic", then the relationship between Σ and the unstable asymptotic behavior of f can be made very precise. Let $\Lambda \subset S^1$ be the set of points which have the property that their orbits do not tend asymptotically to one of the stable periodic orbits. The set Λ contains all of the "chaotic" orbits of f, though excluded from Λ are orbits which might appear chaotic initially for a long period before they become asymptotic to stable periodic orbits. The relationship between Λ and Σ is given by a $1-1$ correspondence ϕ which assigns to each orbit in Λ its *itinerary*: $\phi(x) = \underline{i}$ if $f^l(x) \in I_{i_l}$ for all l. In other words, $\phi(x)$ keeps track, in order, of those arcs I_j which are visited by the orbit of x. It follows immediately from this definition that ϕ has the property that if $\phi(x) = \underline{i}$, then $\phi[f(x)] = \sigma(\underline{i})$ or $\sigma \circ f = \sigma \circ \phi$. This says that, on the symbolic level, σ acts exactly the way f acts on the set $\Lambda \subset S^1$. The Markov property implies that every sequence in Σ is realized by an orbit of f. The map ϕ is onto. Thus Σ and σ together give us a symbolic description of the unstable behavior of f. We can look at Σ instead of Λ when we want to study the qualitative properties of the chaotic behavior of f.

We would like to go back now from this description of the map f on the circle to the forced van der Pol equation. Unfortunately, the excursion cannot be made with the certainty which comes from rigorous analysis. We can go no farther than saying that our description is consistent with all numerical calculations which have been made thus far for the forced van der Pol equation. There are values of the parameters (ε, b, ω) of the forced van der Pol equation for which there are stable subharmonic solutions of periods $3(2\pi/3)$ and $5(2\pi/3)$. Based upon the preceding analysis, we conjecture that for some region in this set of parameters, the remaining unstable trajectories of the system are contained in a set whose cross section looks like the subshift described above (but consisting of sequences whose indices run from $-\infty$ to

$+\infty$) and an isolated completely unstable orbit and nothing more. This flow should be *structurally stable*; the picture of the asymptotic behavior should remain the same for slight changes in the parameter values.

9.6 Discussion

Let us now stand back a bit from all of the examples we have discussed in an effort to place the role of instabilities of natural phenomena into better perspective. These examples should give one an indication of the range of different circumstances in which such instabilities are found. Particularly in confronting the natural world, we are faced frequently with erratic and irregular phenomena within which there is some hope that underlying patterns may emerge. (The weather is a good example of such phenomena.) The models that we have outlined display chaotic dynamics, but whether their chaos is an accurate reflection of the kinds of erratic behavior found in experimental and observational data is an unresolved question in most instances.

Questions of this sort seem inherently difficult for a variety of reasons. One difficulty is practical. Studying the erratic properties of natural phenomena requires substantial amounts of data. Usually, this means, at best, that considerable time and expense are needed to obtain the data and, at worst, that the necessary data are simply unobtainable. The hydrodynamic experiments described earlier in this volume probably represent the best attempts at coping with the problems of obtaining data from unstable phenomena. A second difficulty in comparing difference and differential equations models with data lies in the sensitivity of the dynamics to the choice of initial conditions. Indeed, this property can be regarded as an essential characteristic of "chaos" in the models. The implication of this sensitivity is that a direct comparison between data and model predictions may not be appropriate. Over moderately long time spans, the sensitivity to initial conditions in the model may destroy completely the ability to come close to matching the data. Statistical tests for comparing model and data are needed, but choosing and using these tests complicates matters further. A third difficulty in dealing with chaotic models has already been mentioned. In many examples, there are systems of partial differential equations which lie intermediate between the natural phenomena and finite dimensional systems of differential or difference equations. There are substantial mathematical problems in passing from one level to the other. All of these issues have been most successfully addressed for the instabilities studied in this volume, those found in the transition to turbulence in hydrodynamics. Obtaining results for other phenomena which are as satisfying as the hydrodynamic results remains for the future.

Despite the primitive state of affairs in dealing with irregular natural phenomena, there are some grounds for optimism about the utility of dynamical systems theory as a tool for understanding these phenomena. One of the observable features of experiments is the dependence of qualitative aspects of the outcome upon changes in parameters. Qualitative changes in experiments resulting from changes in parameters are mirrored in the bifurcation theory of models. Bifurcation theory of dynamical systems studies the qualitative changes in asymptotic behavior which occur as the system is deformed. The recurring patterns of bifurcation which are experimentally observed in many phenomena can be understood in their simplest form in the geometric setting of dynamical systems. The theme which one hopes to pursue consistently is that the typical bifurcation behavior which appears in models depending upon a small number of parameters can be studied in the setting of low dimensional dynamical systems. This expedites both the analysis and visualization of the different possibilities. In "good" situations, techniques are available for transferring these results to infinite dimensional settings involving partial differential equations.

A final remark about the role of these chaotic models is more philosophical than the previous ones. There are limits to our ability to reproduce experimental observations from mathematical models. Formulating questions which are well posed in the sense that their answers vary continuously with parameter changes is a difficult task when both model and data are chaotic. The inherent limits on our abilities to do so are likely to be severe. Many questions which at first glance seem reasonable may be ill posed. Studying the wilder features of chaotic dynamical systems seems to be one of the few means available of addressing these limits in terms which are sensitive to the special features of individual phenomena.

Acknowledgements. This research was partially supported by National Science Foundation Grant MSC77-04112-A01.

References

9.1   F. H. Busse: Magnetohydrodynamics of the earth's dynamo. Annu. Rev. Fluid Mech. **10**, 435–462 (1978)

9.2   A. E. Cook, P. H. Roberts: The Rikitake two-disc dynamo system. Proc. Camb. Philos. Soc. **68**, 547–569 (1970)

9.3   K. A. Robbins: A new approach to subcritical instability and turbulent transitions in a simple dynamo. Math. Proc. Camb. Philos. Soc. **82**, 309–325 (1977)

9.4   M. W. Hirsch, S. Smale: *Differential Equations, Dynamical Systems, and Linear Algebra.* (Academic Press, New York 1974)

9.5   H. Haken: Analogy between higher instabilities in fluids and lasers. Phys. Lett. **53** A, 77–78 (1975)

9.6   L. F. Olsen, H. Degn: Chaos in an enzyme reaction. Nature **267**, 177–178 (1977)

9.7   R. A. Schmitz, K. R. Graziani, J. L. Hudson: Experimental evidence of chaotic states in the Belousov-Zhabotinskii reaction. J. Chem. Phys. **67**, 3040–3044 (1977)

9.8   O. E. Rössler, K. Wegmann: Chaos in the Zhabotinskii reaction. Nature **271**, 89–90 (1978)

9.9   J. J. Tyson: The Belousov-Zhabotinskii reaction, *Lecture Notes in Biomathematics*, Vol. 10 (Springer, Berlin, Heidelberg, New York 1976)

9.10 J.J.Tyson: On the appearance of chaos in the Belousov reaction. J. Math. Biol. **5**, 351–362 (1978)

9.11 A.T.Winfree: Spiral waves of chemical activity. Science **175**, 634–636 (1972)

9.12 N.Kopell, L.N.Howard: Pattern formation in the Belousov reaction. Lect. Math. Life Sci. **7**, 201–216 (1974)

9.13 R.J.Field, R.M.Noyes: Oscillations in chemical systems. IV. Limit cycle behavior in a model of a real chemical reaction. J. Chem. Phys. **60**, 1877–1884 (1974)

9.14 R.M.May: Simple mathematical models with very complicated dynamics. Nature **261**, 459–466 (1976)

9.15 T.Li, J.A.Yorke: Period three implies chaos. Am. Math. Mon. **82**, 985–992 (1975)

9.16 J.Guckenheimer, G.Oster, A.Ipaktchi: Dynamics of density dependent population models. J. Math. Biol. **4**, 101–147 (1977)

9.17 M.L.Cartwright, J.E.Littlewood: On nonlinear differential equations of the second order: I. The equation $\ddot{y} + k(1 - y^2)\dot{y} + y = b\lambda k \cos(\lambda t + a)$, k large. J. London Math. Soc. **20**, 180–189 (1945)

9.18 S.Smale: Differentiable dynamical systems. Bull. Am. Math. Soc. **73**, 747–817 (1967)

9.19 M.Levi: Memoirs Am. Math. Soc. **32**, No. 244 (July 1981)

9.20 N.Levinson: A second order differential equation with singular solutions. Ann. Math. **50**, 127–153 (1949)

9.21 J.Guckenheimer: Symbolic dynamics and relaxation oscillations. Physica 1D, 227–235 (1980)

10. Recent Progress

F. H. Busse, J. P. Gollub, S. A. Maslowe, and H. L. Swinney

With 1 Figure

10.1 Introductory Comments

In the few years since the publication of the first edition of this book there has been much progress in understanding the transition to weak turbulence (see [10.1–3] for reviews of recent developments). Several routes to turbulence, including period doubling and intermittency, have been identified in experiments and have become fairly well understood through the application of renormalization group ideas. The strange attractors described by Lanford in Chap. 2 have been directly observed in experiments on Rayleigh-Bénard convection [10.4], baroclinic systems [10.5], surface waves [10.6], and Couette-Taylor flow [10.7]. Immediately beyond the onset of nonperiodic flow the strange attractors have a fractional dimension that is, at least in small aspect ratio systems, quite small, indicating that the dynamical behavior should be describable by models having only a few degrees of freedom (see the discussions of such hydrodynamic chaos in Chaps. 2 to 4).

Another area that is now receiving much attention concerns wavelength selection and the development of spatial patterns. References to pattern selection, hydrodynamic chaos, and other recent developments may be found in the following sections.

10.2 Rayleigh-Bénard Convection

Our knowledge about the onset of nonperiodic behavior in Rayleigh-Bénard convection has increased considerably in the past few years, especially from the experimental point of view. Reviews of new results that appeared at about the time of the first edition of this book were given by *Ahlers* [10.8], *Busse* [10.9], and *Gollub* [10.10]. Here we mention only a few selected topics.

10.2.1 Routes to Chaos in Convection

It had been known for a number of years that transitions to chaotic states in convection in small aspect ratio boxes resemble those found in simple dynamical systems described by a few coupled nonlinear ordinary differential equations. But the first quantitative comparison became possible when the *Feigenbaum* [10.11] sequence of period doublings was observed in convection in small boxes [10.12–15]. In particular, the experiment by *Giglio* et al. [10.14] showed five period doublings and fair agreement with the Feigenbaum theory. It should be emphasized, however, that the realization of the period doubling route to chaos depends sensitively on the parameters of the experiment, and other routes to chaos are frequently found even in small aspect ratio convection boxes. For example, see *Gallub* and *Benson* [10.13].

Another route to chaos that has been studied recently involves intermittent chaotic bursts at irregular intervals. *Dubois* et al. [10.16] have been able to show that this process is described accurately by one-dimensional Poincaré return maps.

The transition to chaos via a quasi-periodic state consisting of two incommensurate oscillations has been predicted to have universal properties analogous to the period doubling route. In particular, the power spectrum at the onset of chaos is expected to have a characteristic self-similar structure. An experimental test of these predictions has recently been reported by *Fein* et al. ([10.17] and references therein). Some evidence for spectral self-similarity was found ; however, the experiment was extremely delicate and conclusive results await future work.

Finally, *Malraison* et al. [10.4] have succeeded in measuring the dimension of the attracting set in a chaotically convecting fluid. The observation that the dimension was small (2.8 ± 0.1 in one case) provides strong evidence that weakly-turbulent small-aspect-ratio convection can be described by a low-dimensional strange attractor.

10.2.2 Pattern Evolution and Defects in Large Aspect Ratio Convection Layers

Convection layers with aspect ratios of the order of 20 or larger permit the realization of a large manifold of nearly equivalent modes of convection. The patterns consist mainly of curved rolls with various kinds of defects, including dislocations like those of a crystal lattice and disclinations where the local orientation of the rolls is singular. There is a strong tendency for rolls to align perpendicular to lateral boundaries. The evolution of convection patterns is extremely slow, often longer than the horizontal thermal diffusion time. The evolution process is dominated by the motion of defects in the patterns. Several groups have recently reported detailed experimental studies of convection patterns [10.18–21].

Theoretical efforts to explain the process of pattern evolution and wave-number selection have centered on the use of two-dimensional model equations that are much simpler than the full hydrodynamic equations and yet correctly describe many of the observed phenomena near threshold. A theoretical description of the problem in the spirit of the small amplitude evolution equation of Newell and Whitehead is given in the recent book by *Haken* [Ref. 10.22, Chap. 9]. We note especially the analysis of *Cross* [10.23] based on the Swift-Hohenberg equation; related numerical computations by *Greenside* et al. [10.24]; and the work of *Pomeau* et al. [10.25] on dislocation motion. *Cross* et al. [10.26] have also shown that the presence of lateral boundaries affects the entire flow even in a large layer. *Siggia* and *Zippelius* [10.27] studied theoretically the dynamics of the motion of isolated dislocations in the roll structure. They also showed [10.28] that the amplitude equation must be supplemented by an equation for the large-scale vertical component of vorticity when convection with stress-free boundaries in a fluid of finite Prandtl number is considered.

10.2.3 Other Time-Dependent Phenomena

In large aspect ratio convection at moderate Prandtl number ($1 < P < 10$), time-dependence on the scale of the vertical rather than horizontal thermal diffusion time is caused by the skewed varicose instability. This time-dependence is chaotic [10.18] and may be identified with the onset of (weak) turbulence. This effect has also been seen in numerical experiments [10.27], but is not really well understood. The numerical simulation of three-dimensional convection has progressed considerably [10.29–30]. Even though the computations are still limited by a small periodicity interval in the horizontal dimensions, computational simulations offer an important complement to the experimental measurements because of their flexibility and high information content.

Time-dependence can be induced even near threshold by time-periodic modulation of the imposed temperature gradient or heat current. Experiments and a satisfactory model which is a generalization of the Lorenz equations were reported recently by *Ahlers* et al. [10.31].

10.2.4 Convection with Magnetic Field

The added parameter of the magnetic field strength in convection layers of electrically conducting fluids yields an even richer spectrum of phenomena than is observed in non-magnetic convection. An oscillatory onset of convection and subcritical finite amplitude onset are possible in the presence of a vertical magnetic field. The various types of supercritical bifurcations can be analyzed mathematically by a five mode model [10.32], which appears to capture qualitatively all features of the exact solutions, in contrast to the non-magnetic Lorenz model of convection. Justification for the two-dimensional analysis is partially provided by the results of the analysis of stability of convection rolls in the presence of a magnetic field [10.33–34].

Experimentally, the case of an imposed horizontal magnetic field has been investigated in detail [10.15, 35]. The availability of two experimentally controllable parameters (temperature gradient and magnetic field strength) facilitates the study of the bifurcation sequences. Various routes to chaos were detected in different regions of the parameter space.

10.3 Instabilities and Transition in Flow Between Concentric Cylinders

An increasing variety of phenomena are being found in studies of the Couette-Taylor systems as a function of the radius ratio of the cylinders, η; the aspect ratio, $\Gamma = h/d$; the boundary conditions at the ends of the annulus; and the Reynolds numbers R_i and R_o associated with the rotation rates Ω_i and Ω_o of the inner and outer cylinders, respectively,

$$R_i = \Omega_i a d / v, \qquad R_o = \Omega_o b d / v. \tag{10.1}$$

The phase diagram of *Andereck* et al. [10.40] in Fig. 10.1 shows the ranges of different flow regimes observed in a system with a radius ratio equal to 0.883 and aspect ratio equal to 30. Each flow state was approached along the following path: R_o was increased quasistatically from zero to its final value, and then R_i was increased quasistatically from zero to its final value [10.39–40]. Other paths in control parameter space leading to the same final values of η, Γ,

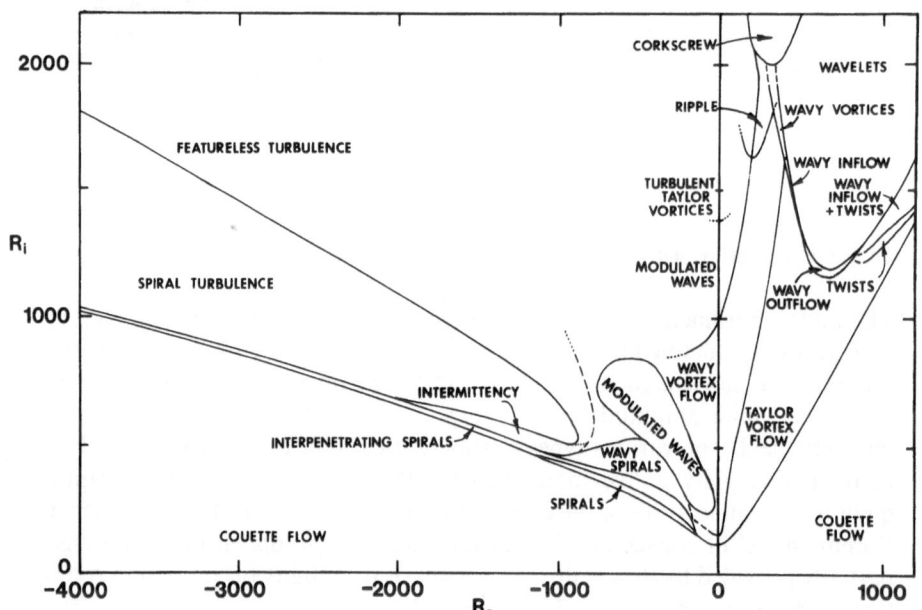

Fig. 10.1. Flow regimes observed in flow between independently rotating cylinders with radius ratio 0.883 and aspect ratio 30. (Adapted from [10.40])

R_o, and R_i, and end conditions lead to *many* other flow states; hence the flow is highly nonunique [10.39, 40, 43, 44, 60, 92]. Moreover, even if the same path were followed in Raynolds-number space (increasing first R_o, then R_i), the phase diagram could in general be quite different from that in Fig. 10.1 if η [10.54, 66, 67] or Γ (Sect. 10.3.1) were different from the values for which Fig. 10.1 was obtained.

The occurrence of a wide variety of phenomena in flow between rotating concentric cylinders has led to an experimental and theoretical interest in this problem that is now greater than at any time in the century since *Mallock* [6.1, 2] and *Couette* [6.3] conducted studies of this system. We will briefly describe some of the areas of recent interest. References to these studies [10.36–100] are arranged alphabetically. Almost all of these studies concern flows with the inner cylinder rotating and the outer cylinder at rest. Very little work has been done with the outer cylinder rotating.

10.3.1 Taylor Vortex Flow and Finite Length Effects

The effect of finite cylinder length on the onset and growth of Taylor vortices, discussed in Sect. 6.3, is being explored in experiments and theory [10.43, 44, 62, 63, 67, 69, 71, 79, 81, 82, 84–87, 96]. One approach to the analysis of finite-length effects is through amplitude equations such as equations (6.28, 33, and 34). The attractiveness of amplitude equations is their universal character near a transition; they should identify features that are common in diverse systems. For the concentric cylinder problem amplitude equations are providing insight into the effects of end walls on transition [10.47, 62, 69, 90], the propagation of vortex fronts [10.36, 45, 70, 95], and wavelength selection [10.50, 55].

An elegant experiment on wavelength selection by *Cannell* et al. [10.50] has demonstrated that in a system with a gentle taper of the inner cylinder radius, which produces a reduction in the relative Reynolds number $\varepsilon = (R - R_c)/R_c$ from $\varepsilon > 0$ to $\varepsilon < 0$, there is a selection of a unique axial wavelength as predicted theoretically [10.131]. In contrast, boundaries such as rigid ends which allow wavelength adjustment to occur by the creation or destruction of a vortex pair, give a wide range of stable states.

Amplitude equations should also be useful in understanding the variation of vortex wavelength with axial position [10.38, 86] and, in the small gap limit, the onset of wavy vortex flow [6.27; 10.64].

Some studies have attempted to minimize end effects by making the aspect ratio very large. Other recent work has examined the other extreme, aspect ratios as small as ~ 0.5 [10.51, 71, 79–82]. Then the ends have a major effect – the flow, even for $R \ll R_c$, is no longer approximated by the Couette solution (6.1), but, on the other hand, the problem (at least at small R) is amenable to numerical simulation with realistic end boundary conditions [10.71].

Other recent papers that discuss Taylor vortex flow include [10.41, 43–44, 47, 53, 57–59, 65, 76–79, 81, 96]. Other recent papers that discuss finite-length effects include [10.43–44, 62–63, 67, 69, 84, 87, 96].

10.3.2 Wavy Vortex Flow and Other Periodic and Multi-Periodic Flows

Experiments on wavy vortex flow have examined the range of stability of wavy vortex flow states with different numbers (m) of azimuthal waves as a function of axial wavelength, aspect ratio, and Reynolds number [10.37, 38, 68, 84]. Numerical simulations of wavy vortex flow have predicted that the onset depends very sensitively on radius ratio for radius ratios near 0.75 [10.65]; recent experiments [10.83] corroborate this prediction. Numerical simulations have been extended well beyond the onset of wavy vortex flow, where the flow is highly nonlinear [10.74, 75]; the predicted angular velocities of the waves are in excellent agreement with experiment [10.67]. Other papers that discuss wavy vortex flow include [10.45, 46, 56, 69, 82, 86, 89, 90, 100].

Quasiperiodic flows characterized by velocity power spectra with two incommensurate frequencies [10.53, 61] have been shown to correspond to two traveling azimuthal waves in most cases examined [10.92], although flows have also been observed that have one fundamental frequency corresponding to a traveling azimuthal wave and another one (or more) corresponding to nonpropagating modes [10.100]. Flows with more than two incommensurate frequencies have been observed [10.53, 60, 68, 72, 73], but they are apparently relatively rare.

10.3.3 Chaos and Turbulence

As mentioned in Sect. 10.1, phase space attractors have been constructed from velocity time series data for Couette-Taylor flow. An attractor is *strange* (chaotic) if nearby the trajectories emanating from nearby points in phase space separate exponentially fast on the average (Chap. 2). The separation rate of nearby trajectories is characterized by the largest *Lyapunov exponent* λ; λ is zero for periodic or multiperiodic flows, while it is (by definition) positive for a chaotic flow. For a chaotic system any uncertainty in the specification of the initial state, no matter how small, will rapidly grow to be the size of the attractor; since the state of a flow can never be determined with infinite precision, long-term future behavior is unpredictable, even though the Navier-Stokes equation is deterministic.

The largest Lyapunov exponent was found to be zero for wavy vortex flow and modulated wavy vortex flow, but λ was found to become positive above the onset of nonperiodic flow (which is marked by the appearance of broadband spectral noise; see Sect. 6.5.2) [10.7, 48, 49].

It is perhaps not surprising that the behavior of weakly turbulent (chaotic) flow is inherently unpredictable. But what is surprising is that the fractal (noninteger) dimension of the phase space attractors is small ($\lesssim 6$), even well beyond the onset of nonperiodicity [10.5–7, 48, 49]. This suggests that these chaotic flows should be describable by low-dimensional deterministic models with only a few variables [10.1–3, 72–73]. This may not by the case, however, for very large aspect ratio systems which contain defects or dislocations in the flow patterns [10.56].

Some experiments have been conducted at Reynolds numbers far beyond the onset of nonperiodicity [10.42, 53, 88]. The three-dimensional turbulent flows are found to contain large organized structures [10.42, 53]; in fact, the Taylor vortices even persist for $R \gtrsim 600 R_c$ [10.94].

10.3.4 Theory and Numerical Analysis

Nonlinear stability theory cannot be extended very far beyond the primary instability, but realistic numerical simulations have been made of Taylor vortex flow and wavy vortex flow [10.51, 58, 65, 66, 71, 74–78, 97–99], and supercomputers make it possible in some cases to extend the simulations to fairly large Reynolds numbers [10.74–75]. These numerical studies together with symmetry considerations [10.52, 59, 61, 91, 93] should lead to new insights into the variety of flows exhibited by the Couette-Taylor system and into the mechanisms of the instabilities.

10.4 Shear Flow Instabilities and Transition

Three-dimensional shear flow instabilities, both linear and nonlinear, are topics of continuing active research. A particularly interesting instability mechanism termed "direct resonance" is described in [10.101]. This mechanism involves a resonance between modal solutions of the Orr-Sommerfeld equation for the vertical velocity and forced modes of the vertical vorticity equation. One of the many possible applications may be to provide a subcritical growth mechanism for flows not possessing a linear neutral curve such as pipe flow (Sect. 7.4.1).

In related earlier work on the linear viscous initial-value problem for plane Couette flow [10.102], directly resonant modes were identified. It seems that in shear flows the requisite conditions are typically satisfied for waves that are slightly damped on a linear basis. The consequences in the nonlinear theory [10.101] are potentially most significant for multi-mode disturbances, including waves that are "standing" in the spanwise direction.

Although most of the recent work on the linear initial value problem (Sect. 7.3) has been for viscous flows, *Brown* and *Stewartson* [10.103] have extended and corrected previous work on inviscid stratified shear flows. Of particular interest is their discovery of a subtle error in [7.36] which, when corrected, eliminates the algebraically growing modes noted in Sect. 7.3.1.

The nonlinear critical layer theory (Sect. 7.4.2) has been employed in a study of *Smith* and *Bodonyi* [10.104] of neutral modes in a boundary layer developing in a favorable pressure gradient. Different scalings result from the Blasius case, where $\bar{u}''(0) = 0$. This paper also addresses complexities arising when the disturbance amplitude is large enough so that nonlinear effects are significant in the viscous wall layer (Fig. 7.5). The same authors have found [10.105] that nonaxisymmetric neutral modes with nonlinear critical layers exist in pipe flows; by contrast, only damped waves exist in the linear theory.

A numerical study by *Orszag* and *Patera* [10.106] presented results that appear to be significant in explaining the subcritical transition of channel flow. It was found that infinitesimal oblique wave disturbances can grow exponentially for Reynolds numbers greater than 1000 when the basic steady flow has superimposed upon it a quasiperiodic two-dimensional wave of finite amplitude. After some amplification of the oblique wave, random behavior is observed and it is conjectured that this is indicative of breakdown to turbulence.

Vortex methods provide an interesting alternative to the more widely used finite-difference and Galerkin numerical schemes. By employing vortices with a core structure, singularities in the potential vortex model are eliminated. Although such methods have been proved convergent, the convergence is slow near boundaries. A modification that remedies this difficulty has been introduced recently by *Chorin* [10.107], namely, to use vortex-sheet segments in the region adjacent to a boundary. Both two- and three-dimensional versions of the method have been used to simulate boundary-layer transition, and some interesting results have been presented in [10.107].

Of the numerous valuable experimental studies reported recently, two will be discussed briefly here. First, an investigation of a turbulent-mixing layer subject to periodic two-dimensional forcing has been described in [10.108]. The mechanism for generating the forced oscillation was a flap at the end of a splitter plate initially separating the two streams. Despite the large Reynolds number of the flow (on the order of 10^4), coherent structures were observed and the ideas on vortex pairing discussed in Sect. 7.5.1. are relevant. In a certain parameter range the forced oscillation resonates with the mixing layer with one result being that vortex interaction and, consequently, mixing-layer growth, are suppressed. Surprisingly, a negative Reynolds stress is produced, implying that energy from the turbulence goes back into the mean flow.

The second experimental investigation cited here [10.109] deals with observations of a turbulent spot generated in a laminar boundary layer by injecting fluid though a small hole downstream of the leading edge of a flat plate being towed through water. The spot was visualized by means of a fluorescent dye which becomes visible when excited by a light source, in this case a 5 W argon laser. Some quite spectacular photographs are shown which describe the growth of the spot. While growth was due to turbulent entrainment in the vertical direction, in the spanwise direction it was observed that instabilities were induced by the spot and these, in turn, formed new turbulence. The spot itself was reportedly composed of random eddies which interacted only very weakly.

Finally, a welcome addition to the stability literature is the text of *Drazin* and *Reid* [10.110]. Many topics touched upon briefly in Chap. 7 herein are treated in great depth, particularly the asymptotic theory and numerical solution of the Orr-Sommerfeld equation. The final chapter is a well-written survey of current nonlinear stability theories.

10.5 Instabilities and Chaos in Other Systems

Double diffusive instabilities, discussed in the chapter on geophysical fluid dynamics (Sect. 8.6), have recently been elegantly reviewed by *Huppert* and *Turner* [10.111], and baroclinic instabilities (Sect. 8.5) have been discussed thoroughly in a review by *Hide* [10.112]. The book by *Drazin* and *Reid* [10.110] contains (in Sects. 44 and 45) brief accounts of the linear stability theory that is the basis for Sects. 8.3 and 8.5.

A clear example of a hydrodynamic strange attractor has been found in studies of surface waves excited on a fluid layer by small amplitude vertical oscillation of the container [10.6]. The chaos was shown to arise from competition between two distinct normal modes of the surface oscillation. A four-variable phenomenological model described most of the experimental data, including the route to chaos and the dimension of the attractor.

There has been a veritable explosion of interest in chaos in nonhydrodynamic systems. Four systems that have been extensively investigated in experiments and have been found to be described well (at least in some parameter ranges) by low-dimensional models are the following: the Belousov-Zhabotinskii reaction [10.113–117], chick heart cells [10.118–121], semiconductors and Josephson junctions [10.122–126], and lasers and optical cavities [10.127–130].

10.6 Conclusion

The brief discussions and bibliography contained in this chapter clearly show that the subject of instabilities and the transition to turbulence continues to evolve rapidly.

References

10.1   T. Tatsumi (ed.): *Turbulence and Chaotic Phenomena in Fluids.* Proc. Kyoto IUTAM Conference, 1983 (North Holland, Amsterdam 1984)
10.2   P. Bergé, Y. Pomeau, Ch. Vidal: L'Ordre dans le chaos (Hermann, Paris 1984)
10.3   D. Campbell, H. Rose (eds.): Order in chaos. Physica **7** D, Nos. 1–3 (1983);
        N. Abraham, J.P. Gollub, H.L. Swinney: Testing nonlinear dynamics. Physica **11** D, 252 (1984);
        H.L. Swinney, J.P. Gollub: Characterization of hydrodynamic strange attractors. Physica D, to appear (1985)
10,4   B. Malraison, P. Atten, P. Bergé, M. Dubois: Dimension of strange attractors: an experimental determination for the chaotic regime of two chaotic systems. J. Phys. Lett. (Paris) **44**, L 897 (1983)
10.5   J.D. Farmer, J. Hart, P. Weidman: A phase space analysis of a baroclinic flow. Phys. Lett. **91** A, 22 (1982);

10.5   J. Guckenheimer, G. Buzyna: Dimension measurements for geostrophic turbulence. Phys. Rev. Lett. **51**, 1438 (1983)

10.6   S. Ciliberto, J. P. Gollub: Pattern competition leads to chaos. Phys. Rev. Lett. **52**, 922 (1984); Chaotic mode competition in parametrically forced surface waves. J. Fluid Mech., to appear (1985)

10.7   A. Brandstater, J. Swift, H. L. Swinney, A. Wolf, J. D. Farmer, E. Jen, J. P. Crutchfield: Low dimensional chaos in a hydrodynamic system. Phys. Rev. Lett. **51**, 1442 (1983)

10.8   G. Ahlers: Onset of convection and turbulence in a cylindrical container, in *Systems Far From Equilibrium*, ed. by L. Garrido and J. Garcia (Springer, Berlin, Heidelberg 1980)

10.9   F. Busse: Transition to turbulence in thermal convection with and without rotation, in *Transition and Turbulence*, ed. by R. E. Meyer (Academic, New York 1981)

10.10 J. P. Gollub: Recent experiments on the transition to turbulent convection, in *Nonlinear Dynamics and Turbulence*, ed. by G. I. Barenblatt, G. Iooss, and D. D. Joseph (Pittman, New York 1983)

10.11 M. J. Feigenbaum: Quantitative universality for a class of nonlinear transformations. J. Stat. Phys. **19**, 25 (1978)

10.12 A. Libchaber, J. Maurer: Une experience de Rayleigh-Bénard de géometrie réduite: multiplication, accrochage, et démultiplication de fréquences. J. Physique **41**, 13 (1980)

10.13 J. P. Gollub, S. V. Benson: Many routes to turbulent convection. J. Fluid Mech. **100**, 449 (1980).

10.14 M. Giglio, S. Musazzi, U. Perini: Transition to chaotic behavior via a reproducible sequence of period-doubling bifurcations. Phys. Rev. Lett. **47**, 243 (1981)

10.15 A. Libchaber, C. Laroche, S. Fauve: Period doubling cascade in mercury, a quantitative measurement. J. Physique Lett. **43**, L 211 (1982)

10.16 M. Dubois, M. A. Rubio, P. Bergé: Experimental evidence of intermittencies associated with a subharmonic bifurcation. Phys. Rev. Lett. **51**, 1446 (1983)

10.17 A. P. Fein, M. S. Heutmaker, J. P. Gollub: Scaling at the transition from quasiperiodicity to chaos in a hydrodynamic system. Physica Scripta (to appear)

10.18 J. P. Gollub, A. R. McCarriar, J. F. Steinman: Convective pattern evolution and secondary instabilities. J. Fluid Mech. **125**, 259 (1982); J. P. Gollub, A. R. McCarriar: Convective patterns in Fourier space. Phys. Rev. A **26**, 3470 (1982); M. S. Heutmaker, P. N. Fraenkel, J. P. Gollub: Convection patterns: Time evolution of the wavevector field. Phys. Rev. Lett. **54**, 1369 (1985)

10.19 V. Croquette, M. Mory, F. Schosseler: Rayleigh-Bénard convective structures in a cylindrical container, J. Physique **44**, 293 (1983); A. Pocheau, V. Croquette: Dislocation motions: a wavenumber selection mechanism in Rayleigh-Bénard convection. J. Physique **45**, 35 (1984)

10.20 J. A. Whitehead: The propagation of dislocations in the Rayleigh-Bénard rolls and bimodal fow. J. Fluid Mech. **75**, 715 (1976)

10.21 V. Steinberg, G. Ahlers, D. S. Cannell: Pattern formation and wave-number selection by Rayleigh-Bénard convection in a cylindrical container. Physica Scripta (to appear)

10.22 H. Haken: *Advanced Synergetics, Instability Hierarchies of Self-Organizing Systems and Devices*, Springer Ser. Syn., Vol. 20 (Springer, Berlin, Heidelberg 1983)

10.23 M. C. Cross: Ingredients of a theory of convective textures close to onset. Phys. Rev. A **25**, 1065 (1982)

10.24 H. S. Greenside, W. M. Coughran, Jr., N. L. Schryer: Nonlinear pattern formation near the onset of Rayleigh-Bénard convection. Phys. Rev. Lett. **49**, 726 (1982)

10.25 Y. Pomeau, S. Zaleski, P. Manneville: Dislocation motion in cellular structures. Phys. Rev. A **27**, 2710 (1983)

10.26 M. C. Cross, P. G. Daniels, P. C. Hohenberg, E. D. Siggia: Phase-winding solutions in a finite container above the convective threshold. J. Fluid Mech. **127**, 157 (1983)

10.27 E. D. Siggia, A. Zippelius: Dynamics of defects in Rayleigh-Bénard convection. Phys. Rev. A **24**, 1036 (1981)

10.28 E. D. Siggia, A. Zippelius: Pattern selection in Rayleigh-Bénard convection near threshold. Phys. Rev. Lett. **47**, 835 (1981)

10.29 G.Grötzbach: Direct numerical simulation of laminar and turbulent Bénard convection. J. Fluid Mech. **119**, 27 (1982)

10.30 J.B.McLaughlin, S.A.Orszag: Transition from periodic to chaotic thermal convection. J. Fluid Mech. **122**, 123 (1982);
J.H.Curry, J.R.Herring, J.Loncaric, S.A.Orszag: Order and disorder in two- and three-dimensional Bénard convection. J. Fluid Mech. **147**, 1 (1984)

10.31 G.Ahlers, P.C.Hohenberg, M.Lücke: Externally modulated Rayleigh-Bénard convection: experiment and theory. Phys. Rev. Lett. **53**, 48 (1984)

10.32 E.Knobloch, N.O.Weiss, L.N.DaCosta: Oscillatory and steady convection in a magnetic field. J. Fluid Mech. **113**, 153 (1981)

10.33 F.H.Busse, R.M.Clever: On the stability of convection rolls in the presence of a vertical magnetic field. Phys. Fluids **25**, 931 (1982)

10.34 F.H.Busse, R.M.Clever: On the stability of convection rolls in the presence of a horizontal magnetic field. J. Méc. Théor. Appl. **2**, 495 (1983)

10.35 A.Libchaber, S.Fauve, C.Laroche: Two-parameter study of the routes to chaos. Physica **7 D**, 73 (1983)

10.36 G.Ahlers, D.S.Cannell: Vortex-front propagation in rotating Couette-Taylor flow. Phys. Rev. Lett. **50**, 1583 (1983)

10.37 G.Ahlers, D.S.Cannell, M.A.Dominguez-Lerma: Fractional mode numbers in wavy Taylor vortex flow. Phys. Rev. Lett. **49**, 368 (1982)

10.38 G.Ahlers, D.S.Cannell, M.A.Dominguez-Lerma: Possible mechanism for transitions in wavy Taylor vortex flow. Phys. Rev. A **27**, 1225 (1983)

10.39 C.D.Andereck, R.Dickman, H.L.Swinney: New flows in a circular Couette system. Phys. Fluids **26**, 1395 (1983)

10.40 C.D.Andereck, S.S.Liu, H.L.Swinney: Flow regimes in a circular Couette system with independently rotating cylinders, to be published (1985)

10.41 I.P.Andreichikov: Branching of the secondary modes in the flow between cylinders. Fluid Dynamics **12**, 38 (1977). Translated from Izvestiya Academic Nauk SSSR, Mekhanika Zhicbkosti i Goza, No. 1, 47 (1977)

10.42 A.Barcilon, J.Brindley: Organized structures in turbulent Taylor-Couette flow. J. Fluid Mech. **143**, 429 (1984)

10.43 T.B.Benjamin, T.Mullin: Anomalous modes in the Taylor experiment. Proc. R. Soc. London A **377**, 221 (1981)

10.44 T.B.Benjamin, T.Mullin: Notes on the multiplicity of flows in the Taylor experiment. J. Fluid Mech. **121**, 219 (1982)

10.45 H.Brand, M.Cross: Phase dynamics for the wavy vortex state of the Taylor instability. Phys. Rev. A **27**, 1237 (1983)

10.46 A.Brandstater, U.Gerdts, A.Lorenzen, G.Pfister: Structure and dynamics of non-stationary Taylor-vortex flow, in *Nonlinear Phenomena at Instabilities and Phase Transitions*, ed. by T.Riste (Plenum, New York 1981)

10.47 A.Brandstater, G.Pfister, E.O.Schulz-DuBois: Excitation of a Taylor vortex mode having resonant dependence of coherence length. Phys. Lett. **88** A, 407 (1982)

10.48 A.Brandstater, H.L.Swinney: Distinguishing low dimensional chaos from random noise in a hydrodynamic experiment, in *Fluctuations and Sensitivity in Nonequilibrium Systems*, ed. by W.Horsthemke and D.Kondepudi, Springer Proc. Phys. 1, (Springer, Berlin, Heidelberg 1984) pp. 166–171;
Low-dimensional strange attractors in Couette-Taylor flow (to be published) see also [10.7]

10.49 A.Brandstater, J.Swift, H.L.Swinney, A.Wolf: A strange attractor in a Couette-Taylor experiment, in *Turbulence and Chaotic Phenomena in Fluids*, ed. by T.Tatsumi (North-Holland, Amsterdam 1984), p.179

10.50 D.S.Cannell, M.A.Dominguez-Lerma, G.Ahlers: Experiments on wave number selection in rotating Couette-Taylor flow. Phys. Rev. Lett. **50**, 1365 (1983)

10.51 K.A.Cliffe: Numerical calculations of two-cell and single-cell Taylor flows. J. Fluid Mech. **135**, 219 (1983)

10.52 P.Chossat, G.Iooss: Primary and secondary bifurcations in the Couette-Taylor problem. Preprint (1984)

300   F. H. *Busse* et al.

10.53   G.Cognet, A.Bouabdallah, A.A.Aider: Laminar-turbulent transition in Taylor-Couette flow, in *Stability in the Mechanics of Continua*, ed. by F.H.Shroeder (Springer, Berlin, Heidelberg 1982) p. 330

10.54   J.A.Cole: The effect of cylinder radius ratio on wavy vortex onset, in *Third Taylor Vortex Flow Working Party Meeting* (Université de Nancy, 1983) p. 1.a1

10.55   M.Cross: Wavenumber selection of soft boundaries near threshold. Phys. Rev. A **29**, 391 (1984)

10.56   R.J.Donnelly, K.Park, R.Shaw, R.W.Walden: Early nonperiodic transitions in Couette flow. Phys. Rev. Lett. **44**, 987 (1980)

10.57   P.M.Eagles, K.Eames: Taylor vortices between almost cylindrical boundaries. J. Eng. Math. **17**, 263 (1983)

10.58   H.Fasel, O.Booz: Numerical investigations of supercritical Taylor vortex flow for a wide gap. J. Fluid Mech. **138**, 21 (1984)

10.59   M.Golubitsky, I.Stewart: Symmetry and stability in Taylor-Couette flow. Preprint (1984)

10.60   M.Gorman, L.A.Reith, H.L.Swinney: Modulation patterns, multiple frequencies, and other phenomena in circular Couette flow. Ann. N.Y. Acad. Sci. **357**, 10 (1980)

10.61   M.Gorman, H.L.Swinney: Spatial and temporal characteristics of modulated waves in the circular Couette system. J. Fluid Mech. **117**, 123 (1982);
M.Gorman, H.L.Swinney, D.A.Rand: Doubly periodic circular Couette flow: Experiments compared with dynamics and symmetry. Phys. Rev. Lett. **46**, 992 (1981)

10.62   R.Graham, J.A.Domaradzki: Local amplitude equation of Taylor vortices and its boundary condition. Phys. Rev. A **26**, 1572 (1982)

10.63   P.Hall: Centrifugal instabilities of circumferential flows in finite cylinders: The wide gap problem. Proc. R. Soc. London A **384**, 359 (1982)

10.64   P.Hall: The evolution equations for Taylor vortices in the small gap limit. Phys. Rev. A **29**, 2921 (1984)

10.65   C.A.Jones: Nonlinear Taylor vortices and their stability. J. Fluid Mech. **102**, 249 (1981)

10.66   C.A.Jones: On flow between counter-rotating cylinders. J. Fluid Mech. **120**, 433 (1982)

10.67   G.P.King, Y.Li, W.Lee, H.L.Swinney, P.S.Marcus: Wave speeds in wavy Taylor vortex flow. J. Fluid Mech. **141**, 365 (1984)

10.68   G.P.King, H.L.Swinney: Limits of stability and irregular flow patterns in wavy vortex flow. Phys. Rev. A **27**, 1240 (1983)

10.69   A.Lorenzen, G.Pfister, T.Mullin: End effects on the transition to time-dependent motion in Taylor experiment. Phys. Fluids **26**, 10 (1983)

10.70   M.Lücke, M.Mihelcic, K.Wingerath: Propagation of Taylor vortex fronts into unstable circular Couette flow. Phys. Rev. Lett. **52**, 625 (1984)

10.71   M.Lücke, M.Mihelcic, K.Wingerath, G.Pfister: Flow in a small annulus between concentric cylinders. J. Fluid Mech. **140**, 343 (1984)

10.72   V.S.L'vov, A.A.Predtechenskii, A.I.Chernykh: Bifurcation and chaos in a system of Taylor vortices: A natural and a numerical experiment. Sov. Phys. JETP **53**, 562 (1981)

10.73   V.S.L'vov, A.A.Predtechensky: On Landau and stochastic attractor pictures in the problem of transition to turbulence. Physica 2 D, 38 (1981)

10.74   P.S.Marcus: Simulation of Taylor-Couette flow. I. Numerical methods and comparison with experiment. Fluid Mech. **146**, 45 (1984)

10.75   P.S.Marcus: Simulation of Taylor-Couette flow: II. Numerical results for wavy vortex flow with one travelling wave. J. Fluid Mech. **146**, 65 (1984)

10.76   R.Meyer-Spasche, H.B.Keller: Computations of the axisymmetric flow between rotating cylinders. J. Comp. Phys. **35**, 100 (1980)

10.77   R.Meyer-Spasche, H.B.Keller: Numerical study of Taylor-vortex flows between rotating cylinders, I and II. Appl. Math. Repts. Caltech (1978 and 1984).
R.Meyer-Spasche, H.B.Keller: Some bifurcation diagrams for Taylor vortex flows. Appl. Math. Rept., Caltech (1984)

10.78   G.Frank, R.Meyer-Spasche: Computation of transitions in Taylor vortex flows. J. Appl. Math. Phys. (ZAMP) **32**, 710 (1981)

10.79 T. Mullin: Mutations of steady cellular flows in the Taylor experiment. J. Fluid Mech. **121**, 207 (1982)

10.80 T. Mullin, T. B. Benjamin, K. Schätzel, E. R. Pike: New aspects of unsteady Couette flow. Phys. Lett. **83** A, 333 (1981)

10.81 T. Mullin, G. Pfister, A. Lorenzen: New observations on hysteresis effects in Taylor-Couette flow. Phys. Fluids **25**, 1134 (1982)

10.82 T. Mullin, T. Brooke Benjamin: Transition to oscillatory motion in the Taylor experiment. Nature **288**, 567 (1980)

10.83 K. Park: Unusual transition sequence in Taylor wavy vortex flow. Phys. Rev. A **29**, 3458 (1984)

10.84 K. Park, G. L. Crawford: Deterministic transitions in Taylor wavy vortex flow. Phys. Rev. Lett. **50**, 343 (1983)

10.85 K. Park, G. L. Crawford, R. J. Donnelly: Determination of transition in Couette flow in finite geometries. Phys. Rev. Lett. **47**, 1448 (1981)

10.86 K. Park, G. L. Crawford, R. J. Donnelly: Characteristic lengths in the wavy vortex state of Taylor-Couette flow. Phys. Rev. Lett. **51**, 1352 (1983);
K. Park, K. Jeong: Stability boundary of Taylor vortex flow. Phys. Fluids **27**, 2201 (1984)

10.87 K. Park, R. J. Donnelly: Study of the transition to Taylor vortex flow. Phys. Rev. A **24**, 2277 (1981)

10.88 B. Perrin: Apparition d'un mode periodique dans le domaine turbulent d'un écoulement de Couette cylindrique. J. Phys. Lett. (Paris) **43**, L 5 (1982)

10.89 G. Pfister, U. Gerdts: Dynamics of Taylor wavy vortex flow. Phys. Lett. **83** A, 23 (1981)

10.90 G. Pfister, I. Rehberg: Space-dependent order parameter in circular Couette flow transitions. Phys. Lett. **83** A, 19 (1981)

10.91 D. Rand: Dynamics and symmetry, predictions for modulated waves in rotating fluids. Arch. Rat. Mech. Analysis **79**, 1 (1982)

10.92 R. S. Shaw, C. D. Andereck, L. A. Reith, H. L. Swinney: Superposition of traveling waves in the circular Couette system. Phys. Rev. Lett. **48**, 1172 (1982)

10.93 M. Shearer, I. C. Walton: On bifurcation and symmetry in Bénard convection and Taylor vortices. Stud. Appl. Math. **65**, 85 (1981)

10.94 G. P. Smith, A. A. Townsend: Turbulent Couette flow between concentric cylinders at large Taylor numbers. J. Fluid Mech. **123**, 187 (1982)

10.95 P. Tabeling: Dynamics of the phase variable in the Taylor vortex system. J. Phys. Lett. (Paris) **44**, L 665 (1983)

10.96 I. C. Walton: The transition to Taylor vortices in a closed rapidly rotating cylindrical annulus. Proc. R. Soc. London A **372**, 201 (1980)

10.97 H. Yahata: Temporal development of the Taylor vortices in a rotating fluid III. Prog. Theor. Phys. **64**, 782 (1980)

10.98 H. Yahata: Temporal development of the Taylor vortices in a rotating fluid IV. Prog. Theor. Phys. **66**, 879 (1981)

10.99 H. Yahata: Temporal development of the Taylor vortices in a rotating fluid V. Prog. Theor. Phys. **69**, 396 (1983)

10.100 L. H. Zhang, H. L. Swinney: Nonpropagating oscillatory modes in Couette-Taylor flow. Phys. Rev. A **31**, 1006 (1985)

10.101 D. J. Benney, L. H. Gustavsson: A new mechanism for linear and nonlinear hydrodynamic stability. Stud. Appl. Math. **64**, 185 (1981)

10.102 L. H. Gustavsson, L. S. Hultgren: A resonance mechanism in plane Couette flow. J. Fluid Mech. **98**, 149 (1980)

10.103 S. N. Brown, K. Stewartson: On the algebraic decay of disturbance in a stratified linear shear flow. J. Fluid Mech. **100**, 811 (1980)

10.104 F. T. Smith, R. J. Bodonyi: Nonlinear critical layers and their development in streaming-flow stability. J. Fluid Mech. **118**, 165 (1982)

10.105 F. T. Smith, R. J. Bodonyi: Amplitude-dependent neutral modes in the Hagen-Poiseuille flow through a circular pipe. Proc. R. Soc. Lond. A **384**, 463 (1982)

10.106 S. A. Orszag, A. T. Patera: Secondary instability of wall-bounded shear flows. J. Fluid Mech. **128**, 347 (1983)

10.107 A. J. Chorin: Vortex models and boundary layer instability. SIAM J. Sci. Stat. Comput. **1**, 1 (1980)

10.108 D. Oster, I. Wygnanski: The forced mixing layer between parallel streams. J. Fluid Mech. **123**, 91 (1982)

10.109 M. Gad-el-Hak, R. F. Blackwelder, J. J. Riley: On the growth of turbulent regions in laminar boundary layers. J. Fluid Mech. **110**, 73 (1981)

10.110 P. G. Drazin, W. H. Reid: *Hydrodynamic Stability* (Cambridge Univ. Press, London 1981)

10.111 H. Huppert, J. S. Turner: Double diffusive convection. J. Fluid Mech. **106**, 299 (1981)

10.112 R. Hide: High vorticity regions in rotating thermally driven flows. Met. Mag. **110**, 335 (1981)
Also in L. Bengtson, J. Lighthill (eds.): *Intense Atmospheric Vortices* (Springer, Berlin, Heidelberg, 1982) p. 313

10.113 J. L. Hudson, J. C. Mankin: Chaos in the Belousov-Zhabotinskii reaction. J. Chem. Phys. **74**, 6171 (1981)

10.114 J. C. Roux: Experimental studies of bifurcations leading to chaos in the Belousov-Zhabotinskii reaction. Physica **7** D, 57 (1983)

10.115 J. C. Roux, R. H. Simoyi, H. L. Swinney: Observation of a strange attractor. Physica **8** D, 257 (1983)

10.116 R. H. Simoyi, A. Wolf, H. L. Swinney: One-dimensional dynamics in a multi-component chemical reaction. Phys. Lett. **49**, 245 (1982)

10.117 C. Vidal, J. C. Roux, S. Bachelart: Experimental study of the transition to turbulence in the Belousov-Zhabotinskii reaction. Ann. N. Y. Acad. Sci. **357**, 377 (1980)

10.118 M. R. Guevara, L. Glass: Phase locking, period doubling bifurcations and chaos in a mathematical model of a periodically driven biological oscillator: A theory for the entrainment of biological oscillators and the generation of cardiac dysrhythmias. J. Math. Biol. **14**, 1 (1982)

10.119 M. R. Guevara, L. Glass, A. Shrier: Phase locking, period-doubling bifurcations and irregular dynamics in periodically stimulated cardiac cells. Science **214**, 1350 (1981)

10.120 M. C. Mackey, L. Glass: Oscillation and chaos in physiological control systems. Science **197**, 287 (1977)

10.121 L. Glass, M. R. Guevara, A. Shrier, R. Perez: Bifurcation and chaos in a periodically stimulated cardiac oscillator. Physica **7** D, 89 (1983)

10.122 E. Ben-Jacob, Y. Braiman, R. Shainsky: Microwave-induced "devil's staircase" structure and "chaotic" behavior in current-fed Josephson junctions. Appl. Phys. Lett. **38**, 822 (1981)

10.123 B. A. Huberman, J. E. Crutchfield, N. H. Packard: Noise phenomena in Josephson junctions. Appl. Phys. Lett. **37**, 750 (1980)

10.124 R. F. Miracky, J. Clarke, R. H. Kock: Chaotic noise observed in a resistively shunted self-resonant Josephson tunnel junction. Phys. Rev. Lett. **50**, 856 (1983)

10.125 J. Testa, J. Peres, C. Jeffries: Evidence for universal chaotic behavior of a driven nonlinear oscillator. Phys. Rev. Lett. **48**, 714 (1982)

10.126 S. W. Teitsworth, R. M. Westervelt, E. E. Haller: Nonlinear oscillations and chaos in electrical breakdown in Ge. Phys. Rev. Lett. **51**, 825 (1983)

10.127 H. M. Gibbs, F. A. Hopf, D. L. Kaplan, R. L. Shoemaker: Observation of chaos in optical bistability. Phys. Rev. Lett. **46**, 474 (1981)

10.128 K. Ikeda, H. Daido, O. Akomoto: Optical turbulence: Chaotic behavior of transmitted light from a ring cavity. Phys. Rev. Lett. **45**, 709 (1980)

10.129 R. S. Gioggia, N. B. Abraham: Anomalous mode pulling, instabilities and chaos in a single-mode, 3.39 micron HeNe laser. Phys. Rev. A **29** (1984)

10.130 F. T. Arecchi, R. Meucci, G. Puccioni, J. Tredicce: Experimental evidence of subharmonic bifurcations, multi-stability, and turbulence in a Q-switched gas laser. Phys. Rev. Lett. **49**, 1217 (1982)

10.131 L. Kramer, E. Ben-Jacob, H. Brand, M. C. Cross: Wavelength selection in systems far from equilibrium. Phys. Rev. Lett. **49**, 1891 (1982)

Subject Index

Page numbers given in *italics* refer to the basic explanation of the entry. Only introductory citations are given for oft-mentioned subjects.

Topics in Applied Physics Founded by Helmut K. V. Lotsch